Human Evolution

Genes, Genealogies and Phylogenies

Controversy over human evolution remains widespread. Howev
the Human Genome Project and genetic sequencing of many other species have provided myriad precise and unambiguous genetic markers that establish our evolutionary relationships wit other mammals. *Human Evolution* identifies and explains these identifiable rare and complex markers, including endogenous retroviruses, genome-modifying transposable elements, gene-disabling mutations, segmental duplications, and gene-enabling mutations. The new genetic tools also provide fascinating insights into when, and how, many features of human biology arose: from aspects of placental structure; vitamin C-dependence and trichromatic vision; to tendencies to gout, cardiovascular disease and cancer.

Bringing together a decade's worth of research and tying it together to provide an overwhelming argument for the mammalian ancestry of the human species, this book will be of interest to professional scientists and students in both the biological and biomedical sciences.

GRAEME FINLAY is Senior Lecturer in Scientific Pathology at the Department of Molecular Medicine and Pathology, and Honorary Senior Research Fellow at the Auckland Cancer Society Research Centre, University of Auckland, New Zealand.

Human Evolution

Genes, Genealogies and Phylogenies

GRAEME FINLAY

Department of Molecular Medicine and Pathology,
Auckland Cancer Society Research Centre,
University of Auckland, New Zealand

CAMBRIDGE
UNIVERSITY PRESS

University Printing House, Cambridge CB2 8BS, United Kingdom

Cambridge University Press is part of the University of Cambridge.

It furthers the University's mission by disseminating knowledge in the pursuit of education, learning and research at the highest international levels of excellence.

www.cambridge.org
Information on this title: www.cambridge.org/9781107040120

First published 2013
3rd printing 2014

Printed in the United Kingdom by Clays, St Ives plc

A catalogue record for this publication is available from the British Library

Library of Congress Cataloguing in Publication data
Finlay, Graeme, 1953–
Human evolution : genes, genealogies and phylogenies / Graeme Finlay, Department of Molecular Medicine and Pathology, Auckland Cancer Society Research Centre, University of Auckland, New Zealand.
pages cm
Includes bibliographical references and index.
ISBN 978-1-107-04012-0 (hardback)
1. Human evolution. 2. Human population genetics. 3. Evolutionary genetics. 4. Genetic genealogy. I. Title.
GN281.F54 2013
599.93'8–dc23 2013015863

ISBN 978-1-107-04012-0 Hardback

Contents

Preface

Histories are subject to different interpretations. We would expect biological history to conform to this variety of understandings. But the strange thing is that the very existence of biological history is denied in some quarters. This field of science has acquired a 'more than scientific' aura to it. People argue about it as if it were an ideology. Vast resources, including a lot of goodwill, have been expended in the debate. To have achieved this notoriety, we must conclude that biological history (or evolutionary biology) is widely misunderstood. But the evidence for it is there; and a vast volume of fresh genetic data has been added recently. Such data are compelling.

This is a history book, and for two reasons. It attempts to describe, in a very limited and situated sense, a spectacular period in the history of science. Its timeframe covers, with somewhat fuzzy edges, the first decade of the twenty-first century. This is the period during which the human genome sequencing project has been elaborated to ever increasing degrees of detail, and during which myriad fascinating insights into the biological basis of our humanness have been revealed.

Secondly, it describes the evolutionary history of our species, as inscribed in great detail in our genomes. The DNA that we carry around as part of our bodies is an extraordinary library of genetic information. But it is more than simply a blueprint for the human body plan; it also carries, inscribed in its base sequence, a record of its own formative history. Multiple other mammal and vertebrate genomes have also been sequenced over the last decade or so, and this means that we have access to their histories too. When our genomic history is laid out, side by side with those of other species, particular discrete changes in the historical records can be identified in our genome and in the genomes of cohorts of other species. We can thus infer, unambiguously and with a great deal of confidence, that most of our genetic history has been shared with the genetic histories of other primates and, more inclusively, other mammals. Our evolutionary history is well documented.

Molecular evolution is at least as old as the work of Alan Wilson, who used molecular data to infer evolutionary relationships between organisms as long ago as the 1960s. Phylogenetic analyses of DNA and protein sequences have also been used to generate evolutionary trees. Such approaches require expertise in statistics and computation, and require specialist treatments. However, the novel and intuitively appealing approaches surveyed in this book are based, in general, on the identification of particular complex mutations. These arise in unique events. When any such mutation is found in multiple species, it is only because it has been inherited from the one ancestor in which the mutation arose. These are thus very powerful signatures of phylogenetic relatedness.

Along the way, we find out many fascinating things about our biology. We discover that our genome is an entire ecosystem in which semi-autonomous units of genetic material play out their own life cycles. We discover why some people have violent allergic reactions to eating certain animal products. We find out why we must have vitamin C in our diets, whereas other organisms lack this requirement. We learn of the basis of our tendency to suffer from gout. We find clues as to why humans may be particularly cancer-prone. We discover how three-colour vision arose. Indeed many processes through which new genetic functionality has been generated have been laid bare.

Everything that is presented herein is in the public domain. Anything that I have not reported accurately, or that calls for further elaboration, can be fully checked against the source literature. To me, as a cell biologist, the wonder of our DNA-inscribed history is that it requires no logic other than that which is fundamental to all genetics. (Perhaps if I were a palaeontologist, the study of fossils would be just as intuitively compelling! But I am not a palaeontologist and I suspect that far fewer people are knowledgeable about fossils than are knowledgeable about the basic mechanisms of heredity.) I believe that the logic of this book will be widely available, although it will require a modicum of biological literacy.

I am very grateful to my superiors in the University of Auckland and the Auckland Cancer Society Research Laboratory, Professors Peter Browett and Bruce Baguley, for allowing me the space and time to work on this book. I thank many senior colleagues who have provided kind and helpful advice: Professor Bill Wilson and Associate Professor Philip Pattemore, Associate Professor Andrew Shelling, Professors Wilf Malcolm, Richard Faull, Malcolm Jeeves and John McClure. Theological input has come from the late Dr Harold Turner, as well as Dr Bruce Nicholls and Dr Nicola Hoggard-Creegan. I am hugely indebted to personnel at the Faraday Institute for Science and Religion, St Edmunds College, University of Cambridge, including Dr Denis Alexander, for sharing their erudition and for their encouragement.

I am deeply grateful to the editorial staff at Cambridge University Press and Out of House Publishing for their unvarying courtesy, patience and helpfulness. It has been a pleasure to work with and learn from them.

I am also grateful to those who have given me scope to work out ideas and evolve ways of expressing them. In particular, I thank the editors of the Paternoster Press periodical *Science and Christian Belief*, and the multi-author book *Debating Darwin: Is Darwinism True & Does it Matter?* (2009). They have allowed me to explore, and reflect upon, earlier phases of an explosively expanding scientific field.

Prologue

Charles Darwin did not discover biological evolution. The concept had been brewing in people's minds for decades and Darwin grew up in an ambience of evolutionary speculation. His own grandfather, Erasmus, who died seven years before Charles was born, had ventured the possibility that all warm-blooded animals had evolved from a single ancestor. Erasmus undoubtedly had a great influence on his grandson through family links and his book *Zoonomia.*

In the first half of the nineteenth century, many biologists propounded the idea that humans had evolved from single-celled microbes. The physician-turned-biologist Robert Grant embraced evolutionary ideas from both Erasmus Darwin and the French evolutionary theorist Lamarck (who had proposed that organisms generated adaptive responses when presented with environmental challenges, and that these were heritable). Grant, in turn, passed these ideas on to the young Charles Darwin when he was studying medicine at Edinburgh. Grant then moved to University College London where he continued to popularise evolutionary thinking.

A book promoting the idea that humans evolved from simple ancestors (*Vestiges of the Natural History of Creation*) was published in 1844. It was published anonymously, but was later revealed as the work of a journalist, Robert Chambers. It was derided by its reviewers, but remained hugely popular during the rest of the nineteenth century. The philosopher Herbert Spencer (who coined the term 'survival of the fittest') also wrote on themes of human and social evolution. Spencer contributed to the wider intellectual environment of receptivity to evolutionary ideas. These works prepared popular thinking for Darwin's *Origins* when it was finally published in 1859 [1].

I DARWIN'S SCIENCE

Darwin was the first to offer a plausible *mechanism* for evolutionary development [2]. In this he was closely followed by Alfred Russel Wallace, who had spent time exploring the Amazonian and South East Asian rainforests. The outline of this scheme, known as *natural selection*, is elegantly simple.

- Resource limitations will always prevent a population from increasing at the rate that it is potentially capable of. In every generation, the individuals that become parents are a subset of the individuals that were born into that generation.
- The individuals of a species vary in many features. When a population is presented with environmental challenges or opportunities, the individuals endowed with variations that enable them to best tolerate or exploit those conditions will have a better chance of producing offspring. Parents are a *selected* group.
- Offspring tend to inherit their parents' characteristics. Features conferring reproductive success will become progressively more widely represented or more strongly developed in the population. Continuously changing conditions will drive the continuous modification of the biological features possessed by populations.

Darwin drew parallels between natural selection and the *artificial selection* performed by breeders of domesticated plants and animals. The characteristics of cereals and fruits, and of dogs and horses, are progressively altered as breeding is limited to those individuals that display the characters people desire. A spectacular example (not known to Darwin) is the way in which humans transformed the grass teosinte into maize in a few thousand years. The kernels of teosinte are few (no more than a dozen per ear), attached to long stalks and protected by a hard case. The kernels of maize are many, attached to a cob (peculiar to maize) and unprotected. A large number of genes underwent selection during the transformation from teosinte to maize [3]. Dramatic as these effects are, the particular features established by selective breeding are retained only as long as the appropriate selective pressures are applied.

Darwin identified another source of selection known as *sexual selection*. Male and female individuals of a species are often highly distinctive. The sexual dimorphism of the Indian peafowl is a classical example. In such cases, the factor driving evolutionary change is a behavioural one: choice by potential mates. The genes favoured in the case of the peacock are genes for glamour, not for usefulness.

Darwin developed many other insights that have been validated subsequently. He promoted the idea of common descent, ultimately represented by the image of a single tree of life. He perceived that an authentic taxonomic system simply reflects the branching patterns of this tree, and that extant species are a mere sample of all those that have existed, because of the wholesale extinction of linking intermediate species. He accounted for the geographical distributions of species in terms of patterns of adaptive radiation, according to which organisms evolve to take advantage of all available habitats.

He developed the concept of the vastness of time required for evolution. He accepted that the concept of gradual evolutionary change encompasses stepwise innovations, anticipating the discovery of punctuated equilibrium in the late twentieth century. Other areas of Darwin's prescience included the concerted evolution of mutually interacting species (*co-evolution*). He recognised that complex interactions occur between species (the economy of nature), and so anticipated ideas that would find their place in the science of ecology.

Darwin compiled a huge volume of evidence supporting his evolutionary paradigm. Such evidence featured comparative anatomy, physiology and behaviour, the illuminating – but necessarily incomplete – fossil record, the geographical distributions of plants and animals, and analogies with artificial breeding. These approaches have been the staple of evidential discussion (almost) to the present day [4]. The cumulative evidence for evolution was impressive, but inherently circumstantial. No-one had seen a wing evolve.

But the idea of natural selection faced one huge hurdle. Darwin knew no genetics. He did not know how heredity worked. He and most of his contemporaries considered that hereditary information was somehow distilled from throughout the parents' bodies and imprinted on to the appropriate sites of the developing embryo. This system of inheritance entailed that distinctive parental characteristics would be blended in their offspring. Such blending of inherited features engendered an unfortunate consequence. Useful adaptations would be diluted out with each succeeding generation, and ultimately lost. This was argued cogently on mathematical grounds by Fleeming Jenkin in the late 1860s.

Blending inheritance presented what appeared to be an intractable problem to Darwin's theory. As he wrestled with it, he reverted increasingly to the idea that environmental challenges could induce adaptive features in organisms, and that these were transmissible to the next generation. To get around the problem of blended inheritance, he suggested that environmental conditions might affect all the individuals in a population in a concerted manner. For much of his life, Darwin was more a Lamarckian than a Darwinian [5].

2 GENETICS ARRIVES ON THE SCENE

In the early 1900s, Gregor Mendel's work was rediscovered. It provided a first hint of the existence of units of inheritance that would later be known as genes. The answer to the problem of blending inheritance is that inheritance is quantised. Darwinian evolution only became established in the 1920s with the synthesis of natural selection and genetics. But the biochemical substance that acted as the repository of genetic information remained unknown until 1944. In that year, the material of inheritance was shown to be a constituent of cells, called DNA. People had not thought DNA particularly interesting up until that time.

In 1953, James Watson and Francis Crick proposed a model of the chemical structure of DNA, and revealed how it could embody genetic information. A DNA molecule contains myriad chemical units called *bases*, arranged in linear sequence, which are information-bearing. Watson and Crick showed how DNA could be faithfully copied and transmitted from generation to generation. And their model revealed – at last! – how DNA could undergo structural changes that would account for heritable (and non-blending) variation. Changes in the chemical units (and information content) of DNA would be transmitted from parents to their children, and thence to succeeding generations.

An important corollary of the heritability of DNA variants is that particular novelties in genetic information identify organisms connected by descent. DNA constitutes a record of family relationships. Indeed, the genetic information inscribed in DNA is an archive of long-term (evolutionary) histories. But a digression is first necessary. This book is written for biologists, and for people in medical and allied sciences who are familiar with biological concepts. But, hopefully, it will be read by all sorts of interested people – teachers, students, pastors and theologians – and so the conventions used to depict the nature of genetic information should first be reviewed.

The DNA double helix is an icon of biology. DNA consists of two helical strands, each of which consists of a backbone from which projects a succession of bases. There are four different bases, designated A (adenine), T (thymine), G (guanine) and C (cytosine). Each base hanging off one backbone interfaces with a base hanging off the opposite backbone. But size and shape considerations mean that A must pair with T, and G must pair with C. In a moment of exhilarating intuition, Watson perceived how this arrangement underlies the mechanism of heredity. Genetic information is inscribed in the order (or *sequence*) in which the bases occur. If the two strands of a DNA molecule (each backbone with its bases) are separated, the base pairing rules ensure that each is able to direct the synthesis of a new strand with its ordered complement of bases. One double helix generates two identical double helices. When cells divide, the DNA of the parent cell is duplicated and an identical copy bequeathed to each daughter cell.

Conceptually, we can unwind the double helix to produce a ladder in which the rungs are the base pairs. By convention, we read the base sequence of the top strand, as set out for the hypothetical sequence below, from left (designated 5′) to right (designated 3′). The bottom strand is read in the opposite direction. If we are thinking about gene sequences, the top strand is called the *coding* or *sense* strand (again, conventionally), because this is the sequence that specifies the order in which amino acids are added to make proteins.

Coding strand: 5′-CATATTACATAGGA-3′
Non-coding strand: 3′-GTATAATGTATCCT-5′

The most economical way of depicting genetic sequence is to present the coding strand, CATATTACATAGGA. We do not need the 5′ or 3′ signs, because we know it reads from left to right; nor do we need to write out the complementary base sequence, because we know that A, T, G and C must specify T, A, C and G as their respective complements. It is in this minimalist form that genetic sequences may be portrayed.

3 THEOLOGICAL RESPONSES TO DARWIN

Humanity had formulated no plausible scientific theory to account for the development of new species (including humans) and the diversity of life forms until Darwin. In the absence of scientific knowledge, the default position had been to account for *physical* realities (the adaptations and diversity of organisms) by using *metaphysical* concepts. It was sufficient to say that living species possess their particular constellations of characteristics because God made them that way. But such reasoning transgresses category boundaries.

The Darwinian revolution exploded this long-held conflation of concepts. The spectacular diversity of life was for the first time explained in physical cause-and-effect terms. The development of evolutionary theorising simply illustrated the dictum that scientific questions require scientific answers. Theologians had to rethink the relationship between the God whom they perceived as being at work in human history, and physical or biological mechanisms. The question of whether the cosmos was *creation* had to be accepted (or rejected) on the basis of considerations other than scientific ones.

Theologians had to recognise that the biblical concept of 'creation' referred to *ontological* origin (God creates all things at all times), not *temporal* origin (God creates particular things at particular times) [6]. A biblical creator had to be understood as the cause of everything but scientifically the explanation of nothing [7]. Such a creator could not be conceived as a component of, or an alternative to, any scientific formulation. No process – and certainly no aspect of cosmic or biological history – could be out of bounds to empirical investigation. The created order had an authentic evolving history [8], and such histories were open to empirical investigation, and on their own terms.

Many Christians accommodated their thinking to Darwin's new scientific paradigm. Darwin agreed with the Reverend William Whewell, Master of Trinity College, Cambridge (and inventor of the word *scientist*), that in the material world, 'events are brought about not by insulated interpositions of divine power, exerted in each particular case, but by the establishment of general laws' (1859). The Reverend Charles Kingsley (later Professor of History at Cambridge) articulated similar sentiments: it is 'just as noble a conception of Deity, to believe that he created primal forms capable of self-development' as to believe that God had to make a fresh act of intervention to fill every taxonomic gap (1859).

Darwin was religiously agnostic but advocated strategies of reconciliation. He did not see how evolution should shock the religious feelings of anyone. His chief supporter in America was the Christian, Asa Gray (Professor of Natural History at Harvard). They shared the conviction that evolution was 'not at all necessarily atheistical' (1860). Towards the end of his life, Darwin rejected (in private correspondence) any reason why the disciples of religion and of science 'should attack each other with bitterness' (1878). He stated that it was absurd to suggest that a man could not both have an ardent faith in God and be an evolutionist (1879) [9].

Such perspectives have been restated in the years since Darwin wrote. For example, the judge summarising the comprehensive *Kitzmiller* vs *Dover* legal case (2005) affirmed that 'the theory of evolution represents good science, is overwhelmingly accepted by the scientific community' but that it 'in no way conflicts with, nor does it deny, the existence of a divine creator' [10]. Historians marvel at the irony that Darwin's characteristic courtesy, irenicism and openness to accommodation have dissolved into acrimonious polarisation [11].

Many Christians refused to embroil the *Genesis* creation stories in conflicts with the emerging results of empirical research. To do so would denigrate Scripture [12]. Benjamin Warfield, a giant of American theology and a forerunner of the fundamentalist movement (d. 1921), argued that there was no reason why any part of Scripture, including the creation stories of *Genesis*, should be considered incompatible with biological evolution [13]. Warfield represented a tradition of conservative biblical scholars in America who urged Christians to refrain from interpolating theology into biology [14]. Their theological understanding that all reality is divinely ordered, legitimated an untrammelled mechanistic science.

Archaeological research showed that the *Genesis* creation stories were best understood against the background of Ancient Near Eastern creation stories. The *Genesis* accounts portrayed Israel's distinctive perspective on the nature of God and on people's place in the world. They were composed in the literary forms of the day, and assumed ancient cosmological understandings, but possessed radically new content: the distinctiveness of Israel's God. This God was order-conferring, rational, faithful, and declared creation to be resoundingly good. *Genesis* contained no science, but introduced a law-instituting God who made science possible [15]. Theological leaders who have gladly accepted the scientists' description of biological history, as they concern themselves with the theologians' description of human history, include J R Stott, J I Packer, Tom Wright and Richard Bauckham [16]. Christian theology does not require evolution denial.

But many people never made the transition to the new science. They persisted in the category error of regarding physical concepts (scientifically formulatable mechanism) and metaphysical concepts (divine agency) as mutually exclusive alternatives. Evolution became an obsession, a threat to be resisted. Part of the problem is that Darwinism itself became overlaid with metaphysical disputes, which could not be resolved through appeal to its scientific character.

Darwinism *as science* entails the random generation of variation screened by lawful natural selection, leading to biological adaptation and diversification. But when this mechanism is asserted to be either purposive or non-purposive, Darwinism is changed into a *metaphysical* consideration. Such deliberations may be properly carried out, but not as a *scientific* activity. For science is blind to the concept of purpose. Whether the process of natural selection entails no purpose (as a materialist might suppose) or is a means to an end, such as a creature that expresses the image of God (as a Christian might suppose) are equally metaphysical *interpretations*. Neither teleology nor a denial of teleology should be accepted as an integral component of a scientific understanding.

This confusion is illustrated by Charles Hodge, Principal of Princeton Theological Seminary (1851–78) and an older colleague of Warfield. He is renowned for his statement 'What is Darwinism? It is atheism!', which has been a rallying cry for opponents of evolution ever since. However, Hodge was not in principle opposed to either evolution or natural selection. His hostility was based upon the (metaphysical) belief that biological adaptations reflected design, and was directed to the (metaphysical) denial of teleology that was often imposed upon evolutionary science. His particular understanding of 'design' invoked the deistic metaphor of the 'divine watchmaker' popularised by William Paley (d. 1805). Hodge provides no reason to reject biological evolution. But his mingling of religious and scientific terminology, leading to an unnecessary conflict of ideas, should motivate us to distinguish between Darwinism *as science* and various *metaphysical extrapolations* from that science [17].

Confusion reached fever pitch in the 'Monkey Trial' at Dayton, Tennessee (1925). A young teacher, John Scopes, was taken to court for contravening a statute forbidding the teaching of evolution in public schools. William Jennings Bryan, a Christian and high-profile Democrat politician, acted as a counsel for the prosecution. Bryan technically won his case, but was humiliated in the process. He failed to recruit scientists as expert witnesses to present the case against evolution. He was ridiculed for relying on the writings of George McCready Price, who lacked scientific training, and whose crusade against evolution was inspired by the Seventh Day Adventist prophetess, Ellen White. Bryan was forced to concede that the world was much older than Price's strictly literalistic interpretation of *Genesis* would allow. The event revealed that Creationists were hopelessly divided [18].

Religion had taken on science and science had triumphed. Or so it seemed. But George McCready Price was to become the pioneer of today's biblical literalists. And the textbook that Scopes used [19], which contained an innocuous section on biological evolution, was laced with ideology. It was explicitly racist – white people were the apex of the evolutionary tree. It was pervasively eugenicist – the underclass of society were parasites who would be exterminated had they been animals. The undefined 'feeble-minded' should not be allowed to breed. Thus it was that both the anti- and pro-evolution camps transgressed the boundaries of scientific evolutionary theory, seeking to exploit its findings for non-scientific purposes. The way forward is to respect the integrity of scientific methodology, and distinguish evolutionary theory from more widely ranging world-view questions.

4 INTERPRETATIONS OF EVOLUTION TODAY

Science post-Darwin has shown that metaphysical interpretations of nature cannot disregard evolutionary biology. For those who approach the issue from a Christian perspective, any credible reflection on whether biology may be interpreted as embodying purpose (it is debatable whether 'design' is even a biblical concept) must engage with the reality of our evolutionary past. Evolutionary biology is often interpreted as destroying any sense of cosmic purpose, but there are possibilities of interpretation that are compatible with evolution as the unfolding of a story.

The role of chance in evolution tended to erode Darwin's belief in God; the lawfulness of the universe tended to sustain it [20]. But it is widely recognised that the *blend* of chance variation followed by lawful selection is a remarkably fruitful strategy for generating biological innovation. Such strategies have been adopted by software engineers (in genetic algorithms) and by molecular biologists (in directed evolution) [21]. Current theological approaches perceive purpose [22] in the way these polarities of contingent chance and lawful necessity co-inhere with such anthropic fruitfulness [23]. The gift of chance (or freedom) generates novelty. The gift of necessity (or lawfulness) directs that novelty along specifiable paths. This synergy is evinced in the way in which biological innovations arise multiple times given the same challenges [24] and in the ubiquity of evolutionary convergence [25]. Perhaps physical reality is so constituted that creatures who discuss God and evolution are a destination inherent within the evolutionary process.

Whether (or not) we perceive natural selection as entailing purposiveness is determined more by our metaphysical prejudgements than by the data of biology. For example, the suffering inherent in evolutionary history is a theological issue, and one that finds deep resonances in Christian theology [26]. For Christians, purpose is disclosed not in cosmic or biological history, but in human history, particularly in the phenomenon of Jesus of Nazareth. Christians who seek to controvert evolution should heed theologian Tom Wright's assessment. In terms of biology, 'Darwin put his finger on a massive truth'. But it is inconsistent to oppose Darwin in the name of a fundamentalist reading of *Genesis* if one accepts Spencer's 'survival of the fittest' creed that legitimates the unjust sequestration of wealth and power [27].

We should distinguish the biological data from enveloping metaphysical interpretations that tempt people to transmogrify that data into weapons of religious warfare. Humility rather than dogmatism should prevail. History illustrates how evolutionary biology has been misapplied, repeatedly, in the service of whatever ideology or metaphysical system has been fashionable. An appropriate response from us all is to let science be science.

5 EVOLUTION AND THE GENOME REVOLUTION

In the last few years, the comparative study of genomic DNA sequences from different species has provided a whole new approach for studying phylogenetics and its mechanisms. Genetics was a late arrival to the party but, from my perspective, now constitutes the ultimate evidence for common descent and the definitive way of defining phylogenetic relationships. It is ironic that I should presume to describe this development. I am a cell biologist who has been working in a cancer research laboratory – not a geneticist or an evolutionary biologist.

However, I have spent years studying cancer cells. I have learned that cancers develop, in part, when particular mutations arise. Once a mutation arises in a cell, it is transmitted to all the descendents of that cell. The same complex mutation in the DNA of two or more cells establishes that those cells are related. They inherited that singular mutation from the same ancestor – the one in which the mutation occurred. A cell population descended from a single progenitor is called a clone. Clones and lineages of cells are identified by shared mutations. The same logic can be applied to evolution. Once I appreciated that genetic evidence establishes the clonal nature of oncogenesis (cancer development), I could appreciate the genetic evidence for phylogenesis (species development).

The logic underlying the science of this book may be illustrated as follows. Each year, I conduct a first-year class through the medical school museum to illustrate the nature of diseases that arise from the effects of our environment. These include major types of cancer. The most common of these cancers in sun-loving New Zealand is basal cell carcinoma. One year I was marking the students' reports of the visit, and was struck to read one student's description of basal cell *casanova* – an expression that was singular and therefore memorable. But I subsequently came across two more students who wrote of basal cell 'casanovas'. Here was a singular error shared by three students. Two students must have copied their work from another. I reviewed the three reports closely and confirmed that this was the case.

I have named this the *casanova phenomenon*. It illustrates how singular *shared* spelling mistakes lead to the conclusion that one text is copied from another, or both from the same original. (One might say that the students' reports were clonal.) When singular novelties in DNA – unique genetic 'mistakes' arising through random and often complex events – are shared by multiple cells, we may conclude that *all* those cells are descended from the *one* cell in which the mutation arose. This basic principle is familiar to everyone involved in the study of the clonal progression of cancers, or the clonal development of lymphocytes in immunity (as revealed by antigen receptor gene rearrangements). When singular complex mutations are shared by multiple individuals, then all those individuals are descended from the one individual (indeed the one reproductive cell) in whom that mutation occurred. And if singular mutations were shared by multiple species, then all those species are derived from the one species (indeed the one reproductive cell) in which each of those mutations occurred.

I provide lymphoma cells to students for experiments, secure in the knowledge that cancer cells are not infectious – at least not in humans [28]. Two infectious cancers are known in other species. One of them is transmitted between dogs when they copulate, and is called *canine transmissible venereal tumour* (CTVT). This dog-to-dog contagious tumour occurs in multiple breeds, and is transmissible to wolves, coyotes and foxes. It has spread worldwide over a timescale of thousands of years. CTVT is able to grow in unrelated hosts because the cells have reduced their expression of immunity-provoking proteins (called major histocompatibility antigens). In any one host, CTVT grows only for a few months, and only at the site of infection, because the host's immune system eventually catches up with it and eliminates it. Nevertheless, such transient tumour growth is sufficient to allow transmission during closely timed copulation events [29].

The second infectious cancer is found in the Tasmanian devil, a dog-like marsupial. In 1996, it was discovered that when devils bite each other, they transmit an aggressive cancer, *devil facial tumour disease* (DFTD), which grows on the face, spreads to the internal organs and is rapidly lethal. It is feared that DFTD could drive devils to extinction by mid-century. Extensive studies, cataloguing genetic variants, have indicated where founder populations of the tumour arose, how clones have evolved and how sub-clones have diversified [30].

All the cells comprising each of these contagious tumours are descended from a *single* cancer cell (the most recent common ancestor that may have lived a long time after the tumour first arose). These infectious cancers are *clonal*. All CTVT cells are defined by a unique mutation that probably occurred in the founding cancer cell: the random insertion of a segment of DNA adjacent to the growth-controlling *MYC* gene. All DFTD cells are defined by a set of unique chromosome rearrangements. Such genetic markers arise uniquely, and all cells that now possess them acquired them by inheritance. Common ancestry is established by shared singular mutations. This is the casanova phenomenon again.

These stories are instructive because they establish the common logic of cancer genetics and evolutionary genetics. These tumours are clonal tumours with features also of evolving asexual organisms. All extant cells of each of these single-celled 'organisms' share particular genetic markers, and are the descendants of one ancestral cell.

Genetic markers establish connections in human families. The power of genetic approaches may be illustrated by work that solved the mystery of what happened to the Romanovs, the last royal family of Russia. Tsar Nicholas II, the Tsarina Alexandra, their five children and some members of their staff were gunned down in the Bolshevik revolution of 1918. The graves where they were buried had not been marked and, through most of the twentieth century, no-one knew where they were.

Old stories led to the investigation in 1991 of a location in woodland near the city of Yekaterinburg in the Urals. Bones were recovered from a shallow mass grave. DNA was extracted from them even though they were badly damaged by fire. Molecular analysis indicated that the remains included those from five members of a family – the parents and three daughters – and were consistent with their being from the Russian royal family [31]. But the remains of two children – one of the princesses and Prince Alexei – were missing. Speculation arose that they had survived and some women claimed that they were Princess Anastasia. But in 2007 two more sets of skeletal remains were discovered near the site from which the first group had been disinterred. DNA analysis showed that the more recently discovered bones were from the two missing children [32]. How can we be sure? Four lines of evidence were generated by the DNA sleuthing.

Firstly, standard forensic DNA testing established the sex of the individuals from whom each set of remains was derived. It also showed that two parents and their five children were represented.

Secondly, mitochondrial DNA sequences, which are maternally inherited, placed the remains firmly within the known Romanov genealogy. Tsarina Alexandra was the granddaughter of Queen Victoria, and the skeletal remains attributed to the Tsarina and her children are of Queen Victoria's mitochondrial lineage. Their mitochondrial DNAs have the same sequence as those of several living descendants of Queen Victoria, including Prince Philip, the Duke of Edinburgh. The remains identified as the Tsar's are of

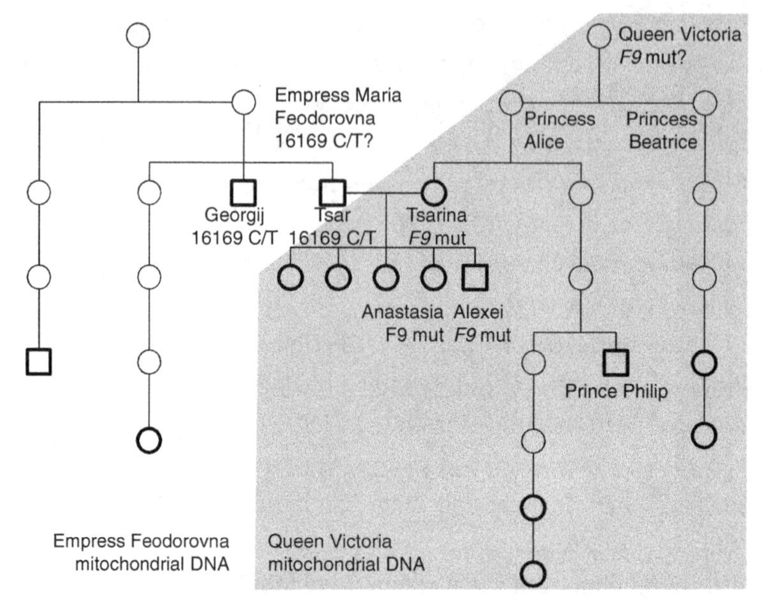

FIGURE P1. DNA IDENTIFICATION OF THE LAST RUSSIAN ROYAL FAMILY
A partial genealogy of the Russian Royal family, depicting females (*circles*), males (*squares*), individuals from whom mitochondrial DNA sequences were determined (*bold outlines*), and the Empress Maria Feodorovna and Queen Victoria mitochondrial sequence types (*background shading*).

the Princess Feodorovna lineage, established by the identity of his mitochondrial DNA with the DNA sequences of several of her living descendants (Figure P1).

But there was one mystery. The remains ascribed to Tsar Nicolas II yielded two populations of mitochondrial DNA molecules, differing at base 16,169. One had the base C and the other had T at this position. The condition in which individuals possess multiple populations of mitochondrial DNA molecules is known as *heteroplasmy*. But no other members of the Tsar's Feodorovna connection possessed the two populations of mitochondrial DNA molecules: all have a T at base 16,169. The suspicion lingered that the DNA sample was contaminated.

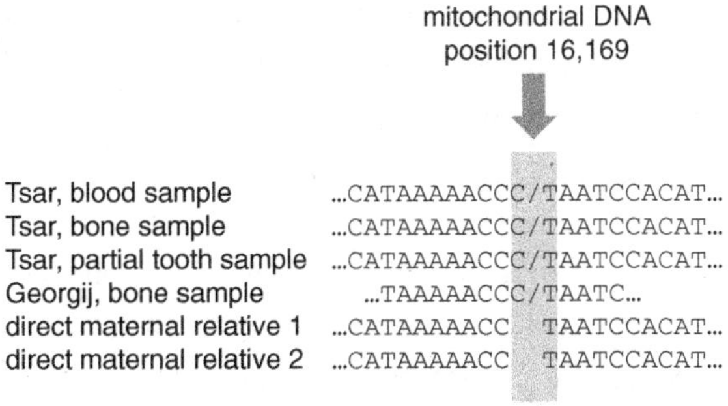

FIGURE P2. A HETEROPLASMIC MARKER ESTABLISHING THE AUTHENTICITY OF THE TSAR'S REMAINS
A small segment of mitochondrial DNA sequence is shown. The *shaded area* shows that the Tsar's and Grand Duke Georgij's tissues contained two populations of mitochondrial DNA molecules, one with a C, and the other with a T, at position 16,169. The population of DNA molecules with the C was lost during transmission to living descendants of the Tsar's mother ('maternal relatives').

To resolve this mystery, the remains of the Tsar's brother, the Grand Duke Georgij, who died in 1899, were exhumed and DNA recovered from a leg bone. The Grand Duke's mitochondrial DNA also showed the same pair of mitochondrial DNA molecules, one of which had a C, and the other a T, at base position 16,169. The heteroplasmy was no longer an embarrassment, but a convincing demonstration of the authenticity of the Tsar's DNA. The issue was settled when DNA from a bloodstained shirt (that the Tsar wore during a failed assassination attempt) showed the same C/T pair of 16,169 markers (Figure P2).

Thirdly, the male-determining Y chromosome is inherited paternally, and Y chromosome markers showed that the remains attributed to the Tsar and Alexei were indeed of the Romanov lineage, again by comparison with living descendants.

Fourthly, a particular disease-causing mutation was identified. Queen Victoria died in 1901. She transmitted to several of the royal families of Europe a mutation that caused haemophilia, although the condition (and its mutation) disappeared without trace after several generations. History has it that Prince Alexei suffered from bouts of severe bleeding. Presumably he had inherited Queen Victoria's haemophilia-causing mutation via his mother.

Alexei's DNA was used to obtain the genetic sequence of two genes known to be mutated in patients with haemophilia. A disabling mutation was discovered in the gene encoding blood coagulation (or clotting) factor IX (the *F9* gene), which resides on the X chromosome. Males have one X chromosome, and Prince Alexei had only a mutated copy of the *F9* gene. Females have two X chromosomes, and the Tsarina and one of her daughters had one normal and one mutated copy of this gene. They were therefore carriers (Figure P1). The identity of Queen Victoria's mutation was discovered from DNA that had lain for 80 years in the damp sod of a temperate forest [33].

Genetic mistakes in old bones connected the Romanovs, demonstrating how mutations can definitively delineate lineages [34]. Genetic markers of the sort used forensically – a mitochondrial DNA mutation manifest as a transient heteroplasmy, and a mutation in the *F9* gene – were used to generate a genealogy. The casanova phenomenon strikes again.

6 THE SCOPE OF THIS BOOK

The following four chapters describe how the casanova phenomenon provides compelling evidence for human evolution and lays out our patterns of relatedness. Each chapter surveys one broad category of genetic marker that is inscribed in our chromosomal DNA. Each class of marker includes myriad instances, each of which acts as a definitive signpost of phylogenetic relatedness.

Retroviruses are a class of viruses that splice their tiny genomes into the DNA of the cells they infect (Chapter 1). Millions of genetic parasites called transposable elements, recognisable as little segments of DNA, are also interspersed collinearly through our genomic DNA. The mode of replication of most of these agents shares some of the strategies used by retroviruses (Chapter 2). The presence of the same inserted piece of DNA in the genomes of two or more cells, organisms or species indicates that those genomes are derived from the one genome into which that piece of DNA was inserted.

Many types of disruptive (*disabling*) mutations are present in our genomes. They are recognisable in derelict genes that have lost the ability to direct the production of functional proteins – which are proteins that are still made by the corresponding gene in other species (Chapter 3). Other mutations have contributed to the acquisition of new genetic function. These are *enabling* mutations (Chapter 4). When particular instances of such mutations are found in the genomes of different species, they demonstrate that all the species that possess them are descendants of one ancestral species – indeed the one ancestral cell – in which the mutation arose.

These molecular signatures inscribed in our DNA constitute definitive evidence that humans and other mammals are descended from common ancestors. It must be stressed that it is the *mechanisms* by which these mutations arise that enable them to act as potent markers of evolutionary relatedness. Familiar molecular transformations are involved. For example, retroviruses and transposable elements are mutagens with precisely defined mechanisms of action. Each marker, spliced into its unique location in the genome, arrived there by an elaborate and interpretable series of biochemical events. The *functionality* of the mutant product is irrelevant with respect to its use as a marker of descent.

I have provided an abundance of examples for two reasons. Firstly, I find each example to be a source of sheer fascination, because of its precise information content and its compelling evidential power. The question of whether large-scale evolutionary change has occurred has been resolved by appeal to a source of historical information that we all carry around with us. Secondly, I want to provide some feeling for the sheer mass of data available. The supreme information-bearing molecule in the known universe, DNA, provides millions of genetic markers for historical reconstruction. If readers find the number of examples excessive, they can move on to the next section.

The research described covers roughly the first decade of this century. This was the time during which the study of the first human genome sequence revolutionised our understanding of human genetics, and provided radically and definitively new ways of documenting evolutionary origins. These issues have been touched on by more-learned authors [35].

I conclude with a consideration of whether the fact of our evolution is in any way a threat to our humanity, or indeed to a spiritual view of ourselves. There will be minimal theological reflection; I have sought to do that elsewhere [36]. It is my hope that this book will calm the misdirected and often lamentably acrimonious controversies over evolution.

1 Retroviral genealogy

I first became involved in cancer research in the early 1980s. It may seem presumptuous that a mere cancer cell biologist should write a book on the definitive evidence for biological evolution, at least as it pertains to our own species. However, it was a background in cancer research that provided useful perspectives – and the *eureka* moments – that enabled me to appreciate the force of the data arising from the field of comparative genomics. And one particular story led me inexorably from cancer biology into evolutionary biology.

The early eighties were heady times for cancer researchers. A revolution was taking place in our understanding of the genetic basis of cancer. Cancer-causing genes called *oncogenes* were discovered. Oncogenes were shown to be derived from normal genes (*proto-oncogenes*) that play vital roles in the regulation of cell proliferation, differentiation and death. During cancer development, proto-oncogenes are damaged by mutations, and their encoded proteins show increased expression, elevated activity and loss of sensitivity to negative regulation. The result is the disruption of cellular regulation and the acquisition of unrestrained patterns of growth. The products of oncogenes undergo *gains* of function that *impel* cancer development.

Concurrently, researchers identified a second class of genes as central players in cancer biology. These were called *tumour suppressor genes* (TSGs), and they were found to play essential roles in restricting cell proliferation and promoting differentiation under normal conditions. They act to counterbalance the effects of proto-oncogenes. Many TSGs are responsible for maintaining the integrity of the genome – often by detecting and repairing DNA damage. During cancer development, TSGs are frequently the target of mutational events that compromise their restraining activities. In contrast to oncogenes, it is the *loss* of TSG functions that *releases* cells down a neoplastic pathway [1].

A third area of discovery was the demonstration that viruses are major etiologic agents in human cancers. The oncogenic roles of viruses had long been debated. But in the 1980s, epidemiological and biochemical evidence implicated oncogenic viruses in 15–20% of human cancers. Hepatitis B virus (HBV) – and later hepatitis C virus – infections were shown to be huge risk factors for liver cancer. Certain types of human papilloma virus (HPV) were implicated in cervical cancer, Epstein–Barr virus in lymphoid cancers and in nasopharyngeal carcinoma in Southern Chinese populations, and Kaposi's sarcoma-associated virus in Kaposi's sarcoma of AIDS patients [2].

Such viruses exert their oncogenic effects by introducing into cells viral genes that act as oncogenes. Some viruses were also found to act as DNA-disrupting (mutagenic) agents. The exponents par excellence of the DNA-disrupting strategy are the *retroviruses*. We need to consider the subversive activities of retroviruses in order to describe their role in oncogenesis – for which they are a major clinical problem in some parts of the world. When we have done this, we will suddenly find ourselves in the world of evolutionary genetics and phylogenesis (the origins of species), complete with definitive answers to the question of whether we have evolved.

1.1 THE RETROVIRAL LIFE CYCLE

Retroviruses cause cancers in birds and mammals. In 1911, Peyton Rous showed that cancers called sarcomas could be transmitted between chickens even when the cancer cells had been pulverised and the lumpy material filtered off and discarded. The filtrate contained a cancer-causing agent, later known as the Rous sarcoma virus. Rous had to contend with widespread disbelief, and had to wait for 55 years before he was awarded the Nobel Prize for Medicine in recognition of his discovery [3].

In the oncological revolution of the early 1980s, retroviruses were shown for the first time to cause disease in humans. Human T-cell leukaemia virus type 1 (HTLV-1) was identified as the causative agent of adult T-cell leukaemia (ATL), an aggressive cancer of lymphocytes that exists in parts of Japan, the Caribbean and Africa. Some 20 million people worldwide may be infected with HTLV-1. Estimates vary as to the proportion of infected people who will ultimately develop cancer (from 0.1% to 5%). HTLV-1 also causes a neurological disease (tropical spastic paraparesis, TSP). This arises from inflammation in the spinal cord, with subsequent nerve damage [4]. Another pathogenic retrovirus is the notorious human immunodeficiency virus (HIV), the cause of AIDS. Because of its toxicity, HIV kills cells rather than causing derangements in their long-term patterns of proliferation. HIV is not believed to directly cause cancers.

Cancer-causing retroviruses pursue their parasitic lifestyle with elegant sophistication. The first step occurs when the infecting virus particle attaches to a cell. It is able to do this because the virus particle displays a protein called the envelope protein (encoded by the retroviral *envelope* or *env* gene) that adheres to a target molecule on the surface of the cell to be infected. This adhesive interaction enables the retroviral membrane to fuse with that of the cell, so that the viral genetic material is delivered into the cytoplasm.

The genetic information of retroviruses is embodied in a molecule called RNA, but retroviruses possess an enzyme that copies (or *transcribes*) the RNA version into a DNA one. The flow of information from RNA to DNA is opposite to that which operates in the genetic expression of cellular organisms. The retroviral enzyme has thus been called a *reverse transcriptase*, and Howard Temin and David Baltimore received the Nobel Prize for its discovery in 1975 (Figure 1.1).

Retroviruses are professional mutagens. The freshly synthesised viral DNA is spliced into the chromosomal DNA of the infected cell. This process is initiated by another virus-encoded enzyme, an *integrase* or *endonuclease*. The enzyme haphazardly selects a *target*

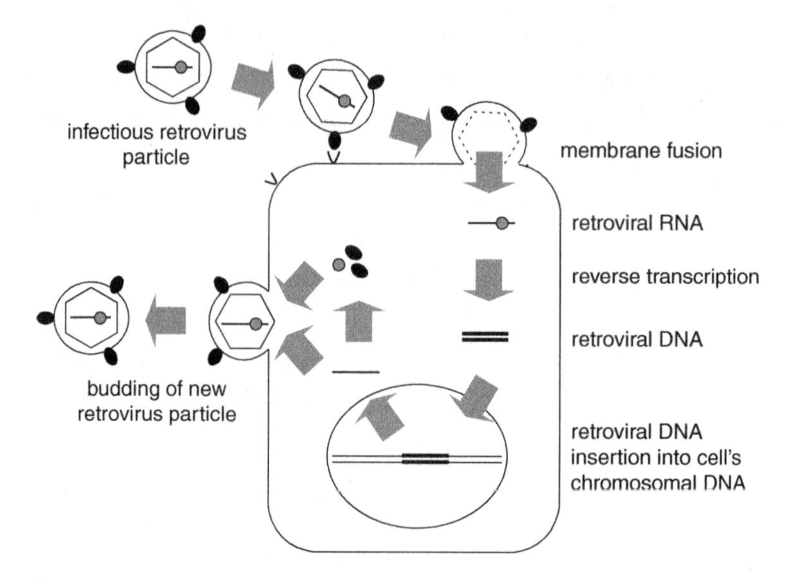

FIGURE 1.1. THE INFECTIOUS CYCLE OF A RETROVIRUS
The retrovirus particle is represented by a *circle* (outer membrane) with envelope protein (*black ovals*), a protein core (*hexagon*), RNA genome (*line*) and associated reverse transcriptase (*grey circle*). The cellular nucleus is indicated by a *large oval* with DNA (*paired thin lines*) and the provirus (*paired dark lines*).

site in the host genome, at which it makes two staggered nicks, four to six bases apart (depending on the type of retrovirus), one nick on each DNA strand. This cleavage event creates a gap in the chromosomal DNA into which the DNA copy of the retroviral genome inserts itself. The integrase has a very loose preference for the bases in the target site. It favours a sequence environment that is rich in A and T bases, and insertion is also favoured in active regions of the genome, in the vicinity of genes.

The final step is to convert the single-stranded lengths of the target site into double-stranded DNA, generating tell-tale *target-site duplications* (TSDs) on either side of the retroviral DNA insert. Cellular enzymes seal the retroviral genome into place (Figure 1.2). The retroviral genome, which is typically 8–10 thousand bases long, has become part of the genome of the cell, and is called a *provirus*.

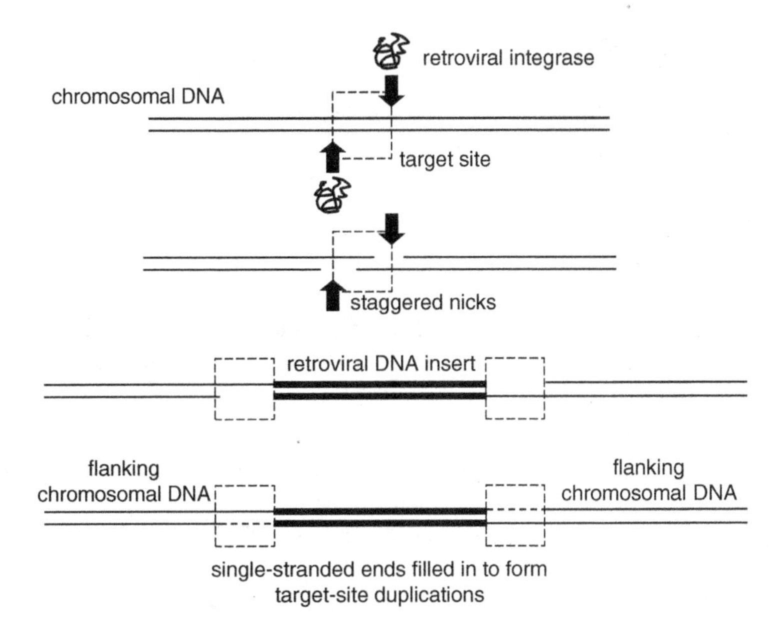

FIGURE 1.2. THE MECHANISM BY WHICH RETROVIRAL DNA IS INSERTED INTO THE CHROMOSOMAL DNA OF A HOST CELL
Target sites and their duplications are depicted by *dashed boxes.*

The insertion of a retroviral genome into that of the infected cell is random with respect to site, and permanently alters the genome of the host cell. In most cases, this will be harmless. In some instances, insertion may compromise the functional integrity of the genome. It may disrupt the regulatory sequences of a gene, for example, with the consequence that genetic function will be compromised. Retroviral insertion thus represents a special type of genetic mutation, and retroviruses are known as *insertional mutagens.* The process by which they splice their genomes into cellular chromosomal DNA is called *insertional mutagenesis.*

The provirus can be recognised by many sequence features. It is bounded by the short (host DNA-derived) target-site duplications, as mentioned above. The provirus itself possesses a large block of duplicated sequence at each end of the virus sequence. These direct repeats may be several hundred to a thousand bases long. They

Table 1.1. *Structural genes common to retroviruses*

Gene	Full name	Function of protein products
gag	group-specific antigen	Packages viral RNA
prt	protease	Processes viral proteins
pol	polymerase	A multi-functional protein with endonuclease, RNA-dependent DNA polymerase (that is, reverse transcriptase) and RNA-degrading activities
env	envelope	A viral membrane protein that mediates viral adhesion to cells; suppresses immunity

are called *long terminal repeats* (LTRs), and they contain the DNA sequence motifs needed to regulate viral gene expression. Situated between the LTRs is a basic set of four structural genes, which are (from left to right) known as *gag*, *prt*, *pol* and *env* (Table 1.1).

The provirus can be transcribed into RNA copies by the actions of cellular enzymes, and these transcripts can be used to direct the synthesis of retroviral proteins. RNA transcripts and new proteins assemble into infectious virus particles that bud off from the cell membrane. The cycle of infection starts all over again. But most significantly, because the provirus has become an integral part of the genome of the cell, it will be inherited by every descendant of the original infected cell, potentially making more viruses over the lifetime of the organism.

1.2 RETROVIRUSES AND THE MONOCLONALITY OF TUMOURS

The presence of such parasitic segments of DNA will usually be innocuous. Much of the genome can tolerate the addition of segments of extraneous DNA. But in rare cases this strategy goes wrong. In the case of HTLV-1, the provirus makes a protein called Tax that has the potential to perturb the mechanisms by which a cell regulates its replication. The disruption of regulatory circuits in an infected cell may cause that cell and its descendants to start dividing in an aberrant way, generating an expanding population of progressively more abnormal cells. In this context of abnormal proliferation, other genetic mutations may accumulate until eventually, decades after the original infection, a lethal leukaemia may become manifest.

Early in the infectious phase, a population of lymphocytes will contain a large number of distinguishable HTLV-1 proviruses. This is because the random nature of target site selection ensures that proviruses are found at myriad different insertion sites. Perhaps every infected lymphocyte will have its own provirus, as defined by the site into which it has inserted. But if one takes (say 50 years after the original infection) a population of leukaemic cells from any one patient, one will find that *every* leukaemic cell possess the *same* HTLV-1 provirus, as defined by one common site of insertion. This demonstrates that *one* original cell with its *singular* provirus initiated a programme of continuous cell multiplication. With time, the expanding clone of cells acquired progressively more abnormal properties until it evolved into a population of cancerous descendants, all of which *inherited* the original, unique, cancer-triggering provirus (Figure 1.3).

Such data demonstrate that ATLs are monoclonal tumours. Surprising as it may seem, the catastrophic leukaemic burden of 10^{10} cells originated from a *single* infected progenitor cell. The particular provirus common to all the cancer cells is the definitive marker of monoclonality. In biological parlance, we may say that the presence of a particular provirus in all the cells of a cancer is formal proof that these leukaemias are monoclonal, derived from a single cell. When we find that a single random genetic 'mistake' is shared by many cells, we may conclude that this 'mistake', and these cells, are copies of the unique original 'mistake' and altered cell. The 'casanova phenomenon' is therefore a thoroughly well-established oncological principle.

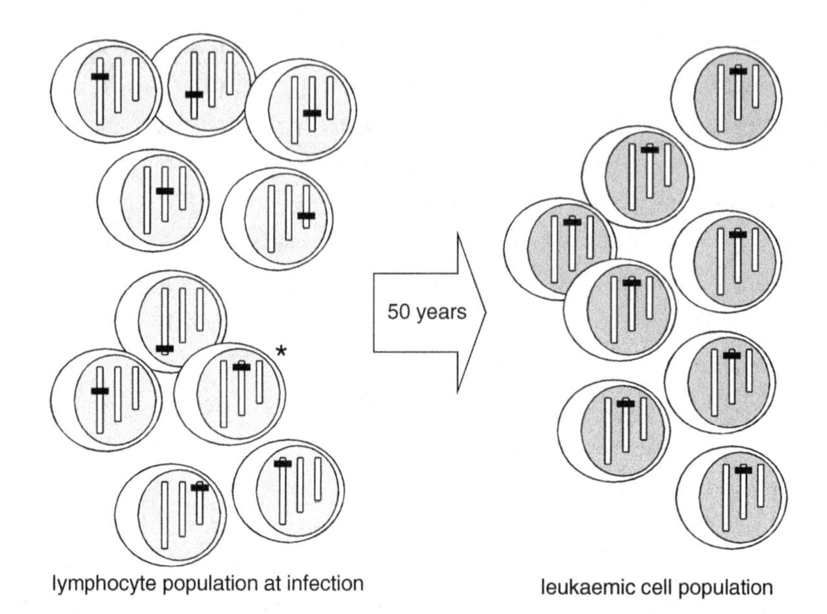

FIGURE 1.3. THE MONOCLONALITY OF HTLV-1-INDUCED TUMOURS
For simplicity, each cell depicted on the left has three chromosomes (*vertical bars*), with an HTLV-1 insertion (*thick horizontal bar*). Many insertion sites are found in the population. The tumour population *on the right* is characterised by one provirus, demonstrating that all the leukaemic cells are descendants of *one* progenitor (marked by the *asterisk*).

Leukaemias from different patients are characterised by distinctive clonal retrovirus insertion sites. In other words, the HTLV-1 provirus is found at a different location in every tumour. A set of target sites from leukaemic and other patients is presented in Table 1.2. In only one case did two HTLV-1 integration events select the same six-base target site (GCTAGG, indicated by asterisks). Any six-base sequence is itself present in the human genome in the order of a million times, and these GCTAGG sites were located in different parts of the genome. It is clear that the chances of finding multiple independent insertions into the same site are pretty remote. Such data establish that HTLV-1 insertion does not strongly favour any particular DNA target site or sequence of bases. The retroviral integrase is pro-

Table 1.2. *HTLV-1 target sites*

Source of DNA	Target site sequence	Ref.
blood cell, TSP patient	ACATTT	5
non-cancer cell, healthy carrier	ACCCGC	5
ATL	ACCTTT	6
non-cancer cell, ATL patient	AGCAAG	5
ATL	CAGCTG	5
blood cell, TSP patient	CATATG	5
ATL	CCATTC	6
non-cancer cell, ATL patient	CCTCTC	5
blood cell, TSP patient	CTGAGG	5
non-cancer cell, healthy carrier	CTGTGG	5
blood cell, TSP patient	CTTGGT	5
ATL	GAATCC	6
non-cancer cell, healthy carrier	GAGAAC	5
ATL	GAGTTG	6
blood cell, TSP patient	GAGAAT	5
ATL	GCATTC	7
non-cancer cell, healthy carrier	GCTTTT	5
non-cancer cell, healthy carrier	GCAACT	5
blood cell, TSP patient	GCTAGG*	5
cerebrospinal fluid, TSP patient	GCTAGG*	5
non-cancer cell, healthy carrier	GGTGTG	5
non-cancer cell, healthy carrier	GTTATA	5
cerebrospinal fluid, TSP patient	TAAAGT	5
blood cell, TSP patient	TAATAG	5
ATL	TAGTTG	5
blood cell, TSP patient	TCAATC	5
blood cell, TSP patient	TCAGTC	5
non-cancer cell, healthy carrier	TCCGCA	5
ATL	TCTTTC	5
non-cancer cell, healthy carrier	TTATGT	5
ATL	TTATTC	5

Note: the asterisks denote the two cases where the target site base sequence is the same.

miscuous in the selection of its chosen substrate. Proviral insertions are largely randomly distributed with respect to DNA site.

Tragic confirmation of the monoclonal nature of retrovirally induced human tumours has been provided by a clinical experiment that went wrong. Children with X-linked severe combined immunodeficiency lack normal immune function because their lymphocytes cannot develop normally. The disease arises because the children inherit a mutant gene that has lost the ability to produce an important signalling molecule (the common γ subunit of the IL-2 receptor). These children are susceptible to infections and, without treatment, die in infancy. A clinical trial was conducted in an effort to rectify the genetic deficiency. Children were treated with a retrovirus engineered to carry the needed gene, in the hope that the missing protein would be expressed and would support normal immune function. Encouragingly, the young patients showed significant improvement in their condition. However, several children developed leukaemias. The malignant cells were found to possess copies of the therapeutic retrovirus in their genomes. Each leukaemia was monoclonal with respect to the viral insertion site, and arose because the therapeutic virus inserted near (and deregulated) the *LMO2* proto-oncogene [8].

It goes without saying that the monoclonality of tumours caused by retroviruses that infect non-human animals (fowl, rodents, cats) is also thoroughly established [9]. An example of one of these retroviral insertion sites is shown in Figure 1.4. It shows a small length of genetic sequence, 26 bases long, from the mouse genome. The six-base sequence ...GTTTGC... (in bold and shaded) represents the target site selected by the retroviral integrase. The upper sequence shows the retroviral DNA insert flanked at each end by the ...GTTTGC... target site sequence, and otherwise neatly spliced into the mouse genome [10]. A unique insertion event in one cell induced an uncontrolled programme of cell division, leading to a proviral copy in each of myriad descendant cells.

We can detour from retroviruses briefly. Several other human cancers arise when bits of viral DNA are insinuated into the genomic DNA of infected cells. No other class of oncogenic virus

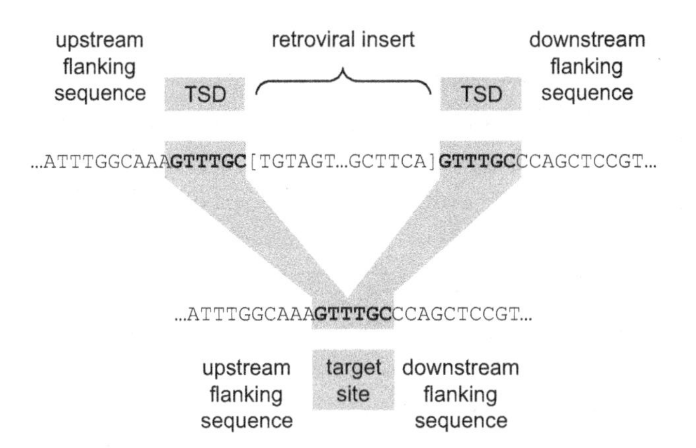

FIGURE 1.4. A RETROVIRAL DNA INSERT IN MOUSE DNA [10]
Sequences represent the original undisturbed target site GTTTGC and the inserted provirus between target-site duplications (TSDs). In this and subsequent figures, target sites and their duplications are in *bold* and *shaded*.

manifests the professional mutagenic sophistication of retroviruses. Nevertheless, the same logic that we have encountered with retrovirus-induced cancers demonstrates the monoclonality of the cancers induced by other classes of viruses.

Several sub-types of HPV cause cervical cancer. The random integration of viral DNA into cellular DNA typically occurs during tumour evolution. In some patients there may be complex patterns of disease, featuring multiple distinct foci of abnormal cells, or multiple tumours that recur over time. The question arises: have the multiple tumours observed in such patients arisen independently, or are they all derivatives of *one* original delinquent cell? If the different tumours have arisen independently, they should all possess *distinctive* viral DNA inserts. But if they are all derived from one cell (that is, if the multiple tumours in a patient are monoclonal), they should all possess the *same* signature viral insert representing *one* originating insertional mutagenic event.

Molecular genetic work has shown that, in most cases, the many tumours arising in one given patient are marked by the *same* insert of HPV-derived DNA. This is illustrated by the clinical history of a patient, in whom surgery for neoplastic cells in the cervix was followed by treatment of a series of abnormal growths arising in the vagina over 12 years. Each of six tissue samples subsequently excised from the patient yielded DNA in which viral and cellular DNA shared the same unique junction point. The series of tumours encountered in different sites of the female reproductive tract were all descendants of one particular cell. This progenitor cell sustained one random viral insertion event and was triggered into an unrestrained and destructive mode of neoplastic growth that gave rise to all the lesions subsequently treated [11].

Similarly, infection with HBV is a major risk factor for developing liver cancer. The viral DNA integrates randomly into liver cell DNA. Multiple tumours are often found in a patient's liver. If these many tumours have the same insert (same bit of viral DNA, same site of the cell genome), then they are derived from the one cell in which the unique insertion event occurred. In many patients with liver tumours, HBV integration sites are common to multiple tumour nodules, and have established that those nodules are of monoclonal origin [12].

In 2008 yet another agent was added to the rogues' gallery of viruses that splice themselves into cellular DNA to exert oncogenic effects in humans. A polyomavirus was shown to be associated with Merkel cell carcinoma, a rare but aggressive tumour of the skin. Again, DNA inserts were common to all tumour cells in a given tumour (but distinct for different tumours), establishing that these tumours too are monoclonal [13].

1.3 ENDOGENOUS RETROVIRUSES AND THE MONOPHYLICITY OF SPECIES

As work on infectious retroviruses gained momentum, a startling discovery was made. Many organisms possess retroviral DNA as an integral part of their genome. In contrast to infectious (or *exogenous*) retroviruses that are transmitted *horizontally* between cells, or between the individuals of a species, many retroviral DNA segments are present as an intrinsic part of the genomic DNA that defines a species, and they are transmitted *vertically* from one generation to the next. They are transmitted in a Mendelian fashion, just as if they were genes, and are known as *endogenous* retroviruses (ERVs) [14].

ERVs enter the genomes of species by infecting germ cells – the cells present in early embryos and in reproductive tissues that generate gametes (eggs and sperm). Once proviral inserts are established in such cells, they are transmittable to future generations. With time, a chromosome (or part of a chromosome) bearing such an insert may increase in frequency (relative to the original, undisrupted length of chromosome) until it replaces the original in the population. At this stage, the ERV becomes *fixed*.

In the early 1980s, ERVs were discovered in the human genome. Their presence was first inferred from the appearance of viral particles that were seen to be budding from cells comprising reproductive tissues, including testicular tumours. These virus particles did not have the capacity to infect other cells, and thus they appeared to be defective. (Later research showed that human ERVs are riddled with inactivating mutations that preclude the production of infectious viruses.) Genetic analysis showed that cells producing these particles possessed messenger RNA molecules encoding the full suite of retroviral genes: *gag*, *prt*, *pol* and *env* [15].

True ERVs are categorised into three major groups: classes I, II and III. Additional retroviral DNA-like units scattered around the human genome also possess long terminal repeats, and are called *LTR retrotransposons*. They lack an *env* gene and therefore are not transmitted between cells. These constitute class IV. True ERVs and LTR retrotransposons (collectively, LTR elements) constitute 8% of the DNA in the human genome. This large fraction of human DNA is distributed around the genome in approximately 400,000 individual inserts with 350 sub-families [16]. Nearly all of these inserts are common to all people on planet Earth. This raises the question of when such lengths of retroviral DNA first entered the genome that we have inherited.

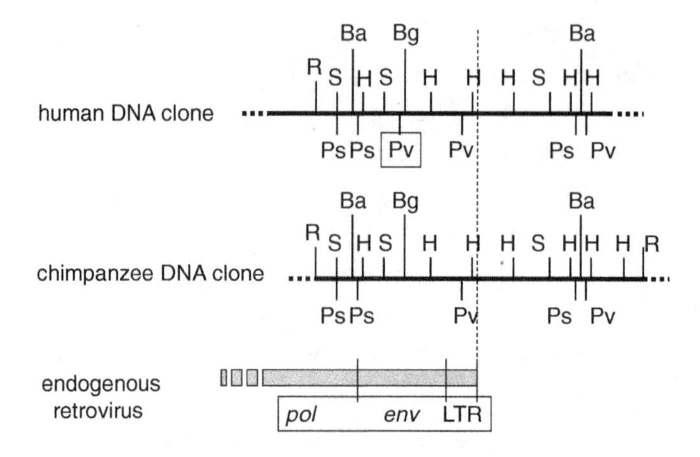

FIGURE 1.5. CLONED LENGTHS OF GENOMIC DNA FROM HUMAN AND CHIMP, WITH AN OVERLAPPING ERV [17]
The *thick horizontal lines* represent lengths of DNA, about 8,000 bases in length. The *vertical lines* represent restriction enzyme-cutting sites, which provide a map of the DNA clones. The one difference between human and chimp is *boxed*.

A 1982 study prepared the way (as far as I was concerned!) for the surprising answer. A length of cloned human chromosomal DNA had been mapped on the basis of restriction enzyme-cutting sites (that provide sequence landmarks along the DNA). An equivalent piece of DNA cloned from the chimpanzee showed almost the same restriction enzyme-mapping sites, indicating that these lengths of cloned DNA were from the corresponding parts of the two genomes. But what is remarkable was that each of these segments of DNA overlapped the sequence of an ERV (Figure 1.5). This finding implied that the ERV in each of the two genomes was inserted at the same location [17]. If indeed it was the same insert (same class of ERV, inserted in precisely the same site with the same target-site duplication, and lying in the same direction), then we would have to conclude that both *species* are descendants of the single progenitor in which this unique insert event occurred. This remarkable conclusion, reflecting the way in which shared proviruses establish the monoclonality of tumours, was forced on me by every instinct inculcated by cell biological experience.

But was the ERV indeed the same one in both species? The definitive answer could only come from DNA sequencing studies, and this pioneering work preceded the high-throughput sequencing revolution. DNA sequencing had not been performed on these cloned lengths of human and chimp genome. The answer was not available. However, this research held out the tantalising prospect that the sequencing of ERV integration sites in related species might provide the definitive answer to the question of whether humans and chimps are monoclonal (as a cell biologist might express it). The word *monophyletic* applies more appropriately to multiple species descended from one ancestor. The distribution of ERVs in the DNA of primate species could provide the ultimate statement on common descent.

Work published in 1999 settled the question of whether shared ERVs could demonstrate human and chimp descent from a common ancestor [18]. This seminal study identified those primate species in which each of six ERVs was present – and defined insertion sites at single-base resolution. The data confirmed that each of these ERVs is shared by humans and chimps. Indeed, each ERV is shared not only by humans and chimps, but also by gorillas and more distantly related primate species (Figure 1.6, white boxes).

- Three of these ERVs were found to be shared by humans, chimps, bonobos (pygmy chimps) and gorillas, but not by orang-utans or other primates. These ERVs entered the primate germ-line in a creature that was ancestral to all the African great apes, but that lived after the orang-utan lineage had diverged from the great ape family tree.
- The other three ERVs were found to be shared by humans, the other apes and Old World monkeys (OWMs) but not the New World monkeys (NWMs). These ERVs had entered the primate germ-line in ancestors common to all the apes and OWMs. The NWMs had already branched out on a separate lineage by this time.

In Figure 1.6, the shape of the primate family tree is presupposed on the basis of other work. But the results of this pioneering ERV study firmly established the reality of the African great apes' ancestral lineage, and of the ape–OWM ancestral lineage.

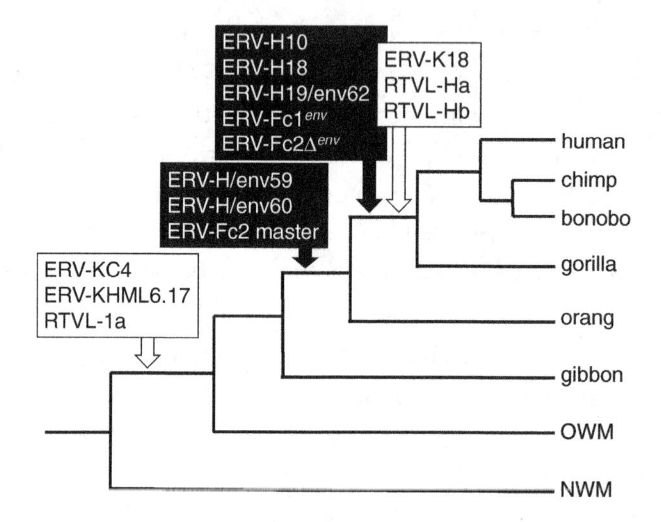

FIGURE 1.6. THE TIMES AT WHICH 14 ERVS ENTERED THE PRIMATE GERMLINE, INFERRED FROM THEIR PRESENCE OR ABSENCE IN THE GENOMES OF PRIMATE SPECIES
White boxes [18]; *black boxes* [19].

These conclusions are unambiguous, unassailable and definitive: strong words in the context of a controversy that has simmered (at least in some quarters) for 150 years. No arcane 'evolutionary' logic was required for this interpretation. The data struck me with compelling force simply because I had been exposed to basic cell biology. The casanova phenomenon was applicable to defining relationships between species, and could demonstrate which species were linked by descent.

More detailed studies of particular ERV classes followed. Class I ERVs include many families of endogenous retroviruses including ERV-H (a large family) and ERV-Fc (a small family with only six members in the human genome). Studies were performed to define the insertion sites of some of these ERVs. DNA sequencing of representative members of these families identified five proviruses that are common to the African great apes (but no other species) and three that are common to all the great apes (Figure 1.6, black boxes) [19].

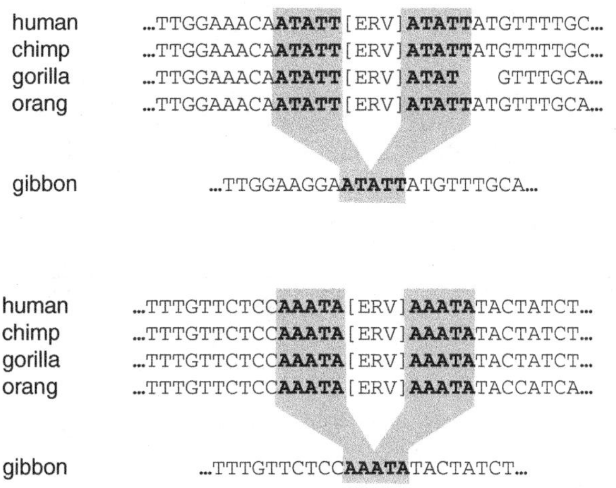

human ...TTGGAAACA**ATATT** [ERV] **ATATT**ATGTTTTGC...
chimp ...TTGGAAACA**ATATT** [ERV] **ATATT**ATGTTTTGC...
gorilla ...TTGGAAACA**ATATT** [ERV] **ATAT** GTTTGCA...
orang ...TTGGAAACA**ATATT** [ERV] **ATATT**ATGTTTGCA...

gibbon ...TTGGAAGGA**ATATT**ATGTTTGCA...

human ...TTTGTTCTCC**AAATA** [ERV] **AAATA**TACTATCT...
chimp ...TTTGTTCTCC**AAATA** [ERV] **AAATA**TACTATCT...
gorilla ...TTTGTTCTCC**AAATA** [ERV] **AAATA**TACTATCT...
orang ...TTTGTTCTCC**AAATA** [ERV] **AAATA**TACCATCA...

gibbon ...TTTGTTCTCC**AAATA**TACTATCT...

FIGURE 1.7. ERVS COMMON TO ALL THE GREAT APES (ERV-H/env59 and ERV-H env60)
From de Parsival *et al.* (2001) [19].

Representative insertion sites are shown for two of these ERVs (Figure 1.7). Both inserts are present in the genomes of humans, chimps, gorillas and orang-utans. These species are collectively known as the great apes and share a common ancestry. The high degree of preservation of the DNA sequences is remarkable. The proviruses are located between five-base target-site duplications (ATATT and AAATA). The gibbon, a lesser ape, retains the undisturbed target site. The gibbon lineage had branched off before the retroviral insert was introduced into the hominoid (or ape) germ-line.

Similar studies have been performed with the Class II ERV-K family, of which there are some 8,000 inserts in the human genome. Most emphasis has been placed on a particular sub-family, designated ERV-K (HML-2). This is an interesting collection of inserts, in that some are found only in humans and are almost intact. These features indicate that they entered the human genome relatively recently – after the human and chimp lineages diverged from their common ancestor [20]. Indeed some of these human-specific ERVs are dimorphic in the human population with respect to presence or

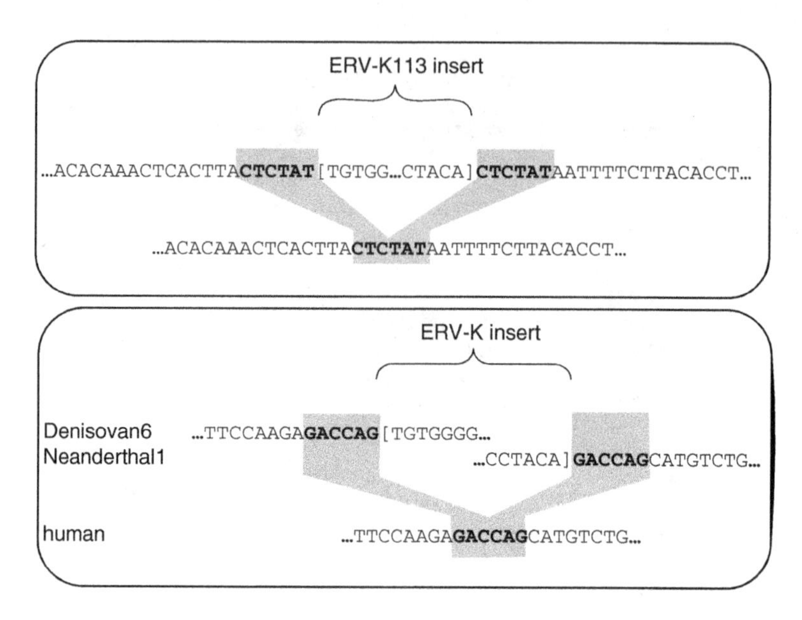

FIGURE 1.8. RECENT ERV-K INSERTS
ERV-K113 (*upper panel*) is dimorphic in the human population [21]; the ERV-K insert (*lower panel*) is found in the DNA from Denisovan and Neanderthal individuals [23].

absence of the provirus, indicating that the insertion events were so recent that only a fraction of the human population has inherited the ERV. For example, the ERV-K113 provirus is present in the genomes of only about 16% of us; the rest of the human population retain the undisturbed target site (Figure 1.8, upper panel) [21]. The ERV-K106 insert, which is fixed in the human population (we all possess it in our genomes), is also very recently acquired. Its long terminal repeats lack mutations – a sign that it was added to the genome relatively recently. Some geneticists have suggested that it arose during the history of anatomically modern *Homo sapiens* [22]. Perhaps infectious (exogenous) retroviruses belonging to this ERV-K clan are still lurking in some geographically isolated human populations.

Further evidence of recent HERV-K activity comes from the study of DNA recovered from the bones of extinct hominins. Fourteen ERVs have been identified in the ancient DNA of Denisovan

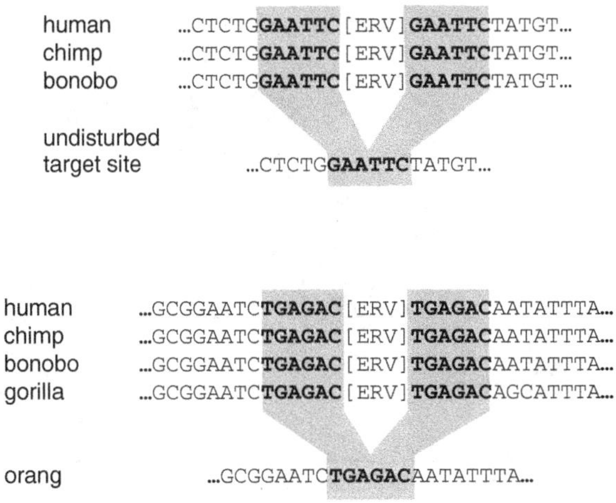

human ...CTCTG**GAATTC** [ERV] **GAATTC**TATGT...
chimp ...CTCTG**GAATTC** [ERV] **GAATTC**TATGT...
bonobo ...CTCTG**GAATTC** [ERV] **GAATTC**TATGT...

undisturbed target site ...CTCTG**GAATTC**TATGT...

human ...GCGGAATC**TGAGAC** [ERV] **TGAGAC**AATATTTA...
chimp ...GCGGAATC**TGAGAC** [ERV] **TGAGAC**AATATTTA...
bonobo ...GCGGAATC**TGAGAC** [ERV] **TGAGAC**AATATTTA...
gorilla ...GCGGAATC**TGAGAC** [ERV] **TGAGAC**AGCATTTA...

orang ...GCGGAATC**TGAGAC**AATATTTA...

FIGURE 1.9. ERVS COMMON TO HUMANS AND CHIMPS (ERV-K105; *UPPER DIAGRAM*) AND TO THE AFRICAN GREAT APES (ERV-K18/K110; *LOWER DIAGRAM*) [18, 20].

and Neanderthal individuals, but they are absent from our genome. Indeed one of these ERVs is shared by these archaic humans (albeit recovered in fragmented form), indicating that Denisovan and Neanderthal populations share a common ancestor that lived after their lineage branched out from ours (Figure 1.8, lower panel) [23].

In contrast, the unique ERV-K105 provirus is present in the human genome, and in those of the two chimp species (Figure 1.9, upper diagram). We must conclude that these species are monophyletic. Neither the ERV nor an undisturbed target site could be found in the genome of the gorilla, which may have undergone a large genetic deletion spanning the site. The time of insertion remains undefined in the case of this ERV.

On the other hand, ERV-K18/K110 (one of those introduced above, see Figure 1.6) is inserted neatly in the genomes of each of the four African great apes (Figure 1.9, lower diagram). As noted, this particular ERV entered the primate germ-line in an ancestor of the African great apes. The orang-utan, the Asian great ape, retains the undisturbed target site [18, 20]. Am I labouring the point?

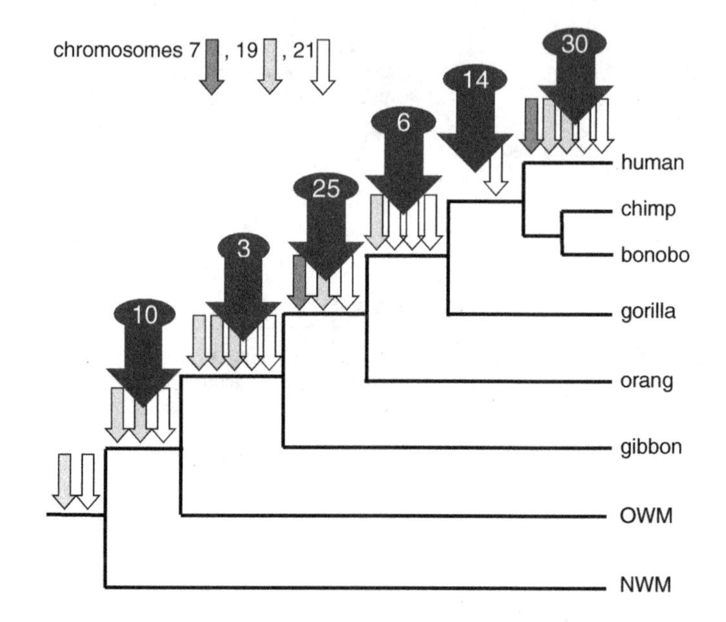

FIGURE 1.10. THE TIMES AT WHICH ERV-K INSERTS ENTERED THE PRIMATE GERM-LINE, BASED ON THE SPECIES DISTRIBUTION OF INDIVIDUAL ERVS
A definitive catalogue of full-length ERV-K (HML-2) inserts in the human genome shows the number (*ovals*) arising at each branch leading to humans [24]. Data for solo LTRs are from chromosome 7 (*dark arrows*), 19 (*light grey arrows*) and 21 (*white arrows*) [25].

Perhaps – but here is an elegant unambiguous demonstration of our evolutionary descent that arises simply from the established and unquestioned principles of medical genetics.

A definitive catalogue of ERV-K (HML-2) inserts that are full-length (or nearly so) has confirmed and extended the validity of the primate phylogenetic tree. The results of this analysis are depicted in Figure 1.10, in which the number of ERVs added to the genome between each bifurcation is indicated in an oval [24]. These studies provide an unambiguous scheme of the relationships of the OWMs and the apes. ERVs of this family have been accumulating in primate genomes on the lineage leading from OWMs to humans, establishing that the Old World primates are monophyletic, all species sharing a particular ERV being descended ultimately from the *single* reproductive cell in which that unique insertion event occurred.

Supporting data were collected independently by another research group, who studied ERV-K inserts on selected chromosomes (Figure 1.10, arrows). Evidence was provided that humans share some inserts also with NWMs [25]. The shape of the family tree revealed by these analyses is congruent with that developed over the years on the basis of a whole range of other criteria. But even if we had never heard of evolution and knew nothing of taxonomy, discovery of the relationships established by patterns of ERV insertions would have compelled us to propose an evolutionary theory of common descent, along the lines that taxonomists have laboured to develop over many years.

ERVs undergo characteristic rearrangements, some of which arise from their distinctive organisation. These rearrangements arise from interactions between long terminal repeats of the *same* provirus, or the exchange of genetic material between *different* ERVs. Each ERV carries a record of its history inscribed in its base sequence. These ERV- and genome-modifying events are outlined below.

A full-length ERV has two long terminal repeats, one at each end. When an ERV is first inserted into chromosomal DNA, its LTRs have the same sequence. If the chromosomal DNA loops back on itself, the two LTRs may align with each other, as depicted in Figure 1.11. When this happens, each of the two lengths of DNA involved may effectively break, and then rejoin with the partner segment present in the alignment. This process is called *homologous recombination*. The result of such an event is that the entire sequence between the breakpoints is looped out of the chromosome and lost, leaving one solitary chimaeric LTR.

Recombination within a *single* ERV occurs in contemporary individuals. An ERV on the Y chromosome contributes sequence content to a gene required for male fertility, the *TTY13* gene. When homologous recombination events occur between the two LTRs of this ERV, the internal content of the ERV, including the embedded

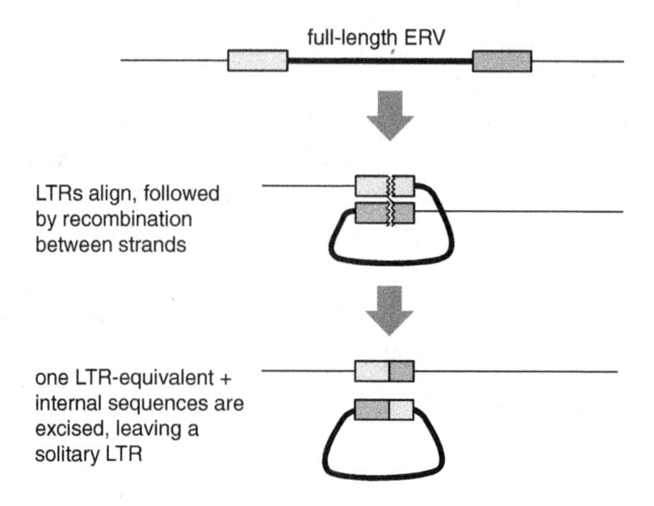

FIGURE 1.11. HOMOLOGOUS RECOMBINATION BETWEEN THE LTRS (*SHADED BOXES*) OF AN ERV

In the *middle diagram* the *jagged lines* indicate breaks in the DNA. The break may be resealed by joining part of one LTR (*light shading*) to part of the other LTR (*dark shading*). The outcome is a solitary LTR and an excised loop of ERV DNA.

portion of the *TTY13* gene, is looped out and lost. The result is inactivation of the *TTY13* gene and male infertility [26].

During evolutionary history, ERVs commonly end up as solitary LTRs. In some cases (such as ERV-K103 and ERV-K113), the human population is polymorphic for an insert: some of us have a complete provirus; others have only a solo LTR. Full-length ERV-K (HML-2) proviruses are outnumbered by solo LTRs by a factor of ten in the human genome [27]. A full-length ERV-H common to all hominoid primates is present as a solitary LTR only in humans [28].

Different proviruses (that is, ERVs found at different places in the genome as a result of independent insertion events) of the same type may also align. In this case, two outcomes may follow the exchange of genetic material. *Equal* homologous recombination generates full-length chimaeric proviruses (Figure 1.12, upper diagram). An extensive amount of genetic material is exchanged between the two interacting lengths of DNA, including flanking

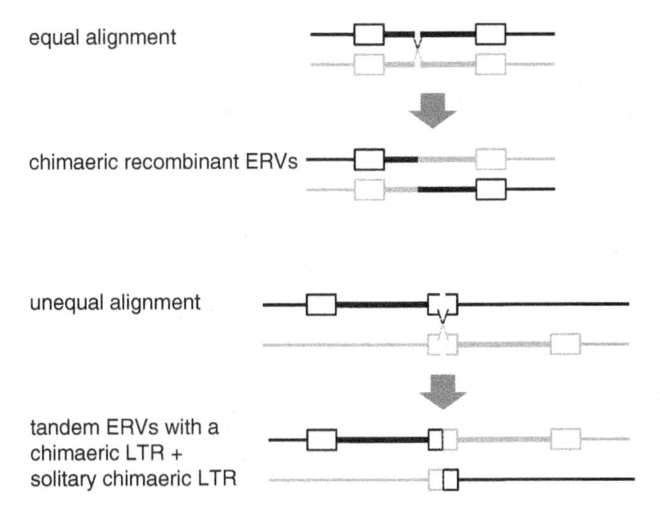

FIGURE 1.12. HOMOLOGOUS RECOMBINATION BETWEEN DIFFERENT ERVS OF THE SAME TYPE
Equal recombination is shown in *upper diagram*; *unequal* recombination in *lower diagram*.

chromosomal DNA that extends for an indeterminate distance beyond the ERV. In the absence of a compensating recombination event, the result would be a chromosome translocation. *Unequal* homologous recombination, say between the downstream (right-hand or 3′) LTR of one ERV and the upstream (left-hand or 5′) LTR of another, leads to very distinctive products. One is a tandemly duplicated, three-LTR proviral structure. The other is a solitary LTR (Figure 1.12, lower diagram).

These processes can also be shown both on the brief timescales of people's lives and on the colossal timescales over which species arise and diversify. Recombination between *different ERVs* on the Y chromosome, in contemporary individuals, results in the deletion of large expanses of intervening genetic material, and of the genes they contain. These events lead to loss of the ability to produce sperm [29]. During evolutionary history, such recombination events have generated extensive exchanges of chromosomal material between distinct loci, with concomitant reorganisation of the genome. One such event has, for example, generated the human-specific tandemly duplicated provirus ERV-K108 [30].

ERVs have been involved in other types of mutational rearrangements. These have generated weird and wonderful derivatives. Such mutational events would be expected to arise as essentially unique happenings, and therefore the presence of such ERV derivatives in multiple species would be a further stratum of evidence that those species are descended from the individual in which the novelty arose. For example, an ERV-H and an ERV-E have been joined together (as the result of a large deletion of genetic material) to form a chimaeric ERV. The deletion extends from the *pol* gene of the ERV-H to just downstream of the left-hand LTR of the ERV-E. This chimaeric ERV is found in humans, chimpanzees and gorillas, and the ERV-H/ERV-E junction point is the same in each species (Figure 1.13). We conclude that humans, chimps and gorillas have inherited that singular ERV from the common ancestor in which the gene deletion event occurred. Multiple copies of this chimaera are also present in each species, indicating that the unique ERV-H/E has been 'copied and pasted' during subsequent history [31].

Another oddity present in our genome is the case in which an ERV-K provirus has undergone a genetic recombination with a cellular gene called *FAM8A1*. The result is a hybrid in which the ERV contains a large fragment of the *FAM8A1* gene in place of a portion of the retroviral gene sequence. As with the ERV-H/E hybrid described above, the chimaeric ERV-K/*FAM8A1* unit has been copied subsequently into a small family of ERVs. Humans share copies of this singular ERV-K/*FAM8A1* chimaera with primates as distantly related as OWMs. The structure could not be found in NWMs, however, indicating that it arose in an ape–OWM ancestor [32].

These examples provide compelling evidence of common descent. But one must ask whether they are representative of the 440,000 LTR elements scattered throughout our genomes. Do anecdotal accounts, no matter how impressive, really tell the whole story? The ultimately rigorous test of the assertion that ERVs establish the truth

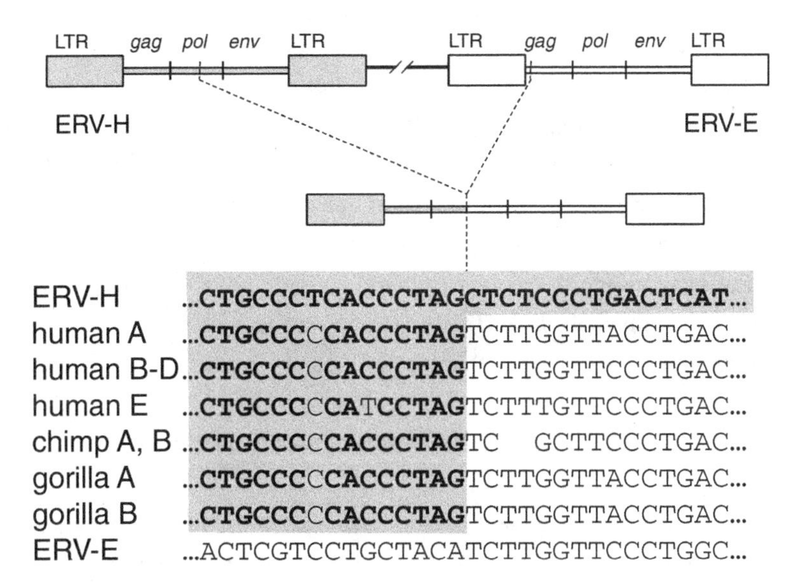

FIGURE 1.13. FORMATION OF A CHIMAERIC ERV BY A DELETION
The junction point is identical in humans, chimps and gorillas in each of several copies [31]. ERV-H sequences (*shaded*); ERV-E sequences (*unshaded*).

of human evolution from remote primate progenitors requires the sequencing of *entire* genomes of multiple species, and a side-by-side comparison of *all* the ERVs residing in them. This would allow every one of the 440,000 ERV and other LTR elements in the human genome to be checked against the equivalent sites of the genomes of other primates.

At the turn of the century, whole-genome comparisons sounded like science fiction. But technological developments have been explosive. The first draft of the human genome sequence was published in 2001 – ahead of schedule and under budget [33]. Analysis of draft sequences of the chimp and bonobo (or pygmy chimp) genomes followed in 2005 and 2012, respectively [34, 35]. Early returns on the gorilla genome [36], and sequence analysis of the orang-utan genome [37] came in 2011, in quick succession. A first draft of the rhesus macaque (an OWM) genome came out in 2007 [38]. And, as already mentioned, sequences of two related archaic extinct humans – the Denisovan [39] and Neanderthal [40] hominins – have been added recently. Many more primate genome sequences are in the pipeline.

If one species had an individualistic collection of ERVs that bore no relation to the ERVs in supposedly related species, then the phylogenetic scheme would crash in a heap. This comparative genomic approach to delineating phylogenetic relationships is inherently very susceptible to falsification – an important criterion for pursuing real science. So what can be said of whole-genome comparisons of ERV content?

I have mentioned that there are four major classes of ERV and ERV-like inserts in primate DNA. In the case of three of them (types I, III and IV), it seems that essentially *all* inserts present in the human genome are shared by chimps and bonobos (Table 1.3). These types of retrovirus had stopped accumulating in the primate germ-line before the human and chimp lineages diverged. Only in the case of the ERV-K family are there human-specific members, and these are approximately 1% of the whole ERV-K complement [35]. We can be confident that even for the ERV-K population of proviruses, the huge majority were inserted into the primate germ-line in individuals that were ancestors of humans and the two chimpanzee species. We can conclude on the basis of over 400,000 inserted markers of monoclonality that humans, chimps and bonobos are descended from common ancestors. Most of this lineage is shared also with gorillas and orang-utans. Full analysis of the orang-utan genome is not yet available. It seems that orang-utans have acquired some additional members of the ERV-E sub-family, but otherwise have inherited the same basic ERV complement that is possessed by humans, chimps and gorillas [37].

Even with the much more distantly related rhesus macaque (an OWM), initial surveys found a high degree of sharing of the ERV population. The one detailed human–macaque comparison currently available involved a selection of those ERVs that have retained both LTRs in both species. This analysis showed that, depending on the

Table 1.3. *ERVs and other LTR elements in the human genome*

		Proportion of ERVs (%) in					
ERV class	Total number in human genome	hum only	bon only	chimp only	bon and chimp only	hum, bon and chimp	hum and mac*
I	105,000	<0.1	<0.1	<0.1	<0.1	>99.9	19
II, ERV-K	4,400	1.3	0.3	<0.1	0.3	99.0	25
III, ERV-L	108,000	<0.1	<0.1	<0.1	<0.1	>99.9	44
IV, MaLR	246,000	<0.1	<0.1	<0.1	<0.1	>99.9	65

Notes: hum, human; bon, bonobo; mac, macaque. Data are largely from the bonobo genome analysis [35]; *full-length LTR elements in the human and macaque genomes [41].

category of ERV, between 19% and 65% of full-length ERVs and LTR elements are shared by these species (Table 1.3). Overall, of 3,781 such well-preserved inserts in the human genome, 1,369 (36%) are present also in the macaque – the same type of ERV at exactly the same location in the respective genomes. This is a colossal weight of evidence for common ancestry of humans and OWMs [41]. One might also suppose that less well-preserved inserts would tend to be older, and that a higher proportion of them would therefore be shared by humans and OWMs.

1.4 NATURAL SELECTION AT WORK: GENES FROM JUNK

Common descent is the defining feature of an evolutionary biology. We have seen how the presence of particular ERVs that become the inherited possession of multiple species provides powerful evidence for this concept. A second tenet that is integral to the Darwinian mode of evolutionary thought is natural selection: the postulate that randomly arising genetic variants may provide new and advantageous functions to the organisms that possess them. Such organisms are enabled to reproduce more efficiently than those that lack the variant, with the result that those variants selectively increase in frequency in the gene pool.

ERVs provide unambiguous examples of the rags-to-riches co-option of randomly accrued DNA sequences. When ERVs first become established in the DNA of their host organism, they typically offer no advantage to that host, and therefore natural selection does not act to preserve them. As a result, ERVs start to accumulate mutations, and degenerate into what appear to be molecular fossils. However, a small proportion of ERVs retain genes with open reading frames, able to specify the production of proteins. Many ERVs also retain genetic regulatory functionality within their long terminal repeats.

1.4.1 ERVs and the placenta

Complete *envelope* (*env*) gene sequences that retain the potential to encode functional envelope proteins have been described in 18 ERVs that reside in the human genome. Most of these ERVs are relative newcomers to the genome. Perhaps their *env* genes have not had time to decay. But some of these ERVs have resided in the primate genome for many millions of years [42]. For example, the unique ERV3 provirus was added to the primate genome in an ancestor of all Old World primates (that is, of apes and OWMs) [43]. Each of the ERV-Pb1 and ERV-V2 proviruses took up residence in the genome of an ancestor of all simian primates (that is, of apes and monkeys) [44]. These viral *envelope* genes have retained coding capacity against all the odds. One can only assume that such intact *envelope* genes were retained under the influence of selective pressure because they have provided useful functions for their host animals.

A particular endogenous retrovirus of the W class (designated the ERV-WE1 provirus) also retains an intact *envelope* gene. ERV-WE1 was spliced into the primate germ-line in an ancestor of the apes and OWMs. The insertion site is shown in Figure 1.14. NWMs, prosimians and a non-primate (the dog) retain the undisturbed, target site (CAAC or similar) [45]. Remarkably, the ERV-WE1 *envelope*

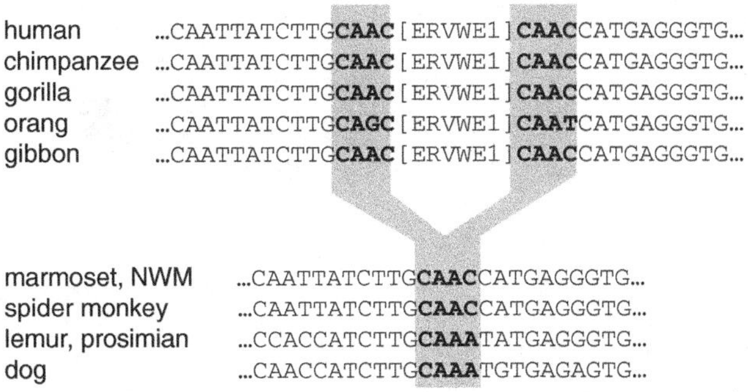

human ...CAATTATCTTG**CAAC** [ERVWE1] **CAAC**CATGAGGGTG...
chimpanzee ...CAATTATCTTG**CAAC** [ERVWE1] **CAAC**CATGAGGGTG...
gorilla ...CAATTATCTTG**CAAC** [ERVWE1] **CAAC**CATGAGGGTG...
orang ...CAATTATCTTG**CAGC** [ERVWE1] **CAAT**CATGAGGGTG...
gibbon ...CAATTATCTTG**CAAC** [ERVWE1] **CAAC**CATGAGGGTG...

marmoset, NWM ...CAATTATCTTG**CAAC**CATGAGGGTG...
spider monkey ...CAATTATCTTG**CAAC**CATGAGGGTG...
lemur, prosimian ...CCACCATCTTG**CAAA**TATGAGGGTG...
dog ...CAACCATCTTG**CAAA**TGTGAGAGTG...

FIGURE 1.14. THE INSERTION SITE OF ERV-WE1 IN PRIMATE GENOMES [45]
Sequence data are not available for OWMs.

gene is active. It is transcribed into messenger RNA in placental tissue. Furthermore, the messenger RNA transcript is translated into a protein [46].

The activity of the ERV-WE1 *envelope* gene has come under the control of an interacting network of signalling proteins. It is regulated by hormones such as corticotropin-releasing hormone, which acts via the intracellular second messenger cyclic AMP [47]. It is regulated also by signals of the so-called *Wingless* pathway, as well as by the PPARγ and RXRα nuclear hormone receptors, and by a suite of regulatory motifs in the local DNA environment. Expression of the retroviral envelope protein has been thoroughly assimilated into the morphogenic (form-generating) control circuitry that operates during embryonic and fetal development [48].

The ERV-WE1 envelope protein is expressed in the outermost layer of the placenta. This tissue acts as a boundary between the blood supplies of the fetus and the mother. It is known as the syncytiotrophoblast because it is a syncytium: an extended structure formed when the membranes of multiple cells fuse to produce a single mega-cell containing large numbers of nuclei. Both the envelope protein and the cellular receptor to which it binds are expressed also on the cytotrophoblast cells that fuse to form the syncytium [49]. It seems that the envelope protein (which enables retrovirus particles

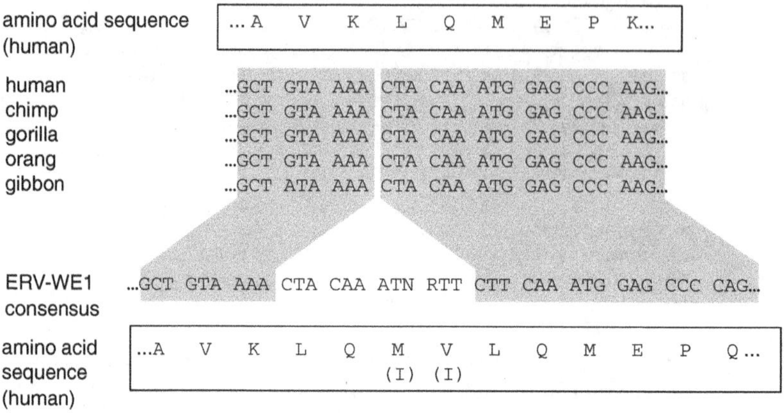

amino acid sequence (human)	...A V K L Q M E P K...
human	...GCT GTA AAA CTA CAA ATG GAG CCC AAG...
chimp	...GCT GTA AAA CTA CAA ATG GAG CCC AAG...
gorilla	...GCT GTA AAA CTA CAA ATG GAG CCC AAG...
orang	...GCT GTA AAA CTA CAA ATG GAG CCC AAG...
gibbon	...GCT ATA AAA CTA CAA ATG GAG CCC AAG...
ERV-WE1 consensus	...GCT GTA AAA CTA CAA ATN RTT CTT CAA ATG GAG CCC CAG...
amino acid sequence (human)	...A V K L Q M (I) V (I) L Q M E P Q...

FIGURE 1.15. A DELETION OF 12 BASES (FOUR AMINO ACIDS IN THE ENCODED PROTEIN) THAT INCREASES SYNCYTIN-1 FUSOGENIC ACTIVITY
The *upper* amino acid sequence (*boxed*; given in one-letter code) is that of the syncytin-1 protein; the *lower* sequence is that of the ERV-W virus envelope protein [51].

to stick to cells) became domesticated following endogenisation to enable cells to stick to each other. Such adherence is a first step in the cell fusion events that generate syncytia. For this reason, the envelope protein has been renamed syncytin-1. To test this hypothesis, the *syncytin-1* gene has been introduced experimentally into cells growing in culture, and expression of the protein promptly generated extensive fusion between cells [50].

The *syncytin-1* gene has sustained various changes during its long residence time in primate genomes. In OWMs, it was inactivated by multiple damaging mutations, and cannot make a functional protein. But in apes, much of the gene has resisted amino-acid-changing mutations, indicating that it has been subject to purifying selection. In other words, gene sequences have been conserved – evidence of functionality. Moreover, the deletion of a short sequence of 12 bases at the 3′ (right-hand) end of the gene results in the loss of four amino acids from the encoded protein (Figure 1.15). This deletion has enhanced the fusogenic activity of syncytin-1 relative to that of the progenitor ERV-W envelope protein – another sign of natural selection. The unique 12-base deletion has been demonstrated in six hominoid species and eight OWM species, and is an independent striking marker of common ancestry [51].

Cell fusion events in the human body are not limited to placental cytotrophoblasts. Syncytia form also in bone and muscle. Bone is a dynamic structure that owes its hardness to a form of calcium phosphate known as hydroxyapatite. Some cells (osteoblasts) increase the mineral content of bone, and other cells (osteoclasts) dissolve the mineral during the remodelling of bone. Osteoclasts are multi-nucleated cells, formed by the fusion of white blood cells known as monocytes. Syncytin-1 and its receptor are expressed as osteoclasts are formed, and these proteins are actively involved in the cell fusion process. Finally, cells called myoblasts fuse to form muscle fibres, and syncytin-1 has been implicated also in this process [52]. All this represents an impressive reformation of manners by an erstwhile pathogenic protein.

ERV-FRD is another ancient provirus. It is present in the genomes of all apes and monkeys, and so has been inherited from a simian ancestor. Its *envelope* gene is also intact and expressed in placental tissue. The ERV-FRD envelope protein (renamed syncytin-2) is expressed in cytotrophoblast tissue, and its receptor in the syncytiotrophoblast layer [53]. Syncytin-2 induces cell fusion, and so may contribute to the formation and maintenance of the syncytiotrophoblast. But syncytin-2 may perform an additional role. Retroviral envelope proteins possess a domain that acts to suppress the immune system. Syncytin-2 may retain this function, and so may contribute to the remarkable phenomenon by which the mother does not reject her immunologically distinct fetus [54]. And on the theme of immune suppression, the envelope proteins from ERV-WE1 and from one or more ERV-K proviruses (also present in cytotrophoblastic cells) may also suppress immunity [55].

It is not easy to acquire experimental support for the hypothesis that retroviral envelope proteins have been recruited (or *exapted*) to perform essential roles in human development. One cannot perform experimental manipulations on syncytin activity in human pregnancies. But, remarkably, species belonging to other mammalian orders have also acquired ERV *envelope* genes that function in placental development. The retroviruses involved are different from those that are found in primates. The ERVs providing the envelope/syncytin proteins are found at different sites of the respective genomes. The ERV genes have been co-opted independently. Other mammals that have acquired their own retrovirally encoded syncytins include mice, guinea pigs (with relatives such as the capybara, chinchilla and Brazilian porcupine), rabbits and also carnivores [56].

ERVs peculiar to sheep and to cattle also encode envelope proteins that are expressed in the placenta. In these cases, however, the exapted proteins appear to function differently from the syncytins [57].

Such non-human organisms are amenable to experimental manipulation. Mice have been generated in which either the *syncytin-A* or the *syncytin-B* gene has been deleted by homologous recombination. Fetuses lacking syncytin-A are severely abnormal. The syncytiotrophoblast layer does not form properly, vascularisation is abnormal, fetal growth is suppressed, and the pups do not develop to term. The absence of syncytin-B has less severe consequences, but syncytialisation is still abnormal. Deletion of both genes results in death of the pups [58]. It may be concluded that a gene that was once part of the infective apparatus of a potentially pathogenic retrovirus (unquestionably junk) is now an essential mediator of placental development (unquestionably part of the riches of a species' genetic endowment).

Sheep demonstrate the interesting situation in which the Jaagsiekte sheep retrovirus exists in both exogenous (infectious) and endogenous (inherited) forms. The *exogenous* forms are frankly pathogenic (oncogenic) viruses of veterinary and economic importance. However, at least one of the *endogenous* versions produces envelope protein at early stages of embryonic development. Scientists have experimentally suppressed the production of this envelope protein in pregnant ewes. This manipulation causes placental abnormalities, leading to abortion of the lamb [59]. The Jaagsiekte retrovirus, currently circulating between infectious and endogenous pools, has given rise to at least one ERV that is the source of a domesticated *envelope* gene now providing an essential function to the sheep. The recruitment of a viral gene into an essential developmental gene is a genetic rags-to-riches story. It exemplifies the efficacy of natural selection [59].

But in fact there are experiments in humans – the tragic experiments of nature – that provide associations between abnormal syncytin expression and placental malfunction. About 6% of human pregnancies are associated with conditions called *pre-eclampsia* (characterised by placental malfunction and high blood pressure) and *intrauterine growth restriction* (failure of the fetus to thrive). These problems arise from abnormalities in placental, particularly syncytiotrophoblast, development. Such placental dysfunction has been correlated with decreased expression of syncytin-1 and syncytin-2, and of other retrovirally derived envelope proteins. The less syncytin-1 and -2 are present, the more severe is the disease [60]. Fetuses with the trisomy 21 (Down's) syndrome also show abnormalities in placental function, aberrant cytotrophoblast fusion and altered patterns of expression of syncytin-2 [61].

It may be concluded that, in at least two cases, ERVs resident in the genomes of simian primates, including humans, have contributed their *envelope* genes to the functioning of their host organisms. Natural selection has retained and tweaked their function, particularly in placental tissue, and they are now essential for our development.

The placenta varies greatly in structure between different groups of mammals. The exaptation of viral genes to form syncytin-1 and -2 indicates that placental development in the simian primates has biochemical features that are dissimilar to those operating in other placental mammals. But the placenta has a history that is much longer than that of the simian primates. The origins of the placenta must have involved some mechanisms that are common to all placental mammals. This early history features ERV-like class IV elements (the LTR retrotransposons).

A small family of such elements within genomes as diverse as those of humans, mice and sheep has been identified. They are exceedingly ancient; long terminal repeats and target-site duplications long ago decayed beyond recognition. These LTR elements (of the Ty3/Gypsy class) lack *env* genes, and it is the *gag* and *pol* genes that have resisted decay through the aeons of their sojourn in mammalian genomes. The persistence of the *gag* and *pol* genes evinces acquired functionality [62].

One such gene, now named *PEG10*, is expressed in various tissues. The mammalian *PEG10* gene even retains a behavioural quirk that is testimony to its LTR retrotransposon provenance. During synthesis of the protein, the *PEG10* messenger RNA requires an adjustment of the ribosome to maintain the appropriate reading frame – just as it does in the case of the original LTR element [63]. *PEG10* function appears to be essential for the development of the placenta. When the gene is inactivated in mice, the placenta fails to develop normally, and the embryo dies. Decreased *PEG10* expression correlates with low birth weight in humans [64]. The gene is not detectable in the platypus, but is present in marsupials and placental mammals. This distribution indicates that the original LTR element inserted into mammalian DNA in an ancestor of marsupial and fully placental (eutherian) mammals [65].

A second retrotransposon-associated gene, *PEG11* (also known as *RTL1*), is expressed in the placenta. Knock-out experiments in mice and clinical syndromes in human beings indicate that abnormal expression of this gene has severe consequences. Over-expression leads to placental hypertrophy, under-expression results in placental hypoplasia, and knock-out is lethal [66]. The gene is present only in placental mammals. Degenerated remnants of the original LTR element are detectable in marsupial DNA, but not in platypus DNA. It appears that the LTR element from which the *PEG11* gene was derived inserted itself into the DNA of a eutherian–marsupial ancestor, but did not retain protein-coding function in the marsupial lineage and, in the absence of any selective pressure, decayed into a barely recognisable genetic relic [67].

The means by which complex organs have evolved are obscure. The placenta is an interesting test case, as it arose only in mammals. The placenta is a rudimentary structure in marsupials, and variable in different groups of eutherians. Thousands of genes are involved in the development of the placenta. In the early stages, these genes are typically ancient (in that they are present widely in eukaryotic organisms) and have roles in basic metabolic processes and growth. Altered regulation of pre-existing genes seems to have been a major factor in early placental development. But during the more advanced stages of placental function, novel genes generated by duplication of existing genes, with subsequent divergence of function, seem to be involved [68]. In addition to these, the exaptation of ERV-derived proteins has had a vital role in the evolution of placental form and function [69].

1.4.2 ERVs that contribute to gene content

ERVs and other LTR elements have contributed to the protein-coding content of genes in ways additional to those discussed above. In the human genome, 50 protein-coding exons are derived from LTR elements. One of these exons is in the *IL22RA2* gene. The product of this gene acts as an inhibitor of a signalling molecule (or cytokine) involved in the control of inflammation. The piece of DNA representing the LTR agent was added to the primate genome in an ape–OWM ancestor, and subsequently underwent an AT → GT mutation (generating a functional donor splice site needed for exonisation) in a great ape ancestor. The functional consequences of exonisation are not known, but the altered protein is made in – surprise! – the placenta [70].

One more example will suffice. Periphilin is an insoluble protein that contributes to the structure of skin. It is produced in keratinocytes as they differentiate, and it contributes to the way they ultimately harden to form the tough (*cornified*) layer that makes us waterproof. An ERV-M resides in the right-hand (3′) end of the gene. This element was inserted into the periphilin gene in an ancestor of the simian primates. It acquired the ability to contribute to gene content in an ancestor of the great apes [71]. It is clear that ERVs have helped to make us what we are. Our retroviral heritage has had a tangible impact on our human shape and physiology.

1.5 NATURAL SELECTION AT WORK: REGULATORY NETWORKS

The apparent complexity of organisms is poorly correlated with the *number* of protein-coding genes they possess. We humans have fewer protein-coding genes than the water flea. What seems to be more significant for complexity is the degree of sophistication of the mechanisms by which those genes are *regulated*. ERVs and other LTR elements possess their own regulatory elements, located in the long terminal repeats. These motifs include promoters (which recruit the enzymes required to transcribe genes), enhancers (which bind regulatory proteins required to modulate the rate of transcription) and polyA sites (which terminate transcription). Just as some *envelope* and *gag* genes have, against all odds, retained coding capacities for proteins that have become essential for our viability, so many regulatory elements in LTRs have resisted degeneration and have been integrated into cellular regulatory networks.

The ERV-9 clan are long-term residents of primate genomes. Essentially all are shared by humans and chimps. Several instances of these, common to all the great apes including humans, possess LTRs that exert regulatory functions on genes that may be more than 40,000 bases downstream [72]. An ERV-L (inherited by all Old World primates) provides a promoter for a gene (*β3GAL-T5*) involved in the synthesis of carbohydrate chains in the colon [73]. A variety of ERVs and LTR elements have also been recruited to control genes that function in the development of the placenta. Table 1.4 shows a

Table 1.4. *Genes expressed in the placenta and controlled by ERVs*

Gene	Protein	Function	ERV type	Age of ERV	Ref.
PTN	Pleiotrophin	Regulates growth and angiogenesis; fetal trophoblast and mesenchyme	ERV-E chimaera	No later than African great ape ancestor	74
INSL4	Insulin-like protein	Regulates early placental syncytiotrophoblast morphogenesis	ERV-K	Old World primate ancestor	75
EDNRB	Endothelin receptor	Constricts blood vessels, raises blood pressure	ERV-E	No later than Old World primate ancestor	76
IL2RB	IL-2 receptor B subunit	Activates T and NK cells (in immunity); fetal trophoblast	THE1D	No later than simian ancestor	77
PRL	Decidual prolactin	Role at maternal–fetal interface?	MER39	Primate–rodent ancestor	78
NOS3	Nitric oxide synthase 3	Dilates blood vessels	ERV-I, LTR10A	Not stated	79

selection of such genes, all of which specify the production of proteins that are involved in signalling between cells, and which are major players in the regulation of cell function.

The regulation of genes that control other genes is of special interest. The famous *TP53* gene is much beloved of cancer biologists because it is a vital tumour suppressor gene. It encodes the p53 protein, which regulates cellular energy usage, and orchestrates responses to stressful conditions. This protein makes the executive decision about whether an injured cell attempts repair, exits the cell division cycle or undergoes suicide. One of the ways by which the p53 protein enforces its decision is to occupy characteristic binding sites on DNA and to activate associated genes. In the human genome, some 1,500 LTRs provide potential p53-binding sites. Indeed more than one-third of all p53-binding sites are located within ERVs. It seems that intensive ERV bombardment of the primate genome, in an era when OWMs were diverging from NWMs, has reconfigured the p53 regulatory network [80].

A relative of p53 is p63. A unique form of p63 is produced in the cells that give rise to sperm. p63 gene transcription is itself initiated in an ERV9 insert, present only in the great apes. The encoded protein acts to induce suicide in progenitors of sperm cells that have sustained genetic damage. Thus a once-disruptive ERV is involved in quality control in the male germ-line [81].

A bioinformatic analysis of the human genome has identified 50,000 ERV promoters at which transcription is initiated. At least 100 of these drive the transcription of nearby genes [82]. The effects of such ERV promoters on gene expression are typically mild, providing subtle influences on gene expression. In some cases, however, ERVs may strongly influence the tissue(s) in which an associated gene is expressed. In these cases, ERVs often activate placenta-specific gene expression [83].

1.6 ARE THERE ALTERNATIVE INTERPRETATIONS OF THE DATA?

The presence of ERVs has established beyond doubt that humans have evolved from ancestors shared with chimps and bonobos. More remotely in time, we share ancestors with gorillas, then (progressively further back) with orang-utans, gibbons, OWMs and NWMs. These are astonishing results that could not have been imagined before the genomics revolution of the last few years. But (at least from the perspective of a cell biologist) the genetics revolution has closed the file on the question of our phylogenetic roots, just as it has closed the file on many criminal investigations that would have been unsolvable a generation ago. Darwin knew no genetics, but it is genetics that have clinched his case. The casanova phenomenon provides an evidential basis for primate evolution of unprecedented detail.

Are there alternative explanations of the data that might undermine these conclusions? In discussion with other people, I have been able to conceive of only three counter-claims through which these findings could be questioned. It might be suggested, firstly, that ERVs insert randomly, but *fortuitously* produce consistent phylogenetic trees. Secondly, that ERVs do not insert randomly, but *necessarily* enter DNA at specified sites that are correlated with other indices of genetic similarity. Thirdly, that ERVs are not retroviral, but are functional sequence motifs inherent to the genomes in which they occur. We will consider these in order.

Firstly, is it plausible to suggest that the retroviral insertion pattern is wholly random but *just happens* to coincide with a pattern indicating that certain species are descended from a common ancestor? A chance pattern of insertions into the genomes of each of a large panel of pre-existing species could never generate the *consistent* phylogenetic tree that is apparent from ERV distributions, and that is also generated on the basis of many other features. Full genome analyses allow the full complement of 440,000 ERVs and LTR elements to be used as phylogenetic markers. And this large number of inserts behaves consistently. More than 99% of these inserts are common to all of the five species of great apes. Occasionally, deletions remove large segments of DNA including the ERVs contained therein, but such deletions can be positively identified. Apart from these occasional events, we may conclude that if we share a particular ERV with NWMs, we will also share it with OWMs. If we share an ERV with OWMs, we will also share it with gibbons and the other great apes. And that is what is observed.

Nevertheless, there are rare exceptions that are discordant with the standard phylogenetic tree (as depicted, for example, in Figures 1.6 and 1.10). The ERV-K-GC1 insert is present in chimps and gorillas but absent in humans (who show the undisturbed target

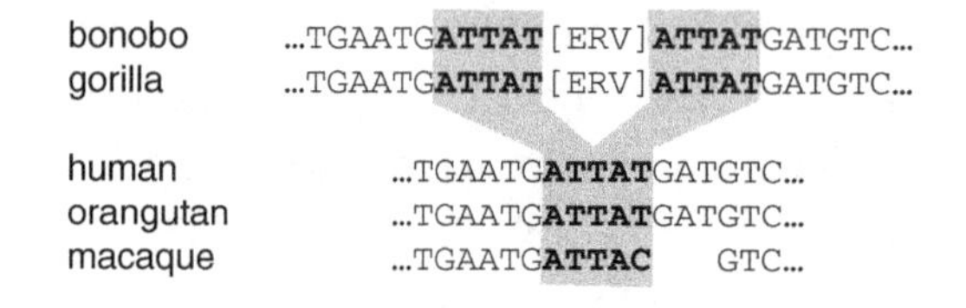

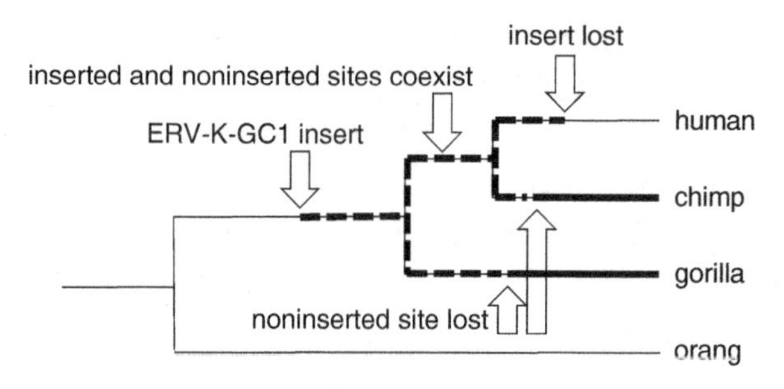

FIGURE 1.16. AN ERV COMMON TO BONOBOS AND GORILLAS (ERV-K-GC1) [84]
Shown are the insertion site (*upper diagram*); and the explanation for the phenomenon (*lower diagram*). The insert is absent (*thin lines*), co-exists with the original undisturbed site (*thick dashes*) or is fixed (*thick lines*).

site along with orang-utans; Figure 1.16, upper diagram) [84]. Such a pattern is not expected, given that humans and chimps share a more recent common ancestor than does either species with gorillas.

The basis for this apparent anomaly is integral to the theory. Every retroviral insertion event generates a unique provirus. If this occurs in the germ-line, the new provirus may be passed on to future generations. Over many generations, copies of the chromosome bearing this particular insert may increase in frequency (usually by random drift) until they totally displace the original (ERV-lacking) chromosome from the population or species. At that point, the ERV is said to be *fixed*, and any species arising from this population will inherit that ERV. This is what is normally observed, and accounts for the unambiguous pattern described hitherto.

But if *speciation occurs rapidly* relative to the time required by an ERV to become fixed, then a parental species may diverge into two (or more) new species at a time when copies of the ERV-containing chromosome constitute only a fraction of the total number of copies of that chromosome. If speciation occurs when an ERV is unfixed (such that the chromosomes with and without the insertion co-exist), then the ERV can be randomly lost or fixed in each diverging lineage. This is known as *incomplete lineage sorting,* and may produce anomalous trees.

The finding with ERV-K-GC1 indicates that this particular insertion event occurred near the time when the human, chimp and gorilla lineages were branching from the ancestral population (Figure 1.16, lower diagram). In this situation, both ERV-integrated and pre-integrated alleles were present as the ancestral population diverged. The integrated allele was lost from the human lineage, but independently fixed in the chimp and gorilla lineages. These data suggest that the gorilla, chimp and human lineages diverged closely in time. This conclusion is confirmed by incomplete lineage sorting of other genetic markers in the African great ape genomes (see later). And the availability of the gorilla genome sequence in 2012 established the reality of incomplete lineage sorting in the African great apes on a genome-wide basis [85]. Thus the ERV-K-GC1 insert breaks the expected pattern in a way that provides further insights to our evolutionary history. Incomplete lineage sorting is not seen at most branching points of our primate history, indicating that the gorilla–chimp–human branching point was an unusually close near-trifurcation.

We now address the second issue: could retroviruses specifically target particular sites of the genome, so that they insert non-randomly, leading to an insertion pattern that merely parallels the closeness with which genetic sequences are related to each other? This suggestion flies in the face of a vast amount of clinical and experimental experience, which stresses that infectious retroviruses are *insertional mutagens.* This genome-disrupting effect cannot be demonstrated with human ERVs, which no longer produce active retrovirus; but in mice, ERVs (including LTR elements) are highly active, and their insertional events are responsible for a staggering 10% of mutations in the germ-line. They also frequently cause mutations in somatic cells, which can result in cancers (particularly leukaemias) [86]. ERVs that actively generate progeny are ruthless mutagens.

Indeed, the random nature of the retroviral insertion event has been exploited experimentally to mutagenise genomes with the purpose of discovering genes that, when damaged, act to drive cancer development [87]. These strategies of gene discovery work only because the retrovirus inserts into chromosomal DNA and alters gene expression *indiscriminately* throughout the genome. Biotechnologists widely use retroviral insertional mutagenesis because this technique provides an unbiased way of targeting unknown genes, of interfering with their activity, of revealing the fact of their existence and of disclosing their function.

The genome is a big place. We have mentioned how children with genetic disease have been treated with therapeutic retroviruses, and how some of them subsequently succumbed to leukaemia. Scientists concerned about the potential of therapeutic retroviruses to cause cancer have wondered whether incoming retroviral DNAs target particular regions of the genome. To investigate the distribution of insertion sites, they mapped the locations of 572 proviruses in blood cells from a group of treated children. They considered that a particular region of the genome was preferentially targeted by inserting retroviruses if any two inserts were found within 30,000 bases of each other; or if three inserts were found within a window of 50,000 bases; or if four inserts were found within 100,000 bases; or if five to nine inserts were found within 200,000 bases [88]. These great expanses of genome indicate that, in living organisms, *relatively* close neighbours can be very distant. Multiple independent insertions into the same sites in different cells are very unlikely.

We have commented on how HTLV-1 target sites lack sequence specificity (Table 1.2). Similarly, ERV-K target sites show a conspicuous lack of consistency. In several studies, no two cases were found to have identical target sites. Even the length of the target site varies, with most being six bases long, but a minority being five-base sequences (Table 1.5). The ERV-K endonuclease is as promiscuous as that of HTLV-1. Shared ERVs do not arise because of specific targeting of particular sequences.

Even though the genome is so vast, is it possible that insertion sites could become saturated, so that later insertions would be forced into a shrinking pool of available sites? The short answer is no: the more ERVs there are in the genome, the more sites there are for entry, because incoming ERVs happily insert into pre-existing ones. One example will suffice. An ERV-H has been shown to lie within the 5′ (left-hand) LTR of an ERV-K. The target site and its duplications (CTAAG) indicate that this was a standard retroviral endonuclease-mediated event, and the identical ERV-within-an-ERV sequence is present in humans, chimps and gorillas, indicating that the ERV-H insertion occurred in an ancestor of these three species [96].

We now come to the third question. Is it possible that ERVs merely *resemble* retroviral DNA but are really distinct entities, perhaps involved *ab initio* in some essential function (such as gene regulation)? We have mentioned those species (such as mice and sheep) in which pathogenic infectious retroviruses are generating endogenised derivatives in an ongoing process. And, conversely, active ERVs are releasing infectious retrovirus. One strain of pedigree mouse has no functional ERVs in its genome. But if antibody-mediated surveillance is suppressed, the swapping of genetic material between these silent ERVs generates active, infectious, lymphoma-inducing retrovirus [97].

Both ancient ERVs and proviruses generated by recent retroviral infection are flanked by telltale target-site duplications. These demonstrate that ancient ERVs and new proviruses were spliced into DNA by the same endonuclease-mediated mechanism. An ERV is not merely a static segment of DNA. Rather, inscribed in its sequence is a record of the biochemical mechanism by which it first became part of the genome. ERVs are highly reliable phylogenetic markers because they arose through retroviral *process*, a stereotypical sequence of genetic events. They display their history.

Table 1.5. *ERV-K target sites*

Identity of ERV-K	Target site sequence	Ref.
ERV-K101	ACCCAG	89
ERV-K109	ATATGC	89
recombinant ERV-K at c11q12	ATCATT	90
cell culture, rejuvenated virus	ATCTCT	91
ERV-K103	ATGGGG	89
ERV-K-GC1	ATTAT	92
ERV-K104	CAGAAC	89
cell culture, rejuvenated virus	CCTGCC	91
ERV-K115	CCTTT	93, 94
cell culture, rejuvenated virus	CTATG	93
ERV-K113	CTCTAT	93, 94
recombinant ERV-K at c3p25	CTTGGT	90
recombinant ERV-K at c3p25	GAAAGT	90
ERV-K105	GAATTC	89
Denisovan6/ Neanderthal1	GACCAG	95
cell culture, rejuvenated virus	GAGGAT	91
recombinant ERV-K at c19p13(b)	GAGGGG	90
recombinant ERV-K at c19p13(b)	GCTGTG	90
ERV-K106	GGCTGG	89
ERV-K102	GGGATG	89
cell culture, rejuvenated virus	GGTGGC	93
ERV-K108	GGTTTC	89
cell culture, rejuvenated virus	GTGCCT	93
cell culture, rejuvenated virus	TACAAC	91
Denisovan2	TACGCC	95
recombinant ERV-K at c19p13(a)	TCCCAG	90
ERV-K110	TGAGAC	89
recombinant ERV-K at c11q12	TGGATT	90
recombinant ERV-K at c19p13(a)	TGTAAT	90
Denisovan3	TTACCA	95

c, chromosome; p and q, short and long chromosome arms, respectively

The ancient ERVs that infected cells at the start of simian history and proviruses generated by recent infection have the same genetic organisation. 'Individual ERVs are present in a form that is indistinguishable from the proviruses that result from retroviral infection of somatic cells' [98]. One must allow, of course, that ERVs get tatty with age, as we all do. Old ERVs and new proviruses possess long terminal repeats (containing regulatory sequences) and the common set of structural genes (*gag*, *prt*, *pol* and *env*). The LTR elements have the same layout but lack internal genes such as the *env* gene. The consistent features of ERV sequences indicate that they are retroviral genomes and function as insertional mutagens *by nature*.

The oldest ERV-Ks and the most recently arising (human-only) ERV-Ks have the same structural organisation. Scientists have compared the sequences of some of the younger, better preserved ERV-Ks to infer the sequence of the original infective retrovirus. They have then used established molecular biological techniques to generate retroviral DNA clones in which all of the identified mutations in the structural genes were removed, so regenerating normal proteins. Revivifying potentially pathogenic viruses that have existed only in fossilised form for aeons is a brave experiment! The rejuvenated retroviruses were found to infect cells in culture, and freshly inserted proviruses formed six-base target-site duplications typical of those of ERV-K [99].

The *envelope* gene of the human-specific ERV-K113 provirus (Figure 1.8) has been experimentally regenerated by removing the mutations that have accumulated in it. The encoded envelope protein has been synthesised. It shows the features typical of a contemporary retroviral envelope protein, including sites required for proteolytic processing and glycosylation [100].

Retroviral gag proteins are required for the maturation of new virus particles. They are cleaved by viral proteases into smaller units (the matrix, capsid and nucleocapsid peptides). The original ERV-K *gag* gene has been recreated in the laboratory, and the gag protein

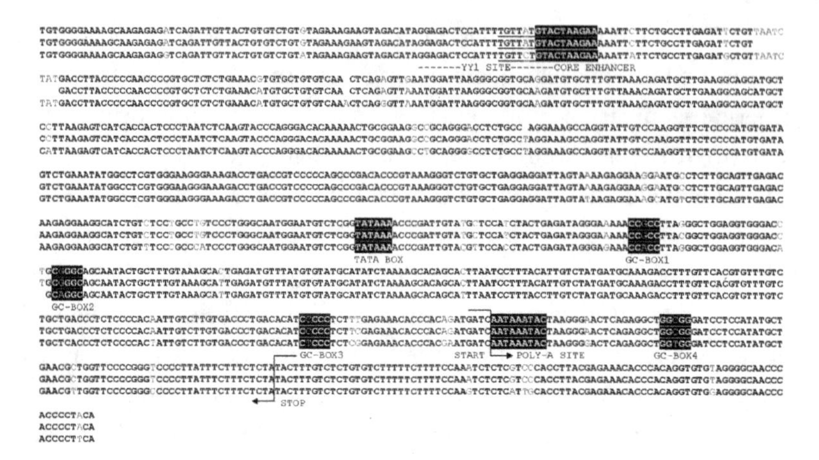

FIGURE 1.17. SEQUENCES OF THE 5′ LTR OF THREE ERVS [102]
The ERVs are (from top to bottom): ERV-Kcon (experimentally rejuvenated); ERV-K106 (human-specific); and ERV-K110 (human sequence, African great ape-specific). Variable bases are indicated by *grey lettering*. Regulatory sequences are *underlined* (glucocorticoid receptor-binding element) or shown against a *dark background* (core enhancer, TATA box, GC boxes and polyA signal). *Bent arrows* indicate transcription start and stop sites.

synthesised from it. Precisely the expected events have been replicated in cultured cells using the artificially regenerated ERV-K *gag* and *prt* (protease) genes [101].

The long terminal repeat sequences contain modules that regulate the transcriptional activity of the provirus. In the case of ERV-K retroviruses, these LTR sequences are about 970 bases long. Figure 1.17 shows the LTR sequences of ERV-Kcon, the experimentally rejuvenated virus that has infectious activity in laboratory experiments; of ERV-K106, a provirus found only in humans and a recent addition to our genome; and of ERV-K110, a provirus shared by all the African great apes (Figures 1.6 and 1.9). Their overall organisation and base sequences are strikingly similar. It is totally implausible to suggest that the human-only ERVs are retroviral but the shared ERVs are some eukaryotic regulatory sequence. If the human-specific ERV-K106 is an authentic retroviral insert, so is the

African great ape-specific ERV-K110. They invaded primate DNA in the same way. And ERV-K110 happened to do so in an African great ape ancestor.

We can conclude that ERVs are authentically retroviral, and retroviruses have always been insertional mutagens. No precursor sequence that might be construed as a 'normal' genetic regulatory element has been identified. ERVs are fossils of ancient retroviral invasions, randomly added to primate DNA, and magnificent markers of evolutionary history. Indeed, the discovery that portions of many ERVs now perform functional roles in our genome is testimony to the opportunistic capacity of natural selection to recruit potentially pathogenic DNA into performing life-sustaining roles. We reiterate that humans share ancestry with chimps, gorillas, orang-utans, lesser apes, OWMs and NWMs.

Retroviruses are specialists at inserting their genetic material into that of cells. Another class of RNA virus, the bornaviruses, locate their genetic material in the nuclei of infected cells, but lack the enzymatic machinery to insert it into the host's chromosomal DNA. However, bornavirus sequences may be inserted fortuitously into cellular DNA. Four such sequences have been found in human DNA – and all are shared with primates as distantly related as the NWMs, indicating that a series of insertion events occurred in the DNA of ancestors of simian primates [103].

More recently, segments of DNA representing a class of viruses known as parvoviruses have also been found to be widely dispersed in animal genomes. In particular, one segment of parvovirus DNA, present in the ninth intron of the *limbin* gene of humans, is also present in the genomes of all other tested primates, and of non-primate mammals as well (carnivores, horses, cattle and dolphins) [104]. The simplest interpretation is that this unit of parvoviral DNA was spliced into a genome that was ancestral to all these mammals. The unscheduled additions of bornavirus and parvovirus gene sequences to primate DNA provide independent confirmation of the power of insertional mutagens to establish phylogenetic relationships.

The same principle is at work in ongoing clinical studies. Human herpesvirus-6 is a virus that causes (generally) mild infections in children. It may insert its genome into the telomeric DNA of chromosomes in germ cells. From thence, the viral DNA is transmitted to future generations. In any one family, all members have the viral genome inserted at the same site. In other words, a cohort of people with the same viral DNA insert will be family members [105].

1.7 CONCLUSION: A DEFINITIVE RETROVIRAL GENEALOGY FOR SIMIAN PRIMATES

The monoclonal nature of leukaemias entails that all the cells in the tumour that share a particular provirus are descended from the one progenitor cell in which that provirus was spliced into the genome. The monophyletic nature of species entails that all the species that possess a particular provirus in their genomes are descended from the one species (indeed the one reproductive cell) in which that particular ERV was spliced into the genome. The one difference is that in a leukaemia, a provirus inserts into a somatic cell and *induces* that cell into an unrestrained programme of cell division. But during evolution, a provirus inserts into a germ cell, which has unlimited proliferative potential (over potentially myriad generations of organisms) as an *inherent* part of its nature.

There are no disagreements regarding the use of unique proviruses to demonstrate that tumours are monoclonal. No-one would suggest that the provirus integration pattern randomly gives the *illusion* that all the cells are descended from one progenitor. Nor would anyone suggest that, in a leukaemia, all the proviruses are in the same site of the cells' genomes because those insertion sites are non-random (that particular sites are specifically selected). No cancer researcher would suggest that the proviruses in a leukaemia are normal cellular regulatory elements, or that they are derived from normal cellular regulatory elements.

In the same way, the distribution of ERVs in multiple species cannot randomly generate a pattern that consistently suggests certain species are descended from a common ancestor. Nor does the consistency arise from selective insertion at particular sites. And ERVs are in no sense normal eukaryotic regulatory elements, or derived from such. ERVs are precisely retroviral sequences containing retroviral genes, spliced into genomes using retroviral strategies.

So there is compelling evidence establishing the monophylicity of simian (anthropoid) primates. ERVs common to all simians – apes including humans, OWMs and NWMs – include ERV-Pb1, ERV-V, ERV-FRD and some with affinities to the ERV-K class. The ultimate experiment is being extended. Sequencing of the genomes of NWMs (such as that of the marmoset) should be expected to provide many further examples of ERVs in NWMs that are *orthologous* – ultimately derived from one unique ancestor – with ERVs found in the human genome.

Many of the ERVs that survive in our genomes have contributed parts of their sequence to provide us with essential functions. All ERVs were viral DNA when they inserted into the genomes of primate cells. But now many of their genes are essential for human development. The *syncytin-1* and *-2* genes are classical examples. And many ERVs have provided regulatory sequences that help control other genes. It has been fashionable in the past to assume that ERVs are simply 'junk'. But biological systems have a knack of appropriating spare bits of DNA and putting them to new uses. Many ERVs have been exapted to provide essential functions.

The ERV-like LTR elements have not featured widely hitherto. Most of these are exceedingly ancient. All are common to humans and chimps, and the great majority to humans and OWMs. The *PEG10* and *PEG11* genes, derived from LTR elements, are orthologous in species as diverse as humans, mice, cows and wallabies. Portions of these elements have been conserved through mammalian evolution because they have come to fulfil essential roles in their hosts.

2 Jumping genealogy

In Chapter 1, retroviruses and LTR retrotransposons were discussed together because they possess long terminal repeats (LTRs): sequences that mediate regulatory functions. These agents generate new copies of themselves using a back-to-front RNA-to-DNA copying step (hence the *retro* prefix). They insinuate these new DNA copies into the genomes of their host cells. And they share the property of being able to replicate in germ cells, so that new copies are transmitted to future generations. Each new ERV and LTR retrotransposon becomes a marker of a family lineage.

But they differ in their life cycles. Retroviruses possess an *envelope* gene that enables them to act as true infectious agents. Retroviruses have the capacity to invade cells that are *different* from the cell that produced them. Each new retrovirus may be domiciled as a provirus in a foreign genome. In contrast, LTR retrotransposons lack *envelope* genes and the capacity for transmission between cells. They are intracellular parasites. They are 'copied and pasted', via an RNA intermediate, to new sites only in the *same* genome.

LTR retrotransposons are not alone in the way they indulge in secretive, genome-modifying activity. Our genomes have been colonised by an extremely diverse and prolific community of partially autonomous segments of DNA. These bits of DNA multiply haphazardly without consideration of the integrity of the genome that harbours them. They are known collectively as *transposable elements* (abbreviated as TEs hereafter), *mobile elements* or (more colloquially) *jumping genes*. They are identified as discrete units of DNA sequence that are present as many copies widely distributed throughout genomic DNA. For this reason, they are also called *repetitive elements*. They clutter up the genome as they replicate, as illustrated by the progressively disrupted sentences below.

Segments of DNA called TEs replicate haphazardly.
Segments of DNA called TEs replicate haphaTEzardly.
SegmTEents of DNA called TEs repTElicatTEe haphaTEzardly.

The pattern by which TEs disperse throughout the genome is similar to that by which dandelions disperse through the lawn. The generation of a daughter TE is an elaborate, lawful process (as is the production, dissemination and germination of a dandelion seed). But TEs (and dandelions) plant their progeny at random in their respective media.

Individual TEs may disrupt genes and cause disease. They may also acquire functional roles. Or their presence may be completely innocuous. Similarly, dandelions may be pests (if they grow on golf courses). They may also be co-opted to a gardener's advantage, to make salad and tea (the leaves), coffee (the roots) or wine (the petals). Or the presence of dandelions may go unnoticed. But any harmful effects or beneficial functions possessed by TEs (or dandelions) have no necessary connection to the fact that, when they insert into genomes (and germinate in lawns), they select the spot on which they alight largely at random. This random process makes TEs brilliant markers of common ancestry, because if two cells, organisms or species have a TE at the same place in their genomes, one may conclude that the randomly germinated, but shared, TE was present in a common ancestor.

TEs are classified into their various families and sub-families on the basis of their archetypal DNA sequences. The two broadest categories of TEs may be distinguished by their strategy of replication [1].

- The first group of TEs replicate via an RNA intermediate and use a reverse transcriptase enzyme by which they generate new DNA copies of themselves. This is the copy-and-paste strategy as already described for retroviruses and LTR retrotransposons. These agents are the *retrotransposons* (sometimes abbreviated to *retroposons*) or *retroelements*. We will focus on these in what follows.

- The second category encompasses the heterogeneous group of *DNA transposons*. These agents do not replicate via an RNA intermediate. Rather, they encode an enzyme known as a *transposase* that enables them to 'cut and paste' themselves, in the form of a mobile segment of DNA, around the genome of the host cell.

Each of these groups may also be subdivided on the basis of their gene content. Some types of TEs possess one or a few genes that are required for their propagation. Such TEs are said to be *autonomous*. Their origins are lost in deep antiquity. They include the famous LINE elements.

Other families of TEs lack genes that encode proteins. They proliferate only by co-opting the proteins made by autonomous elements. These freeloading TEs are said to be *non-autonomous*. Each family of non-autonomous TEs propagates by exploiting the molecular toolkit of a particular family of autonomous TEs. Many non-autonomous TEs originated as copies of conventional genes that do not, of course, transpose, and that typically encode small, non-protein-coding RNA molecules with housekeeping functions. Copies of these progenitor genes accumulated personality-changing mutations that generated descendants with mobile behaviour. Such TEs include the Alu and SVA elements of primates [2].

More than 1,000 different types of TE have been catalogued in the human genome. Analysis of these provides a graphic impression of the vast history over which our genome has been formed. Newly arising TE copies freely insert into older, pre-existing ones. It is thus possible to infer the relative ages of TEs by observing which TEs are found inserted within others. And if *families* of TEs were active at particular periods of the formative history of our genome, then TEs belonging to relatively recent families of TEs should be found inserted into TEs representing relatively old families (Figure 2.1).

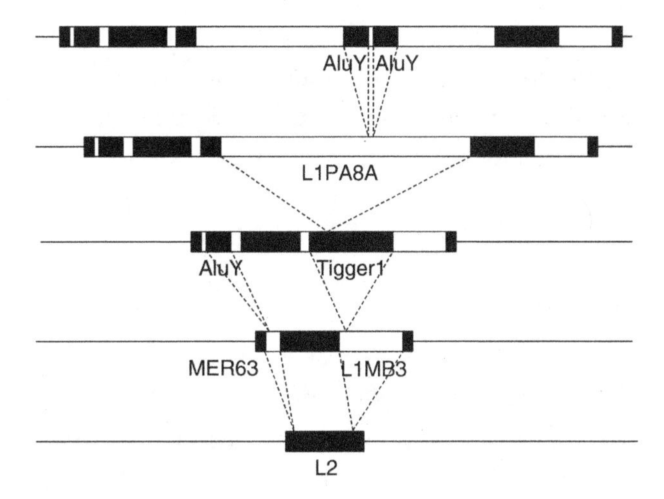

FIGURE 2.1. AN EXAMPLE OF A TE-WITHIN-TE CLUSTER [3]
TEs are represented as *boxes* and the surrounding genomic DNA as a *line*. An ancient L2 element was the first TE present at this locus. Subsequently, the locus was disrupted by MER63 and L1MB3, Tigger1, L1PA8A and AluY elements. Not to scale.

A computational study of the human genome, dubbed *TE defragmentation*, involving 300,000 clusters of TEs-within-TEs and 19% of the genome, has established the relative ages of 360 families of TEs. This approach has been validated by the observation that individual TEs of families identified as being ancient tend to have accumulated many mutations; individual TEs of families found to be recent have accumulated fewer mutations and are much closer to the family 'type' sequence [3]. The successive waves of expansion of hundreds of different TE families demonstrate that our genome is the end result of a vast history of genome formation.

2.1 THE ACTIVITIES OF RETROELEMENTS

Most retroelements in our genome are heavily mutated in comparison to the family archetypal sequence, indicating that they were inserted into the genomes of our forebears a long time ago. They are genomic fossils that have lost the capacity to copy and paste themselves. Their degenerating remains attest to their erstwhile careers as ancient colonisers of genomes. But others have accumulated fewer mutations relative to the family archetype and appear to be recently acquired denizens of genomes. Some of these are still engaged in their genome-modifying pursuits. Three types stand out for their current activity in the human genome.

2.1.1 *LINE-1 elements*

The autonomous LINE-1 elements possess a two-gene set that enables them to complete their copy-and-paste cycles. They are 6,000 bases long if full-length, although the majority of inserts in our DNA are truncated; that is, a variable portion of the 5′ (left-hand) end does not get copied during the haphazard insertion process. More than 500,000 copies are distributed around the human genome, and at least 80–100 of these retain the potential to transpose; that is, to spawn daughter elements. LINE-1 elements comprise a staggering 17% of our DNA [4].

Sequence analysis of LINE-1 elements shows that they may be classified into multiple distinguishable subgroups. Some 25 sub-families are shared with other primates (and are called L1P sub-families, where *L1* stands for *LINE-1*, and *P* stands for *primate*). Most of these sub-families represent a single lineage, in which one sub-family generates the next by the stepwise accumulation of characteristic DNA mutations. More ancient LINE-1 sub-families are present both in primates and widely in other mammals, and are called L1M elements (where *M* stands for *mammal*). And different subgroups of L1M elements may be distinguished and designated as (for example) L1MB and L1ME [5]. The temporal succession of LINE-1 sub-families may be illustrated by the defragmentation study described above (Figure 2.2) [6].

L1PA elements frequently insert into and disrupt L1MB and L1ME elements. L1MB elements frequently disrupt L1ME (but never L1PA) elements. L1ME elements never disrupt elements of either of the other sub-families. The history of our genome, then, has included

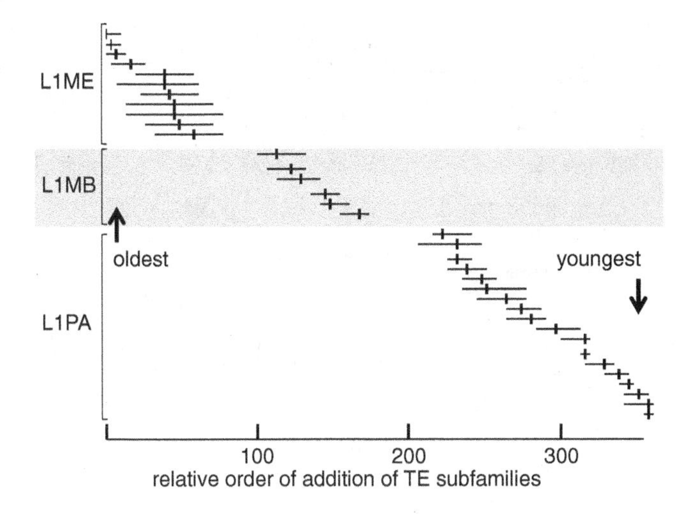

FIGURE 2.2. THE RELATIVE AGES OF 35 LINE-1 SUB-FAMILIES, BASED ON THE HIERARCHICAL WAY IN WHICH SUB-FAMILIES OF TES INSERT INTO EACH OTHER [6]
Numbers on the horizontal axis indicate the temporal order in which TE families were active, from 1 (the oldest studied) to 360 (the most recent). The *horizontal lines* indicate the relative time span over which the sub-family was active; *short vertical lines* indicate the median age of individual elements of each sub-family.

waves of replication of distinct LINE-1 sub-families. Inserts of the L1ME type were the first of the three sub-families to colonise primate genomes, but due to the haphazard way of replicating themselves, eventually ran out of active members and became extinct. Their degenerating inserts lie scattered through the genome. They were followed by an era during which L1MB elements were the active group, and these in turn were superseded by the still-active L1PA elements.

LINE-1 elements replicate (copy and paste) themselves within genomes using a mechanism that is a collaborative effort [7]. Both the host cell and the LINE-1 element contribute enzymes needed for replication. The process starts when the cellular enzyme *RNA polymerase II* transcribes an RNA copy from the parent LINE-1 element (Figure 2.3, step 1). The RNA copy has a long run of 'A' bases added to its terminus. (This is a standard way of tagging RNA transcripts

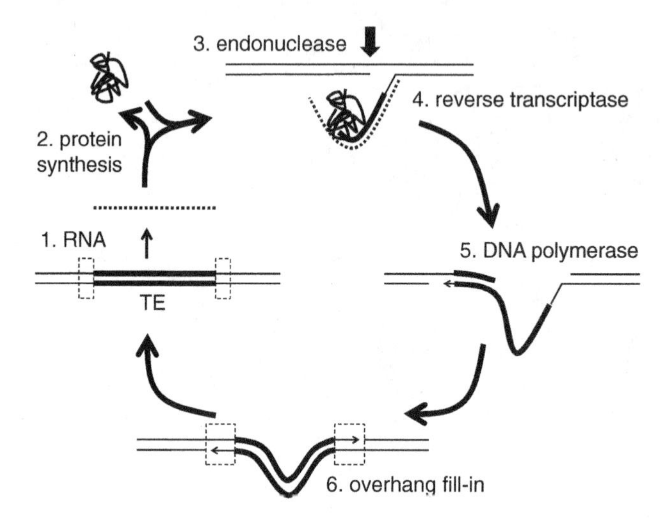

FIGURE 2.3. THE COPY-AND-PASTE CYCLE OF A LINE-1 RETROTRANSPOSON
A resident TE is transcribed into an RNA copy, from which proteins are made (*left-hand side*). A new DNA site is targeted, the DNA cut, a first DNA strand copied from the RNA, and the second DNA strand copied from the first. Representations include double-stranded DNA (*paired lines*), LINE-1 element DNA (*thick lines*), RNA (*dotted line*), and target-site duplications (*dotted boxes*).

in cells, and is called *polyadenylation*. The run of 'A' bases is known as a *polyA tail*.) This RNA is transported into the cytoplasm, where it directs the synthesis of the enzymes required for the TE-specific functions (step 2). In the case of LINE-1 elements, one of these enzymes has DNA-cutting (endonuclease) and reverse transcriptase functions.

The endonuclease selects a target site and generates staggered incisions, one on each strand, separated by several base pairs (the first cut on the lower strand is shown as step 3). The insertion mechanism then proceeds by a process known as *target-primed reverse transcription*.

- The RNA copy of the retroelement base pairs at its 3′ (right-hand) end, by its polyA tail, to a few bases at the lower cut site, so it is positioned to act as a template for RNA-dependent DNA synthesis. The retroelement's reverse transcriptase copies a first DNA strand (the lower strand) from this RNA template (from right to left; step 4).

- A cellular DNA polymerase copies the second (upper) strand of DNA using the newly synthesised lower strand as a template (proceeding from left to right; step 5). By this stage, the original RNA template has done its job and is degraded.
- Cellular enzymes make the final connections between the newly synthesised DNA and the contiguous chromosomal DNA termini (step 6). The result is a new element, permanently spliced into chromosomal DNA. Because the original nicks are staggered (up to 25 bases apart), generating single-stranded lengths on either side of the incoming element, and because these are filled in on both sides of the new element, target-site duplications are formed. Target-site duplications are thus a telltale signature of the covert operation of retroelement endonucleases.
- If the new TE has completed the process successfully, it can then act as a template to start the process all over again.

2.1.2 *Alu elements*

The non-autonomous Alu elements are the second type of jumping gene currently modifying the human genome. These lack the enzymes required for transposition, and consequently copy and paste only when they can commandeer the enzymatic machinery donated by obliging LINE-1 elements. Alu elements are about 300 bases long, and have replicated prolifically to 1,100,000 copies in our genome. The number of those that possess the capacity to transpose is undefined but greatly exceeds the number of active TEs of all other classes [8]. They comprise 11% of our DNA. Alu elements originated from a conventional gene known as the *7SL RNA* gene.

Alu elements are found only in primates. Computational analyses of Alu sequences, as well as Alu-within-Alu analyses, have been used to sub-classify the Alu elements in the human genome, and to work out the relative times at which each sub-family was active. Several intermediate forms bridge the progenitor *7SL RNA* gene and true Alu elements. True Alu elements are classified into AluJ (the oldest), AluS (intermediate in age) and AluY sub-types. These in turn are divisible into more than 200 sub-families. Older Alu element sub-families have generated newer ones by a complex pattern of branching and diversification [9].

2.1.3 SVA elements

A third type of TE, the non-autonomous SVA element, is much less abundant than LINE-1 and Alu elements. However, SVA elements deserve mention because they are also currently active in the human genome. These TEs are a strange hybrid of three different progenitor sequences that appear to have been cobbled together from bits of genetic debris. They consist of a length of sequence derived from an ERV-K, a variable number of copies of a short tandemly repeated sequence, and a fragment of Alu element.

There are approximately 3,000 SVA elements in the human genome. They are represented by six sub-families, four of which are found only in apes, and two of which are human-specific [10].

2.2 RETROELEMENTS AND HUMAN DISEASE

We are generally unaware of the covert operations of these various TEs in our cells, or of the way they undermine the stability of our DNA. But RNA copied from TEs is present in various normal tissues and organs of the human adult. LINE-1 and Alu elements are transcriptionally active in ovary and testis, in stem cells and in nerve cells in the brain. Indeed retrotransposition in brain cells is widespread. LINE-1 and Alu elements frequently disrupt protein-coding genes. The consequences of such somatic cell mutagenesis for normal brain function or for the development of neurological diseases or cancer is not known [11].

RNA copies of Alu elements are present in the retinal pigment epithelium of the eye. The accumulation of these RNA molecules is harmful, and they are usually degraded. As people age, the RNA-degrading mechanisms in retinal cells may become inefficient. Alu transcripts accumulate to undesirable concentrations, activating inflammatory responses, which, over the long term, inflict all sorts of injury upon the surrounding tissue. The result is age-related macular degeneration, the insidious and inexorable loss of vision that affects millions of older people [12]. Perhaps the build-up of Alu RNA will be implicated in the development of other inflammatory diseases of hitherto mysterious aetiology.

The LINE-1 endonuclease causes DNA breaks, which are both toxic for cells and mutagenic. This latter effect implies that they are potentially cancer-causing. The action of this endonuclease can also induce apoptosis, a form of cell death in which damaged cells commit suicide, a response that protects the organism from cells that might manifest potentially disruptive behaviour. In addition, damage mediated by this enzyme can lead to cell senescence, a permanent loss of the ability of cells to replicate. Indeed the LINE-1-encoded endonuclease can wreak havoc on the genome just like other (external) DNA-damaging agents [13].

Our genome is not an ordered or static book of instructions. If we use the book metaphor, our genome should rather be thought of as a book in which particular paragraphs can unpredictably and haphazardly replicate themselves into moveable copies that reinsert into the pre-existing text on any page. It's just too bad if the reinserted paragraphs scramble the existing text. The reader will have to be alert to mysterious paragraphs that are interspersed throughout the original text anywhere in the book, with no regard for context.

The random activities of LINE-1 and their dependent Alu and SVA agents have been studied under artificial conditions in cultured cells. These in vitro experiments have confirmed that the LINE-1-derived endonuclease and reverse transcriptase machinery is needed for the multiplication of Alu and SVA elements [14]. The mechanism by which LINE-1 elements transpose has been investigated also in transgenic mice using engineered versions of the LINE-1 agent. This work shows that the overall strategy of replication is sufficiently well understood to be modified predictably by experimental manipulation. The experimentally determined patterns of new insertion are wholly consistent with prior experience, indicating that these agents act as insertional mutagens [15].

Disease-causing retrotranspositional events have been documented in *somatic* cells; that is, in cells that do not contribute to the germ-line. Such newly arising inserts cannot be transmitted to future generations. The result is *sporadic* (non-inherited) disease. An early example of such an insertion event was found in a case of colon cancer, in which a fragment of a LINE-1 element, 750 bases long, was inserted into a protein-coding exon of the *APC* gene. The result of this insertion was the inactivation of the *APC* gene, a classical tumour suppressor gene that regulates the proliferation and differentiation of colonic epithelial cells, and which is inactivated (by a great variety of mutations) in 80% of all human colon cancers. Loss of *APC* gene function initiates the series of cellular changes that may ultimately lead to colon carcinoma. The insertion site with target-site duplications is shown in Figure 2.4 (top).

When the rearranged *APC* gene in the cancer cell DNA was characterised, only one DNA rearrangement was apparent. The presence of a *single* abnormality in *multiple* cancer cells is consistent with the interpretation that a unique LINE-1 insert (that arose in *one* cell) was characteristic of *all* the neoplastic cells in the resulting cancer. In other words, the multi-million cancer cell population was monoclonal, derived from the one founder cell in which, decades earlier, the LINE-1-mediated mutagenic event had occurred [16].

If cancers can arise as a result of retrotransposition, it seems also that rampant retrotransposition can arise as a result of cancer. Jumping gene activity seems to be rife in the permissive environment of cancer cells, at least in carcinomas, that are derived from epithelial tissues. Perhaps the luxuriant activity of TEs in cancer cells contributes towards the notorious tendency of such cells to generate new, increasingly aggressive variants [17].

Retrotransposition events occur also in *germ-line* cells. Such events generate genomic novelties that are transmitted to succeeding generations. The activities of TEs have been followed in human populations, particularly as large-scale sequencing has enabled researchers to catalogue and compare the TE complements in entire genomes of multiple individuals. It is obvious that in human genomes, a very large number of TE inserts are polymorphic. That is, each of these inserts is present in only a proportion of the human population. Such TEs are responsible for a great amount of the genetic variation that exists between people and that contributes to our diversity. New TEs are being added to the human gene pool at a steady, measurable rate. The rate of accumulation of TEs has been estimated as one new insert in every 100 births for LINE-1 elements, one new insert in every 20 births for Alu elements and one in every 900 births for SVA elements [18].

New retroelements in the germ-line may make their presence known with devastating impact. A new element may insert itself into an important gene, and thereby disrupt or destroy the function of that gene. The retroelement's presence is disclosed by the appearance of a *heritable* genetic disease. Clearly, LINE-1, Alu and SVA elements are insertional mutagens of current clinical significance. Researchers have catalogued TE inserts that are very rare in the human population (allele frequency <5%), which must have been added to the gene pool relatively recently. They have shown that one-third of all such inserts occurred in introns, and 0.9% disrupted exons [19]. It has been estimated that up to 0.3% of all new instances of human genetic disease are caused by these agents. The diseases they have induced include haemophilia, cystic fibrosis, Duchenne muscular dystrophy, Dent's disease, β-thalassaemia and various cancers [20].

An early study implicating the insertion of a TE in heritable disease featured a young man with type 1 neurofibromatosis, a syndrome that includes neurofibromas – benign tumours of nerve sheaths in the peripheral nervous system. An Alu insertion into the *NF1* tumour suppressor gene was found in normal blood cells of the patient (Figure 2.4, second from top). This indicates that the patient had inherited the disrupted *NF1* gene and, with it, the predisposition

insert ...GGAATTAA**GAATAATG** [TGCATGTGTC...AAAAA] **GAATAATG**CCTCCAGT...

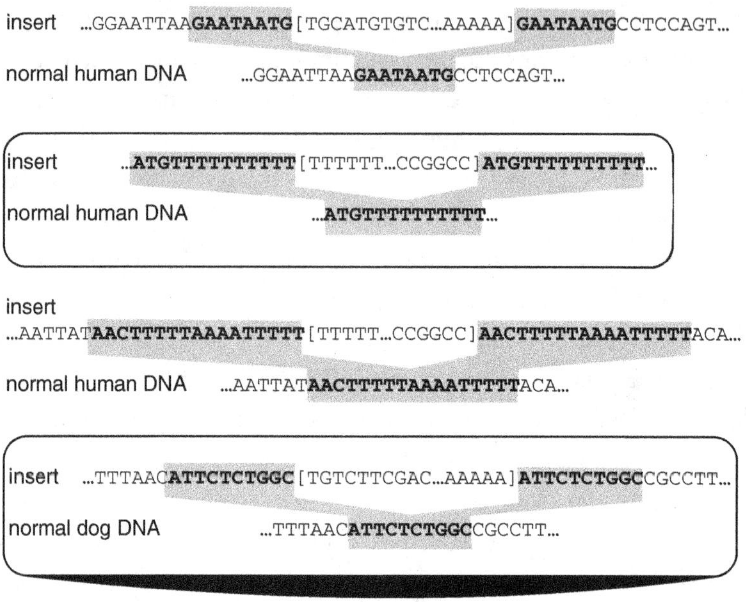

normal human DNA ...GGAATTAA**GAATAATG**CCTCCAGT...

insert ...**ATGTTTTTTTTTT** [TTTTTT...CCGGCC] **ATGTTTTTTTTTT**...

normal human DNA ...**ATGTTTTTTTTTT**...

insert
...AATTAT**AACTTTTTAAAATTTTT** [TTTTT...CCGGCC] **AACTTTTTAAAATTTTT**ACA...

normal human DNA ...AATTAT**AACTTTTTAAAATTTTT**ACA...

insert ...TTTAAC**ATTCTCTGGC** [TGTCTTCGAC...AAAAA] **ATTCTCTGGC**CGCCTT...

normal dog DNA ...TTTAAC**ATTCTCTGGC**CGCCTT...

FIGURE 2.4. INSERTION SITES OF DISEASE-CAUSING TES
From top: a LINE-1 element in the *APC* gene of cells of a colon cancer [16]; an Alu element in the *NF1* gene of a person with neurofibromatosis [21]; an Alu element in the *OPA1* gene in a family transmitting optic atrophy [22]; and a LINE-1 element next to the *MYC* proto-oncogene in CTVT cells [28]. Target sites are in *bold* and *shaded*.

to disease. Neither parent showed the mutation, indicating that the insertion event was a new mutation, perhaps arising in the father's germ cells, and thence transmitted to his son. The presence of this Alu insert disrupted processing of the *NF1* gene transcript. This led to the production of an aberrant messenger RNA, and a scrambled NF1 protein [21].

Some mutations may be well established in families. An eye disease called autosomal dominant optic atrophy involves damage to the *OPA1* gene. In one family possessing this condition, the mutation has been identified as an Alu insertion. Eighteen members of this family, spanning several generations, were shown to have the same mutation. That is, they possess in their genomes the same Alu element (lying in the same orientation), in the same location, with the same target-site duplication (Figure 2.4, third from top). This insert links all the family members who possess it to the one ancestor in which the Alu element was generated [22].

Harmful mutations, including retrotransposon insertions, may spread surreptitiously through unsuspecting communities. They may eventually be present in people who do not realise that they are related. Some people in a community in Japan possess a unique SVA insertion into the *fukutin* gene. The result is the incapacitating condition known as Fukuyama-type congenital muscular dystrophy. This particular mutation was found in 15 of 2,814 Japanese people, but not in any of 969 Chinese or Mongolian people [23]. In northern Japan, a genomic rearrangement in the *HLA-A* region (which controls immune responses), caused by an SVA insertion, was found in three families with no known connections [24]. A community in central Portugal harbours a low-frequency Alu insertion at position 156 of the *BRCA2* breast cancer susceptibility gene [25]. In France, a condition called lissencephaly type II is caused by an Alu insertion into the *POMT1* gene [26].

Each of these mutations arose, perhaps a few thousand years ago, in one individual, and has subsequently been spreading through the population. There are several strands of evidence for the dissemination of such *founder mutations*. The affected populations are limited to circumscribed geographical areas. Each TE insertion is found in a relatively homogeneous genetic environment. That is, it is part of a conserved *haplotype*: since the TE was inserted, the surrounding chromosomal DNA has not had time to be shuffled by recombination. A common haplotype indicates that the TE came from a single source. And of course each mutation is essentially unique: 'an insertion shared between two humans at exactly the same genomic location with identical target site duplications is testimony to an inherited insertion and a common ancestor' [27]. The 'casanova phenomenon' applies to the study of genetic disease.

2.3 RETROELEMENTS AND PRIMATE EVOLUTION

At this point we must address some of the questions that we asked of ERVs. Over what span of time have LINE-1, Alu and SVA elements been accumulating in the genome that we have inherited? In what species other than humans are they found? And if they are found in other species, do we and these other species share particular unique inserts? This is an important issue, because if we do share particular instances with other species, we would have to conclude that we and those other species had inherited each of those inserts from the ancestor in which the shared TE was generated.

2.3.1 LINE-1 elements

This book is about human evolution. But man's best friend is said to be the dog. I introduce this section on LINE-1 elements by considering an anecdote from the canine world.

In the Prologue, I described an infectious cancer, the canine transmissible venereal tumour (CTVT). A number of genetic markers have established that CTVT is clonal: various forms of this tumour, wherever they may be found around the world, have arisen from one progenitor cell. One of these markers is an inserted canine-type LINE-1 element that spliced itself into dog DNA just upstream (to the left) of an important proto-oncogene. It appears that the foreign length of DNA deregulated the proto-oncogene, and contributed to the abnormal growth we recognise as a tumour. The LINE-1 element is bracketed by perfect target-site duplications (Figure 2.4, bottom), and is common to all CTVT cells. It has never been found in any other source of dog genomic DNA [28]. It is a perfect marker of monoclonality (or, if we think of the CTVT as a new obligate parasite of dogs, of monophylicity). In other words, all CTVT cells have been produced from the one cell that sustained the insertion event.

Does our genome contain *particular* LINE-1 elements that are present also in the genomes of other species? An early study identified over 70 relatively young LINE-1 elements in the human genome. The

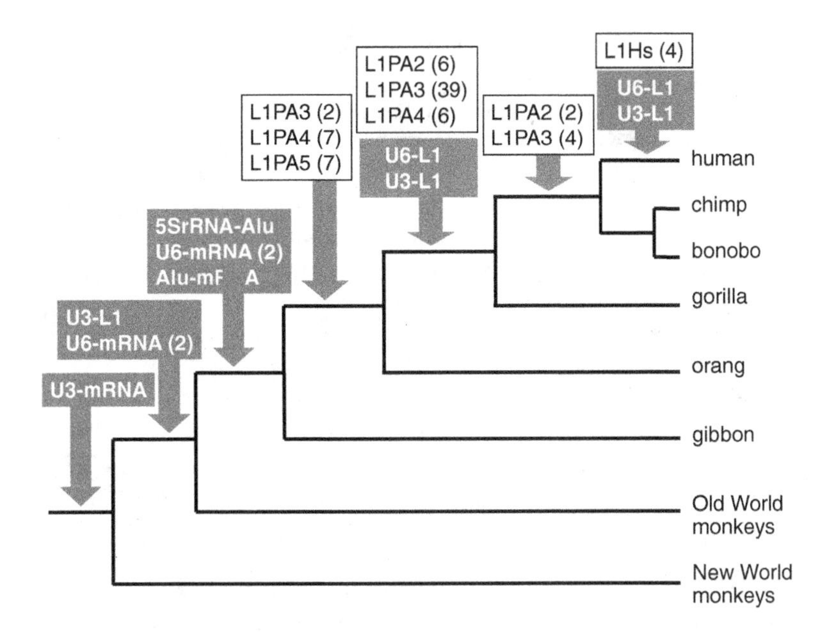

FIGURE 2.5. THE TIMES AT WHICH LINE-1 INSERTS ENTERED THE PRIMATE GERM-LINE, INFERRED FROM THEIR PRESENCE OR ABSENCE IN THE GENOMES OF PRIMATE SPECIES

LINE-1 elements, *open boxes* [29]; chimaeric inserts with both partners indicated, *grey boxes* [31]. *Numbers in brackets* indicate the number of individual inserts mapped.

authors set out to discover whether any of these elements were present in the genomes of other hominoid species. Many cases of shared elements were identified. Humans share many 'young' LINE-1 inserts with chimps, others with chimps and gorillas, and others with chimps, gorillas and orang-utans (Figure 2.5, open boxes) [29]. The shape of the family tree (that is, the pattern of phylogenetic relationships) established using LINE-1 inserts is precisely the same as that obtained using ERV inserts. The presence of particular LINE-1 elements in different primate species establishes that humans, humans and chimps, the African great apes, and the great apes are all monophyletic groups. Each is derived from a lineage of common ancestors.

Whole-genome sequencing provides the ultimate test of common descent. Comparison of all the LINE-1 inserts in the genomes

Table 2.1. *Total and shared TEs in the human genome [30]*

LINE class	Total number in human genome	Proportion of LINE elements (%) in				
		Human only	Bonobo only	Chimp only	Bonobo and chimp only	Human, bonobo and chimp
LINE-1	626,000	0.2	0.1	0.1	0.1	>99
LINE-2	302,000	<0.1	<0.1	<0.1	<0.1	>99.9
LINE-3	37,000	<0.1	<0.1	<0.1	<0.1	>99.9
Alu	731,000	0.7	0.1	0.1	0.1	99
MIR	393,000	<0.1	<0.1	<0.1	<0.1	>99.9

of humans and other primates is required to establish that the inserts selected for study (see above) are representative of the whole population. Of the approximately 600,000 LINE-1 elements in the human genome, all but 2,000 are shared with chimps and bonobos. This means that, as with ERVs, over 99% of LINE-1 elements in our genome have orthologous counterparts in the genomes of the two chimp species (Table 2.1). Similarly, the human genome shares all but 1,860 LINE-1 elements with the gorilla genome. And the orang-utan genome contains only 5,000 LINE-1 elements that are unique to the orang-utan. The rest (a huge majority) are shared.

The only exceptions to the consistent phylogenetic tree are that a number of L1PA2 and L1PA3 elements are present only in gorillas and chimps *or* humans. This reflects incomplete lineage sorting, due to the closely timed branching of the gorilla, chimp and human lineages. It corroborates the evidence provided by ERV-K-GC1 (Chapter 1) that these three groups all diverged over a short time period [30].

In addition, distinct types of LINE element (LINE-2 and LINE-3) are known to be particularly ancient on the basis of the TE defragmentation approach (described earlier). They are present at high copy numbers in primate genomes. Essentially all of these are shared with both chimps (Table 2.1) and OWMs. This means that every individual insert belonging to these families was already resident in primate DNA before the chimp–human and macaque lineages diverged. LINE elements provide evidence of common descent that is as consistent and compelling as that provided by ERVs and LTR elements.

Some LINE-1 elements possess idiosyncrasies that enable them to act as particularly potent markers of phylogenetic relationships. On rare occasions during the LINE-1 retrotransposition process, the reverse transcriptase enzyme acts promiscuously. It disengages from its partner RNA molecule, which it has been copying into the DNA daughter element as is customary, and attaches itself to some other, randomly selected RNA molecule. The reverse-transcribed product is a two-component element that includes portions of the LINE-1 transcript and an innocent bystander RNA. Sometimes two bystander molecules may be involved. They are conjoined at a particular point in the sequence of each molecule – the site at which the template switch occurred. The completed insert lies at a randomly selected point in the genome. Such hybrid elements, generated in a compounded molecular lottery, are exquisitely specific markers of family connections.

Dozens of chimaeric retrotransposed inserts have been characterised in human DNA. Most are shared with other primate species. Examples are shown in Figure 2.5 (grey boxes). The U3-mRNA hybrid insert is shared by humans and all tested species including New World monkeys. This one unique insert demonstrates that humans and NWMs are descended from the unique common ancestor in which the unique insert was generated. We anthropoid (or simian) primates are products of the single reproductive cell in which this freak two-part insert was assembled. We are a monoclonal (in cell biological parlance) or monophyletic assemblage of species [31].

One of these curious hybrid inserts has itself spawned a small family in primate genomes. The common structure consists of a small nuclear RNA designated U6 (or *U6 snRNA*) attached at a unique point to a segment from a protein-coding gene (an intron of a gene encoding a zinc finger protein). Of the seven members of this family, all are shared with chimps and six with orang-utans. And

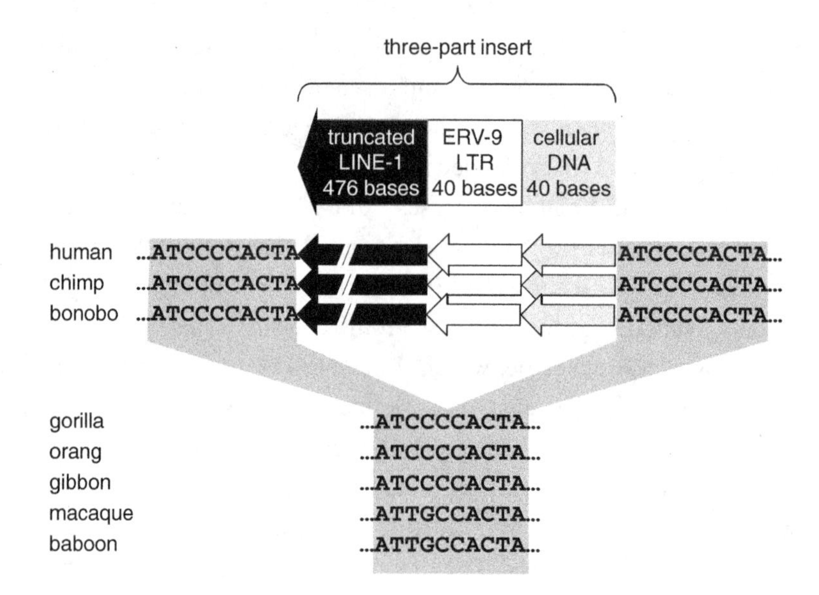

FIGURE 2.6. A TRIPARTITE INSERT IN THE β-GLOBIN GENE LOCUS OF HUMANS AND THE TWO CHIMP SPECIES [33]

two are shared with lemurs, which are prosimians [32]. This indicates that all primate species comprise one great clone.

An even more unusual element exists in the β-globin gene locus. This is a three-component insert, consisting of 476 bases of a LINE-1 element, spliced together with another transcript that is itself a hybrid of an ERV-9 LTR (40 bases) and flanking cellular sequences (also 40 bases). This insert, with its unique target-site duplication, was formed by an unrepeatable sequence of genetic happenstance, a concatenation of singular events, and is found in humans and the two chimp species [33]. Other apes and OWMs retain the undisturbed target site (Figure 2.6). Molecular process has generated a unique genetic fingerprint. Stronger evidence for a human–chimp common ancestor is barely imaginable.

2.3.2 *Alu elements*

The ongoing accumulation of Alu elements in genomes has progressively modified the way in which those genomes function. Alu elements affect gene transcription. High densities of Alu elements also destabilise gene loci, predisposing nearby genes to disruptive rearrangements. Researchers have set out to define the times in history at which certain Alu elements were added to the primate germ-line, leading to innovations in gene regulation or to elevated propensities to develop genetic diseases.

Some Alu insertion events occurred during an epoch of genetic history that was particular to humans. The *ACE* gene, encoding the angiotensin-converting enzyme, has an important role in regulating blood pressure, electrolyte concentrations and cardiovascular function. An Alu insert is found in intron 16 of the *ACE* gene, and this is polymorphic: some people possess the insert, whereas the rest of the human population, as well as chimps and gorillas, retain the undisturbed pre-integration site (Figure 2.7, upper diagram). The ten-base target-site duplications (GTGACTGTAT), and the original target site sequence (GTGACTGTAT) are perfectly conserved [34]. The insert is present in essentially all ethnic groups, albeit at varying frequencies, and so entered the human germ-line before humans were differentiated by current conceptions of race. There is ongoing interest regarding whether the presence of the Alu element affects the risk of developing cardiovascular diseases. Latest results suggest that the possession of this particular Alu insert is associated with lower levels of angiotensin-converting enzyme activity, and with lower blood pressures in children, especially in those who had low birth weights. This element, a relative newcomer to the human genome, may reduce the risk of cardiovascular disease [35].

Another Alu element participates in rare DNA rearrangements involving the *HPRT* gene. Inactivation of this gene is associated with the Lesch–Nyhan syndrome, a condition in children that is characterised by gout, mental retardation and a tendency to self-mutilation. This insert is shared by humans, chimps and gorillas, while orang-utans and OWMs retain the original, undisturbed target site (Figure 2.7, lower diagram). The target site and its duplications (GAATGTTGTGA) are perfectly preserved [36]. This insert dates from an

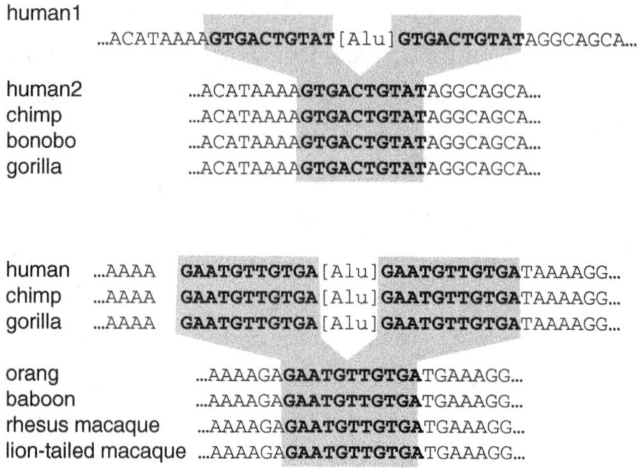

human1
...ACATAAAA**GTGACTGTAT**[Alu]**GTGACTGTAT**AGGCAGCA...

human2 ...ACATAAAA**GTGACTGTAT**AGGCAGCA...
chimp ...ACATAAAA**GTGACTGTAT**AGGCAGCA...
bonobo ...ACATAAAA**GTGACTGTAT**AGGCAGCA...
gorilla ...ACATAAAA**GTGACTGTAT**AGGCAGCA...

human ...AAAA **GAATGTTGTGA**[Alu]**GAATGTTGTGA**TAAAAGG...
chimp ...AAAA **GAATGTTGTGA**[Alu]**GAATGTTGTGA**TAAAAGG...
gorilla ...AAAA **GAATGTTGTGA**[Alu]**GAATGTTGTGA**TAAAAGG...

orang ...AAAAGA**GAATGTTGTGA**TGAAAGG...
baboon ...AAAAGA**GAATGTTGTGA**TGAAAGG...
rhesus macaque ...AAAAGA**GAATGTTGTGA**TGAAAGG...
lion-tailed macaque ...AAAAGA**GAATGTTGTGA**TGAAAGG...

FIGURE 2.7. THE INSERTION SITES OF AN ALU ELEMENT IN THE *ACE* GENE (*UPPER DIAGRAM*) [34] AND THE *HPRT* GENE (*LOWER DIAGRAM*) [36]

ancestor of the African great apes, and its presence suggests the possibility that these species might share a tendency to undergo disease-generating Alu-mediated rearrangements at the *HPRT* locus.

The presence of multiple Alu elements at the *MLL* gene in humans predisposes that locus to damaging rearrangements and to the development of leukaemias. One of these inserts is present in all the great apes. OWMs retain the undisturbed target site (Figure 2.8). This Alu element was spliced into the primate genome in a great ape ancestor. A second Alu element located near the *MLL* gene is present in apes and OWMS, establishing the monophylicity of this wider group of species [37]. Perhaps all species possessing these inserts share a tendency to developing *MLL* aberrations and leukaemias. Data on this question are not available, and high densities of Alu elements near any gene in any species may well contribute to a tendency to delete (or duplicate) parts of genes.

A summary of the data arising from this remarkable research project is shown in Figure 2.9. It indicates the times at which 22 Alu elements, each identified by the associated gene, entered the primate genome en route to humans. And it demonstrates powerfully that

```
human      ...AAAAAAGTAGCC[Alu insert]TAGCCTGTTTCT...
chimp      ...AAAAAAGTAGCC[Alu insert]TAGCCTGTTTCT...
gorilla    ...AAAAAAGTAGCC[Alu insert]TAGCCTGTTTCT...
orang      ...AAAAAAGTAGCC[Alu insert]TAGCCTGTTTCT...

baboon                    ...AAAAAAGTAGCCTGTTTCT...
rhesus macaque            ...AAAAAAGTAGCCTGTTTCT...
lion-tailed macaque       ...AAAAAAGTAGCCTGTTTCT...
```

```
human          ...AGAATTATAATACTTTTTCA[Alu]AATTATAATACTTTTTCAGA...
chimp          ...AGAATTATAATACTTTTTCA[Alu]AATTATAATACTTTTTCAGA...
gorilla        ...AGAATTATAATACTTTTTCA[Alu]AATTATAATACTTTTTCAGA...
orang          ...AGAATTATAATACTTTTTCA[Alu]AATTATAATACTACTTCAGA...
baboon         ...AGAATTATAATACTTTTTCA[Alu]AATTGTAATGCTACTTCAGA...
rhes macaque   ...AGAATTATAATACTTTTTCA[Alu]AATTGTAATGCTACTTCAAA...
lion-t macaque ...AGAATTATAATACTTTTTCA[Alu]AATTGTAATGCTACTTCAGA...

reconstructed target site   ...AGAATTATAATACTTTTTCAGA...
```

FIGURE 2.8. THE INSERTION SITES OF TWO ALU ELEMENTS IN THE *MLL* GENE [37]

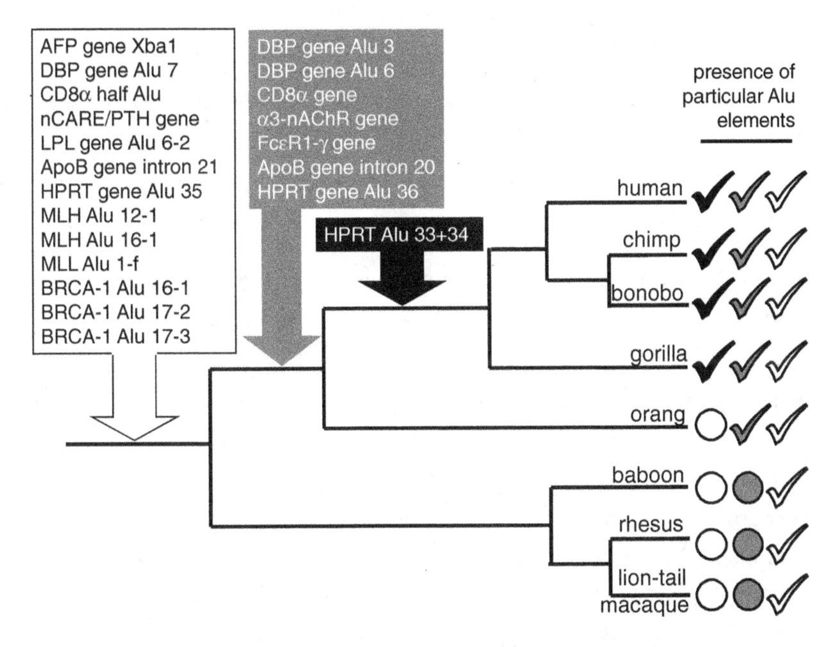

FIGURE 2.9. THE DISTRIBUTION OF A SELECTION OF ALU ELEMENTS IN PRIMATE GENOMES [38]

Ticks and *circles* represent the presence and absence, respectively, of particular inserts. Their *shading* corresponds to the shading of the *boxes* and *arrows*, which indicate when the various Alu elements were added to the primate germ-line.

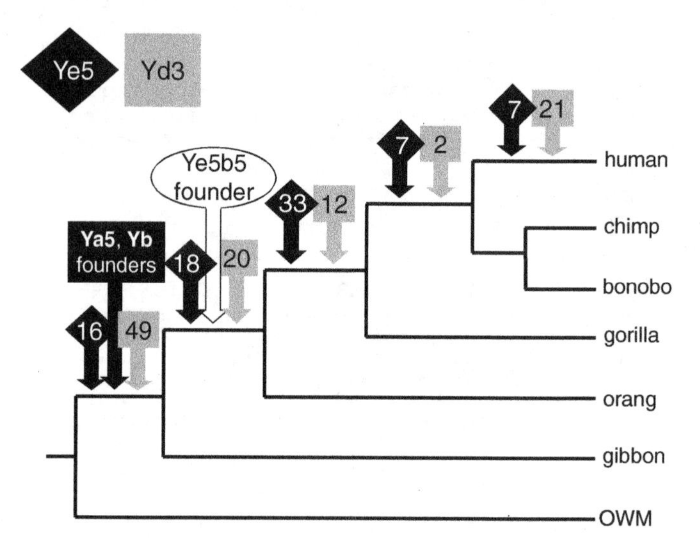

FIGURE 2.10. THE DISTRIBUTION OF A SELECTION OF YOUNG ALU ELEMENTS IN PRIMATE GENOMES [39]
Numerals represent the number of elements that arose at each branch of the phylogenetic tree. The origins of the individual founders of the Ya5, Yb and Ye5b5 sub-families are indicated [41].

the African great apes are monophyletic, as are the great apes, and Old World primates (apes and OWMs) [38].

The whole-genome sequencing era has provided scope for systematic surveys of selected Alu sub-families in multiple species. Two sub-families that have arisen relatively recently (AluYe5 and AluYd3) have expanded during hominoid history, and the distribution of individual inserts in the genomes of different species has provided a definitive outline of hominoid relationships (Figure 2.10). A total of 45 inserts are shared by the African great apes, for example; and each one of these inserts entered the primate germ-line in an African great ape ancestor. Thirty-eight are shared by all the great apes (but by no other species); and 65 by the apes [39].

Only one discordant result was observed: an AluYe5 element present in gorillas and humans but not chimps. The distribution of this element reflects the now-familiar phenomenon of incomplete lineage sorting at the human–chimp–gorilla divergence. We encountered this near-trifurcation with the ERV-K-GC1 provirus, and some L1PA2 and L1PA3 elements. It has been evinced also by an Alu element found in a chromosome 21 survey [40]. However, the single ambiguous element identified in the studies depicted in Figure 2.10 may be compared with seven AluYe5 inserts that are shared by humans and chimps to the exclusion of gorillas, emphasising that humans and chimps represent sister clades, derived from a progenitor that lived after the gorilla lineage diverged.

Two sub-families (AluYa5 and AluYb8) are responsible for most of the new Alu elements that have arisen in the human genome since the human lineage separated from that of the chimp. Unique founder elements of these currently expanding sub-families are present in the genomes of every species of ape, and must have arisen in the genome of an ancestor of all apes (see also Figure 2.10). The single Alu element that founded the small Ye5b5 family is restricted to the genomes of great apes, and generated a few descendants in the orang-utan genome and one in the human genome. Thus elements may remain quiescent for millions of years before they retrotranspose to spawn daughter elements [41].

Whole-genome sequencing has allowed side-by-side comparisons of the Alu elements in human and chimp genomes. Of the 1,100,000 Alu elements in our genome, all but 7,000 are shared with chimps and bonobos. That is, approximately 99% of the Alu elements in our genome are shared with both chimp species, and must have resided in the germ-line of the most recent human–chimp ancestor. Only a few thousand Alu elements in the human genome are absent from the gorilla genome (and vice versa). This indicates that the African great ape ancestor also possessed 99% of the Alu elements that the extant species possess. Given that the orang-utan genome has a mobile element content that is similar to ours, including more than a million Alu elements, and that only 250 Alu elements are unique to orangs, the vast majority of Alu elements must be shared by us and orang-utans, the most distantly related great ape. Early analyses indicate that most Alu elements are shared with OWMs (Table 2.1) [42].

Other non-autonomous families of retrotransposons have also been exhaustively compared by computational surveys of sequenced genomes. The young and still-active SVA elements are present both as human-specific inserts and as inserts shared by other hominoids. In fact, 60% of the SVA elements are unique to the human genome, and 40% are shared with chimps. No SVA elements are found in OWMs, but approximately 100 precursor retroelements, consisting only of the tandem repeat component, are found in OWMs, and most of these are present also in human DNA [43]. The *mammalian-wide interspersed repeat* (MIR) elements are an abundant class of non-autonomous element (dependent on LINE-2 activity for their proliferation). They are also ancient, as essentially all inserts are shared by humans, bonobos and chimps (Table 2.1) [44].

We can gain an impression of the way that TEs have accumulated in primate genomes from representations of the TE landscape around individual genes (*ASPM* and *BRCA1*) in multiple species (humans, chimps, gorillas, orang-utans and macaques). The great majority of TEs associated with these genes – LINE-1, LINE-2, LINE-3/CR1, Alu, MIR, LTR elements and DNA transposons – are common to all five species, demonstrating that the genomes of apes and OWMs share a vast formative history, largely complete by the time of the last common ancestor of apes and OWMs. The only exceptions are a few Alu inserts and a few Alu element deletions following non-allelic homologous recombination events. Now envisage another 21,000 genes, and the vast tracts of intergenic space, all containing a similar patchwork of (relatively old) shared and (relatively recent) lineage-specific TEs, and marvel at the historical record of genome formation thereby disclosed [45].

But several questions remain. *Firstly*, could Alu elements be used to show whether humans, Old World monkeys and New World monkeys have a common ancestor (that is, whether the simian primates are monophyletic)? The answer is an unambiguous 'yes', as exemplified by the Alu insert depicted in Figure 2.11. Some of the target-site duplications on the right of the insert are somewhat frayed

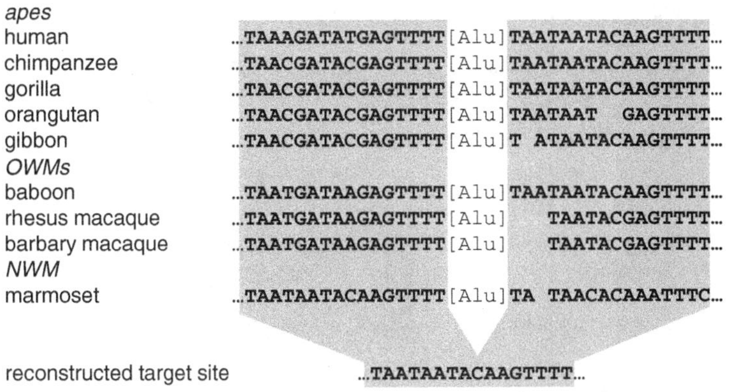

```
apes
human                      ...TAAAGATATGAGTTTT [Alu] TAATAATACAAGTTTT...
chimpanzee                 ...TAACGATACGAGTTTT [Alu] TAATAATACAAGTTTT...
gorilla                    ...TAACGATACGAGTTTT [Alu] TAATAATACAAGTTTT...
orangutan                  ...TAACGATACGAGTTTT [Alu] TAATAAT   GAGTTTT...
gibbon                     ...TAACGATACGAGTTTT [Alu] T ATAATACAAGTTTT...
OWMs
baboon                     ...TAATGATAAGAGTTTT [Alu] TAATAATACAAGTTTT...
rhesus macaque             ...TAATGATAAGAGTTTT [Alu]     TAATACGAGTTTT...
barbary macaque            ...TAATGATAAGAGTTTT [Alu]     TAATACGAGTTTT...
NWM
marmoset                   ...TAATAATACAAGTTTT [Alu] TA  TAACACAAATTTC...

reconstructed target site          ...TAATAATACAAGTTTT...
```

FIGURE 2.11. THE INSERTION SITE OF AN ALU ELEMENT PRESENT IN ALL SIMIAN GROUPS (APES, OWMS, NWMS) [46]

because of the large time span that has elapsed since this Alu insert was spliced into the primate germ-line. Nevertheless, the Alu insert is clearly present at the same unique site in every species tested, from human to marmoset – a NWM. The conclusion is made more compelling because of the number of species included in the study (nine) and the length of the target site (16 bases) [46].

Secondly, what is the relationship between the three major groupings of the primate order? These are represented by the simians (humans, other apes and monkeys) and the two prosimian groups (tarsiers and the loris/lemur group). These three trajectories branched out during the earliest phase of primate history. Each of the three possible sequences of branching for simians, tarsiers and lemurs has been postulated at different times. The issue remained unresolvable until a genome-mining expedition identified four Alu elements that were shared by simians and tarsiers but were absent from the genomes of lemurs, which preserved the unoccupied target site.

The DNA sequence around one of these inserts is shown in Figure 2.12 [47]. In the simians, the right-hand target-site duplication has decayed with the passage of time. The target-site duplications found in the tarsier are highly similar to the left-hand target-site duplication preserved in the simians, and to the unoccupied pre-

human	...AAAA GAACCAGTGGG [Alu] AAAA AAAACA GGG...
macaque	...AAAA CTACCAGTGGG [Alu] AAAA AAAAAA AGG...
douc langur	...AAAA GAACCAGTGGG [Alu] AAAA AAAAAA GGG...
woolly monkey	...AAAAAGAATCAGTTGG [Alu] AAAA AAAAAA GGG...
marmoset	...AAAAAGAATCAGCTGT [Alu] AAAA AAAAAA GGG...
tarsier	...TAAG GAACCAGTGGC [Alu] AAAA GAACCAGTGGG...
ruffed lemur	...ACAG GAAGCAGTGGG...
thick-tailed galago	...AAAA GATTCAGTGGG...
guinea pig	...GAAA GAGCCAGTGG ...
reconstructed target site	...AAAA GAACCAGTGGG...

FIGURE 2.12. THE INSERTION SITE OF AN ALU ELEMENT (ALU-C9) IN SIMIANS (APES, OWMS, NWMS) AND TARSIERS [47]

integration site preserved in the lemur, galago and guinea pig (a rodent). This, and independent work with a suite of TEs (MER45C, FLAM-A, L1MEc, FAM and MLT1C elements) [48], has established that simians and tarsiers are sister clades and monophyletic.

Thirdly, can it be shown whether *all* primate species are descended from a single ancestor? Early searches through available genome sequence databases found several TEs that are shared by all primates from humans to lemurs: these include a LINE-1 element, a MER element and a *free left Alu monomer* (a FLAM element, which represents an intermediate stage in the evolution of the *7SL RNA* gene into a fully fledged Alu element) [49].

Later systematic searches though expanding sequence data sets found that 4 of 1,569 Alu elements residing in the human genome were present also in lemurs, suggesting that 0.25% of Alu elements in the human genome are orthologous with elements in the lemur genome [50]. A search for monomeric Alu elements (ancient progenitors of the typical dimeric Alu element) revealed that 29 of 1,404 inserts were shared by all primate groups, including lemurs and bushbabies [51]. And two very ancient LINE elements, designated L3b and Plat_L3, have been shown to be the common possession of each of 18 primate species, including lemurs [52]. We conclude that all living primates are descended from the one reproductive cell in which each of the particular transposition events occurred.

Fourthly, are colugos (flying lemurs) properly classified as primates? The classification of colugos reflects a long history of uncertainty. However, the lack of Alu elements and of other DNA markers identified above has established that they are not true lemurs and are not primates. Colugos in fact belong to the order Dermoptera. We have defined the limits of the primates [53].

2.3.3 *Retroelements and phylogeny: validation*

The DNA extractable from a hair or a drop of blood embodies the digitised genetic information that delineates the route of our evolutionary history via populations of extinct primates. But one issue must first be addressed: what is the likelihood that two TEs of the same type would *independently* insert into the same DNA site in different species? The independent acquisition of a particular character in multiple lineages is called *homoplasy*. This phenomenon compromises the power of evolutionary studies based on morphology. Similar structures can evolve independently in distantly related organisms, because similar selective pressures generate similar adaptations by convergent evolution. When this happens, it can be inferred quite erroneously that the multiple organisms have inherited their shared traits from the same ancestor. Analogously, independent insertions into the same site (*insertion homoplasy*) would limit the use of TEs as markers of common ancestry. Some types of TEs do possess mechanisms for selecting certain genomic sites for insertion. However, LINE-1 elements and their non-autonomous hangers-on (Alu and SVA elements) are the TEs of relevance to us. They have sloppy insertion-site preferences [54].

Can insertion *hotspots* generate insertion homoplasy, and so account for the distribution of TEs in primate species? Computational surveys of TEs that have inserted into pre-existing TEs indicate that, in general, new insertions occur randomly into DNA. However, a large number of insertion hotspots have been identified in the human genome. Such hotspots are sites into which new insertions occur at a higher frequency than would be expected on a purely random basis. Hotspots are sites into which new insertions occur *preferentially* but not *exclusively*. It follows that the presence of hotspots cannot constrain different species to independently generate the same pattern of insertions. For example, Alu elements show a tendency to insert into pre-existing Alu elements to the right of a sequence of 'A's at base 133. But this tendency cannot generate consistent insertion patterns. Firstly, an incoming Alu element (say in a hominoid ancestor) would have a million resident Alu elements to choose from. Secondly, the Alu hotspot is only one of many hotspots present in TEs. And, thirdly, hotspots comprise only a fraction of all potential insertion sites [55].

Consider an analogy: the location of car accidents. Such events occur randomly due to factors such as excess speed, alcohol-induced misjudgements and driver inattention. If the location of every car accident that had happened in Cambridge over the last decade was marked on a map, certain locations would be overrepresented. These sites are hotspots for car accidents. They might include certain intersections and blind corners. But accidents that occur at hot-spots are still accidents. If you had two maps of Cambridge, and each map provided the same locations for a hundred car accidents, you would conclude that the maps are *copies* of the same set of data. The existence of hotspots cannot independently replicate an identical distribution of accidents in each of two (or more) consecutive years. Similarly, TE insertions at hotspots remain a subset of all insertions. They are essentially random events, and cannot account for the consistent pattern of shared insertions between species.

Hotspot considerations may be illustrated by insertional mutagenesis of the *NF1* gene in neurofibromatosis type 1. A search for mutations has indicated that 0.4% of all *NF1* mutations are retrotransposition events. In one study, 18 TE insertion mutations were identified in unrelated people, and six of these were clustered in a region of 1,500 bases. Three insertion sites were each used twice. The *NF1* gene represents a remarkable hotspot. But the huge majority of people who do not suffer from neurofibromatosis lack new insertions in this gene; and of people with the syndrome 99.6% appear to lack insertions. The *NF1* locus is a hotspot, but this cannot entail that insertions *must* be found there. And of the three sites that were targeted twice, all could be shown to be independent events for one or more of the following reasons: the target-site duplications were of different lengths, the initial cleavage events were on opposite strands of the DNA, or Alu elements of different sub-families were involved [56]. Study of an insertion hotspot in mice ruled out any homoplasious insertions that might have confounded phylogenetic inferences [57].

The hotspot issue is so important for phylogenetic analysis that exhaustive studies looking for insertion-site homoplasy have been performed. Five hundred LINE-1 elements that belong to sub-families peculiar to the human genome were selected. The insertion sites in the human genome were characterised. The corresponding sites in the genomes of other primate species, from chimps to pro-simians (galagos) were then investigated. The question asked was: in *any* of the other species had *any* LINE-1 element inserted into *any* of the 500 sites? The answer was 'no'. There was not a single case in which a LINE-1 element peculiar to another primate lineage shared the same insertion site as one of the human-specific elements. Insertion-site homoplasy is too infrequent to be detected [58].

Similar studies have been performed on Alu elements. The insertion sites of 'young', human-specific Alu elements have been investigated in a representative selection of other primate species. *In no case* was an Alu element characteristic of sub-families belonging to those other species found in the same site as the human insert. Independent insertions into the same site occur too infrequently to be identified [59]. It has been concluded that 'no instances of insertion homoplasy in hominids have been recovered from the analysis of >2,500 recently integrated human Alu insertions' [60].

Insertion-site homoplasy occurs too infrequently to complicate evolutionary studies.

Two primate species *independently* acquire TEs at the one insertion site only as a rare event, below the level of detection. It follows that the probabilities of independent insertions of TEs into the same sites of three or more species are progressively more remote. We can also test predictions based on these shared insertions. If a TE is shared by humans and (say) one OWM species, then we can predict that this TE originated in a human–OWM ancestor, and will be shared by humans and *all* OWM species. Results to date demonstrate that this is the case. Shared TEs are potent markers of common descent. The only way of denying this is to suggest that transposable elements do not transpose; or that insertional mutagens do not insert or are not mutagenic; or that TEs that are actively transposing at present are fundamentally different entities from TEs of the same families that are shared by multiple species. We would have to posit that the defined classes of TEs are not the entities that all workers in the field consider them to be.

Certain rare events may complicate phylogenetic analyses based on TE insertions. The phenomenon of incomplete lineage sorting has already been encountered. *Gene conversion* is a phenomenon in which a variable length of genetic sequence from one element is substituted for the equivalent length of DNA sequence in a related element. This changes the sequence of internal regions of Alu elements, and so does not alter the Alu presence/absence pattern in species. It is apparent only by DNA sequencing, which also reveals the extent of the converted sequence. It is of little significance to the derivation of phylogenetic relationships. Finally, *exact deletion* of Alu elements occurs at a low frequency. Such deletions occur by homologous recombination between identical target-site duplications that are at least ten bases long [61]. These events do not occur at sufficiently high frequencies to undermine TE-based phylogenies. TEs are thus essentially homoplasy-free characters. They provide the compelling evidence for descent-with-modification of which Darwin could not even have dreamed [62].

Hitherto we have focused on humans as the reference organism and have established the monophylicity of humans with other hominoid primates (apes), with anthropoid primates (apes and monkeys) and ultimately with all primates. This anthropocentric strategy has been necessitated by the research emphasis on humans. It is thus important to stress that the phylogenetic relationships of gibbons, OWMs, NWMs and prosimians [63–6] have also been resolved by the patterns of the distribution of Alu elements particular to those taxa. Each of these groups is monophyletic, and their phylogenetic radiations have been deduced [67].

One spin-off of such work is that panels of selected Alu elements permit the classification of primate tissue without the need for other morphological or biochemical data [68]. If you find a tuft of hair, or a bit of bone, in the back of someone's truck, it is now a straightforward matter to extract the DNA and determine, using selected Alu inserts, what species it represents. Are you dealing with a kidnapper or a poacher? TEs provide unique species markers, and this is a reflection of their power in defining evolutionary relationships.

2.4 MORE ANCIENT ELEMENTS AND MAMMALIAN EVOLUTION

The mammals are divided into three groups. Mammals that are nourished before birth through a placenta connected to their mother's uterus lining constitute the *Eutheria*. This term is derived from the Greek terms *eu-* (true or well-formed) and *theria* (wild animals or beasts). We have placed ourselves, somewhat immodestly, among the *real beasts*. In addition to these are the marsupials (the *Metatheria*; or loosely, the *other beasts*) and egg-laying monotremes (the *Prototheria*; which, it seems, have not attained true beasthood).

The Eutheria include many orders, and hundreds of species, which differ radically in size and shape. What is our evolutionary relationship with this spectacular diversity of real beasts? Considerations of size and shape may be of limited assistance. A more recent, and widely used, method of generating phylogenetic trees is to align DNA sequences from multiple species and computationally determine their degrees of relatedness. These statistical methods can generate misleading results, but increasingly extensive data sets have yielded progressively more reliable phylogenies. Such analyses since 2001 have indicated that eutherian mammals are classifiable into three major divisions, each of which is composed of an astonishing diversity of forms. These have the impressive names listed below [69].

First, there are the *Boreo(eu)theria*, or the *northern real beasts*. Their phylogenetic origins are associated with the ancient boreal (northern) landmass. The northern real beasts themselves fall into two major subdivisions:

- *Euarchontoglires* are of special interest to us, although this clumsy term reflects a haphazard etymological history. The word incorporates the components *eu-* (true), *archonta* (*ancestor* of humans and other primates) and *glires* (Latin for dormouse). This group includes five orders: primates, flying lemurs (Dermoptera), tree shrews (Scandentia), rabbits and their friends and relations, and rodents. For simplicity I will refer to it as the primate–rodent group.
- *Laurasiatheria*, or the *Laurasian beasts*, are believed to have arisen in the ancient Laurasian landmass of North America and Eurasia. These include cattle and whales, horses and rhinos, carnivores, bats, moles and hedgehogs.

Second, *Afrotheria*, or the *African beasts*, are associated with origins in Africa. They include elephants, manatees and aardvarks.

Third, *Xenarthra* are associated with South America. Their defining feature is the possession of *xenos* (strange) *arthra* (joints). These include sloths, armadillos and anteaters.

Should we have any faith in taxonomic divisions that are not readily apparent from the study of anatomical features? Can such arcane statistical approaches, understood only by the expert, really be trusted to provide a valid picture of phylogenetic divisions? Are humans really related to mice? And, if so, are we more closely related to mice than we are to dogs or horses? Here one may think wistfully of the way in which TEs have demonstrated how humans, apes and monkeys have developed as successive sub-clones within the primate clone, and wonder whether a suite of comparable genetic markers could ever resolve the more distant relationships between mammals.

2.4.1 *Euarchontoglires: the primate–rodent group*

We have considered the fact that the human genome possesses over 1,000 types of TE, and that some of these are very ancient. Members of these older classes are heavily mutated and no longer possess the capacity to transpose. These include the LINE-2 elements, which once copied and pasted as autonomous agents, and the *mammalian-wide interspersed repeats* (or MIR elements) that co-opted the LINE-2 enzymatic machinery to transpose. Essentially all LINE-2 and MIR inserts in our genomes were domiciled in the primate genome before the hominoid–OWM ancestor lived. As the genomic revolution gained momentum, TE data started to appear that provided evidence for human and rodent relationships within a primate–rodent clone.

An early study identified two elements (an MLT1A0 and a LINE-1 insert) that are shared by primates (humans, baboons) and rodents (rats, mice) but not by mammals classified as Laurasiatherians (cats, dogs, cows, pigs). The insertion site of the MLT1A0 element is shown in Figure 2.13. The target site and its duplications are recognisable, although they have become ragged with the passage of stupendous amounts of time. Nevertheless some reconstructive work shows that the dog retains the original target site (`GTCAT`). Sequence conservation on both sides of the target site (extending beyond that depicted) is apparent for all eight species. Conversely, two elements were identified that are shared by cats, dogs, cows and pigs, but not by primates and rodents [70]. Collaborating data were at hand. Several MIR elements were identified in the genomes of humans and representative rodents (rats, mice and guinea pigs) [71]. More than 300 MIR elements have been *exonised* – that is, MIR sequences have been co-opted into segments of genes (*exons*) that are destined for assembly

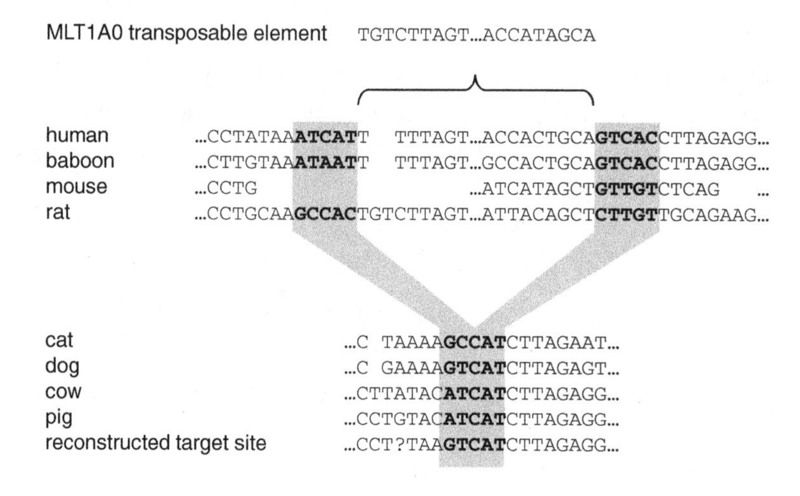

FIGURE 2.13. THE INSERTION SITE OF AN MLT1A0 ELEMENT IN PRIMATES AND RODENTS BUT NOT LAURASIATHERIAN SPECIES [70]

into mature transcripts. Many of those most strongly exonised are shared by primates and rodents [72]. Such MIR elements date from *no later* than the last common ancestor of primates and rodents. These findings vindicate the thesis that Euarchontoglires (the primate–rodent group) and Laurasiatheria are real categories.

Following the publication of the human and mouse genome sequences, computational studies have identified large numbers of TEs that are shared by the two species. In particular, the genomic locations of many old LTR and LINE-1 elements, as well as LINE-2 and MIR elements, are correlated in the two species. In a DNA segment encompassing 1 million DNA bases from the two species, 13 LINE-2 and 30 MIR elements were shown to be shared, or *orthologous,* inherited from the ancestor in which each element entered the DNA. Those TEs that are shared by both species have arisen from insertional mutagenic events and originally lacked functional capacity. It is likely that they have been maintained by selection, suggesting that they now perform useful functions [73].

The branching pattern within the Euarchontoglires has been elucidated. Computational searches have identified five TEs that are shared by primates and tree shrews, and in one case, by the flying lemur. These are all absent from rodents. On the other hand, nine TEs were found to be shared by rodents and rabbits, but are absent from primates [74]. Primates, tree shrews and flying lemurs form one monophyletic group; rodents and rabbits comprise another. But ultimately, we and the mice in the garden share many TEs, each of which arose in the genome of a Euarchontoglires ancestor.

2.4.2 *Boreoeutheria: incorporating the primate–rodent group and the Laurasian beasts*

The next issue that demands attention is whether Euarchontoglires and Laurasiatheria share TE inserts that would necessarily entail their derivation from the same ancestral stock. Early returns for the *PRKAG3* locus in humans, mice and pigs showed an abundance of ancient TEs. Approximately 50% of the LINE-1 elements and 80–5% of LINE-2 elements in the human sequence are found in the same position in the pig sequence. There is also considerable sharing of individual LTR, MIR and MON elements, and of DNA transposons. Each shared element is testimony to a single insertion event that occurred in a Boreoeutherian ancestor of humans and pigs. Of course, many more recently acquired elements are scattered along each genome sequence, and frequently insert into the older ones. Such 'young' TEs in one species always lack similarly located counterparts in the other species, and were inserted into the respective genomes after divergence of the lineages leading to human and pig. The lack of matching partners in the case of the 'new' elements demonstrates the significance of finding matching 'old' elements in the genomes of different species [75].

Our genome contains a large number of DNA segments that are highly conserved among the diversity of beasts that constitute Boreoeutheria. These segments do not have protein-coding function, and are probably gene-regulatory modules. An early survey of such conserved DNA modules, at least 50 bases long, found that more than 10,000 of them originated as parts of TEs. These included LTR,

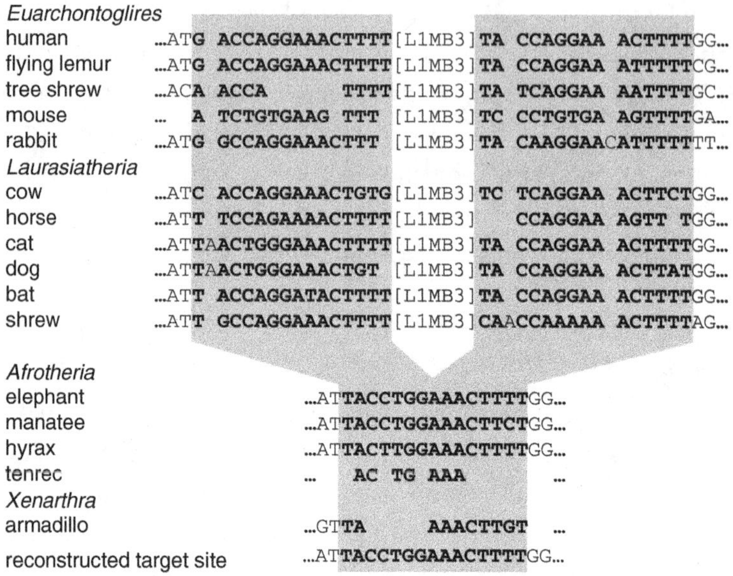

```
Euarchontoglires
human          ...ATG ACCAGGAAACTTTT[L1MB3]TA CCAGGAA ACTTTTGG...
flying lemur   ...ATG ACCAGGAAACTTTT[L1MB3]TA CCAGGAA ATTTTTCG...
tree shrew     ...ACA ACCA       TTTT[L1MB3]TA TCAGGAA AATTTTGC...
mouse          ...  A TCTGTGAAG TTT [L1MB3]TC CCTGTGA AGTTTTGA...
rabbit         ...ATG GCCAGGAAACTTT [L1MB3]TA CAAGGAACATTTTTTT...
Laurasiatheria
cow            ...ATC ACCAGGAAACTGTG[L1MB3]TC TCAGGAA ACTTCTGG...
horse          ...ATT TCCAGAAAACTTTT[L1MB3]   CCAGGAA AGTT TGG...
cat            ...ATTAACTGGGAAACTTTT[L1MB3]TA CCAGGAA ACTTTTGG...
dog            ...ATTAACTGGGAAACTGT [L1MB3]TA CCAGGAA ACTTATGG...
bat            ...ATT ACCAGGATACTTTT[L1MB3]TA CCAGGAA ACTTTTGG...
shrew          ...ATT GCCAGGAAACTTTT[L1MB3]CAACCAAAAA ACTTTTAG...

Afrotheria
elephant                    ...ATTACCTGGAAACTTTTGG...
manatee                     ...ATTACCTGGAAACTTCTGG...
hyrax                       ...ATTACTTGGAAACTTTTGG...
tenrec                      ...   AC TG AAA        ...
Xenarthra
armadillo                   ...GTTA     AAACTTGT   ...
reconstructed target site   ...ATTACCTGGAAACTTTTGG...
```

FIGURE 2.14. THE INSERTION SITE OF AN L1MB3 ELEMENT IN EUARCHONTOGLIRES AND LAURASIATHERIAN SPECIES [79]

LINE and MIR elements as well as DNA transposons [76]. Four years later, a similar analysis of the genomes of 29 eutherian mammals revealed that the number of conserved DNA modules originating from TEs was in fact 280,000. Such randomly accrued TE sequences have been preserved because they have acquired new uses [77].

Multi-species genome database searches have described in detail several well-preserved TEs that corroborate the common ancestry of Euarchontoglires and Laurasiatheria. The insertion site of a L1MB3 element is shown in Figure 2.14. Humans share this particular element with cows, horses, cats, dogs, bats and shrews. The undisturbed (albeit somewhat degenerated) target site is discernable in elephants and armadillos. Humans and bats are descended from the single reproductive cell in which the unique insertion event occurred. People may query *how* the echolocatory system of bats evolved; but there is no doubt that bats evolved from ancestors that lacked this capacity – ancestors that we share [78, 79].

	Species	5′ flank	MIR element	3′ flank
E	human	... AATA**TTTTT**	[AATAGCTACC ... TTGAGCACAG]	**TTACT**T...
	chimp	... AATA**TTTTT**	[AATAGCTACC ... TTGAGCACAG]	**TTACT**T...
	mouse	...GAATG**TTCTA**	[AACAACTGCC ... TATAGCACA]	**ACT**T...
	rat	...GAGTC**CCCCA**	[CACAACT ...	
	rabbit	...TAAGG**CTGAA**	[AACAACTACC ... TAGAACACAT]	**TTA T**T...
L	whale	...TAATG**TGATT**	[AGTGACTCTC ... TTGAGCACAG]	**TTACT** ...
	cow	...AAACA**TGACT**	[AGTGACTCTC ... TTGAGCACAG]	**TTAGT**T...
	pig	...GACGA**GGATG**	[AGGAACTGTC ... AGTTGCA TC]	**TTACT**T...
	horse	...TAATG**TTATT**	[AACAACTGTC ... TAGAGCACAG]	**TTACT**T...
	Asiatic golden cat	...TCATG**TTATT**	[AATAGCTATC ... GAGAGCACAT]	**TTCCC**T...
	Pallas's cat	...TCATG**TTATT**	[AATAGCTATC ... GAGAGCACAT]	**TTCCC**T...
	domestic cat	...TCATG**TTATT**	[AATAGCTATC ... GAGAGCACAT]	**TTCCC**T...
	masked civet	...TGTTG**TTATT**	[AATAACTCTC ... TAGA CACA]	**TTCCC**T...
	grey wolf	...TCATG**TTATT**	[AATGACAATC ... TACAGCACAA]	**TTGCT**T...
	brown bear	...TCGTG**TTACT**	[YATAACTATC ... TAGAGCACAG]	**TTACC**T...
	raccoon	...TCATG**TTATT**	[GATAACTGTC ... TAGAGCACAG]	**TTACC**T...
	marten	...TCATG**TTATT**	[GATAACTGTC ... TAGACCACAG]	**TTACC**T...
	sea lion	...TCATG**TTATT**	[GATAACTGTC ... GAGAGCACAG]	**TTACC**T...
	pangolin	...TGATG**TTACT**	[AATAACTATC ... TAGAACACAG]	**TTAAT**T...
	bat	... TA**TTGTC**	[AGTAACTCTC ... TAGAACACTC]	**TCACC**T...
	mole	...CATGG**AGGAC**	[AGCAAACAGC ... GGGGGCACGG]	**TTGCT**T...
A	elephant	...TACAG**TCATT**	[AATAACCAGC ... TAAATCA G]	**TTACT**T...

possible target site ...TAATG**TTACT**T...

FIGURE 2.15. THE INSERTION SITE OF A MIR ELEMENT IN INTRON 7 OF THE β-FIBRINOGEN GENE [80]
E, Euarchontoglires; L, Laurasiatheria; A, Afrotheria.

2.4.3 *Eutheria*

The search for shared TEs has taken us back to the base of the eutherian radiation. Dare we ask whether all eutherian mammals (that is, the real beasts that nourish their fetuses through a placenta) share an even older subset of TEs? Are we Boreoeutherians derived from ancestors shared with Afrotheria and/or Xenarthra? An answer is at hand. As an illustrative example, an ancient MIR element in an intron of the β-fibrinogen gene has been identified in the genomes of 22 species (Figure 2.15). The target site and its duplications are identifiable as the five-base sequence TTACT. That sequence is retained in one of the target-site duplications of eight species (from each of the three super-orders) and can be obtained by a one-base change in 18 other cases. We share this ancient TE with elephants [80]. Another MIR element that has been recruited into the genetic machinery

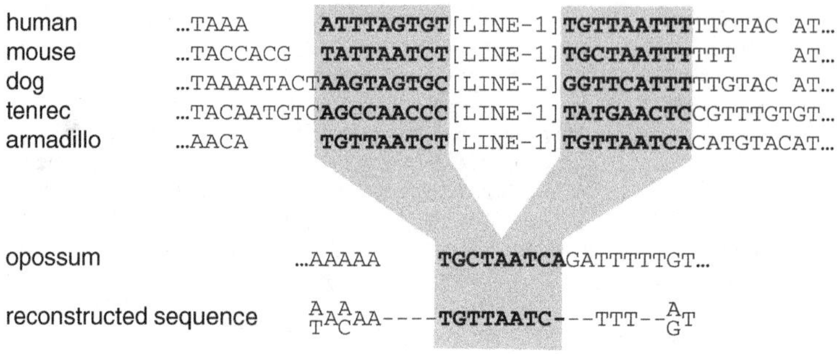

```
human        ...TAAA       ATTTAGTGT[LINE-1]TGTTAATTTTTCTAC AT...
mouse        ...TACCACG    TATTAATCT[LINE-1]TGCTAATTTTTT    AT...
dog          ...TAAAATACTAAGTAGTGC[LINE-1]GGTTCATTTTTGTAC AT...
tenrec       ...TACAATGTCAGCCAACCC[LINE-1]TATGAACTCCGTTTGTGT...
armadillo    ...AACA       TGTTAATCT[LINE-1]TGTTAATCACATGTACAT...

opossum                     ...AAAAA   TGCTAATCAGATTTTTGT...

reconstructed sequence      A A                          A
                            TACAA----TGTTAATC---TTT--T
                                                         G
```

FIGURE 2.16. THE INSERTION SITE OF AN LINE-1 ELEMENT IN REPRESENTATIVES OF EACH OF THE EUTHERIAN SUPER-ORDERS [82]

that boosts expression of the *TAL1* gene (active during blood cell stem cell development) is also shared by humans and elephants (but not marsupials) [81].

A LINE-1 element shared by representatives of all three eutherian super-orders is shown in Figure 2.16. The nine-base target-site duplication has undergone multiple substitutions, but may be reconstructed as TGTTAATC-, the last base being uncertain. The uninterrupted pre-integration site is present in the opossum (a marsupial). This insert is one of at least five identified by the two surveys described above [82]. Humans, mice and armadillos represent an extended and divergent clone. We are all ultimately derived from the one reproductive cell of a eutherian progenitor into the genome of which one particular LINE-1 element insinuated its piece of DNA.

Scientists have developed software that aligns homologous sequences from multiple species and reconstructs the ancestral sequence from which all were derived. They have subsequently scanned these reconstructed sequences for the presence of TEs. This procedure has been performed on a DNA segment of 1,870,000 bases surrounding the *CFTR* gene. (Damage to this gene underlies cystic fibrosis, a disease particularly common in people of European descent.) DNA sequences subject to this analysis were taken from 18 Boreoeutherian species and the armadillo, a representative of Xenarthra. The reconstructed ancestral sequence was found to

```
human                       ...ACATTAAAAATAAAAGACCC[L1MB8]AAAAATTAAAAAGCTACCTT...
chimp                       ...ACATTAAAAATAAAAGACCT[L1MB8]AAAAAGTAAAAAGCTAACTT...
macaque                     ...ACATTAAAAATGAAAGATCT[L1MB8]AAATATTGAAAAGCTGCCTT...
rabbit                      ...ACATTAAAAATGAG      [L1MB8]AAAAATG CAGAGCACTTTT...
dog                         ...ACATTAAAAATGTG AGAGCT[L1MB8]AAATATGAAAGAGATGTCTT...
elephant                    ...ACATTAAAAATGAA ACAGCT[L1MB8]TTTGTAAATTGAGTTGCTCT...
tenrec                      ...ACATTAAAAATGAA ACAGCT[L1MB8]AAAAAGAAGAAAGTTGTTTT...

armadillo                            ...ACGT AAACATGAA AGAGCTGTCAT...
sloth                                ...ACATTAAAAATGAA AGACCTCTCAT...
opossum                              ...TCAGTAAAATTAGA AGTATCTTTCT...
reconstructed target site                 AAAAATGAA AGAGCT
```

FIGURE 2.17. AN L1MB8 INSERT DATING FROM A BOREOEUTHERIA–AFROTHERIA (HUMAN–ELEPHANT) ANCESTOR [86]

possess a large number of very old TEs that had degenerated to such an extent that they were not recognised by computer searches of any one of the sequences on its own. Conventional scans of the human region showed that L2 and MIR elements occupied 46,200 and 34,700 bases, respectively; scans of the reconstructed sequence found an additional 15,300 and 5,600 bases occupied by TEs of these two families. This procedure operates on the assumption that there *is in fact* an ancestral sequence, and its success in identifying TEs that are invisible to conventional analysis establishes its validity. One such marginally discernable TE was spliced into the DNA of an ancestor shared by humans and armadillos [83].

But some unfinished business remains. What is the order of branching of the three super-orders that constitute Eutheria? Inconsistent data had been obtained [84]; perhaps TE analysis had reached its limit. This conundrum initiated a new search for informative TEs that would resolve the deepest branching of Eutheria. And the results were unexpected. All three branching patterns were strongly supported. Many TEs were identified (9 and 25 in two independent studies) that indicated that the Xenarthra lineage branched off first (and that Boreoeutheria and Afrotheria were derived from an ancestor that lived subsequent to this event; see Figure 2.17 for

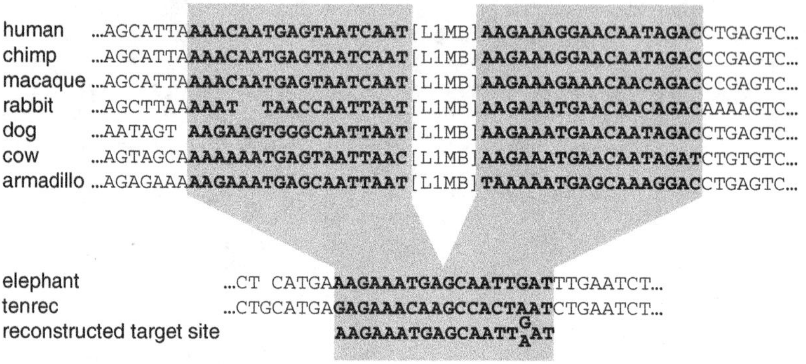

```
human      ...AGCATTAAAACAATGAGTAATCAAT[L1MB]AAGAAAGGAACAATAGACCTGAGTC...
chimp      ...AGCATTAAAACAATGAGTAATCAAT[L1MB]AAGAAAGGAACAATAGACCCGAGTC...
macaque    ...AGCATTAAAACAATGAGTAATCAAT[L1MB]AAGAAAGAAACAACAGACCCGAGTC...
rabbit     ...AGCTTAAAAAT   TAACCAATTAAT[L1MB]AAGAAATGAACAACAGACAAAAGTC...
dog        ...AATAGT AAGAAGTGGGCAATTAAT[L1MB]AAGAAATGAACAATAGACCTGAGTC...
cow        ...AGTAGCAAAAAAATGAGTAATTAAC[L1MB]AAGAAATGAACAATAGATCTGTGTC...
armadillo  ...AGAGAAAAAGAAATGAGCAATTAAT[L1MB]TAAAAATGAGCAAAGGACCTGAGTC...

elephant                   ...CT CATGAAAGAAATGAGCAATTGATTTGAATCT...
tenrec                     ...CTGCATGAGAGAAACAAGCCACTAATCTGAATCT...
reconstructed target site         AAGAAATGAGCAATT(G/A)AT
```

FIGURE 2.18. AN L1MB INSERT DATING FROM A BOREOEUTHERIA–XENARTHRA (HUMAN–ARMADILLO) ANCESTOR [86]

an example). A comparable number of inserts (5 and 22 in the two studies) showed that Afrotheria branched off first (Figure 2.18). And a similar number (8 and 21) indicated that the Boreoeutherian lineage was the first to diverge [85, 86].

We should recognise such anomalous patterns as indicating incomplete lineage sorting. A considerable number of ancient TEs were in a polymorphic state when the pioneers of the three superorders started to go their separate ways. This indicates that the basal radiation of the eutherian mammals was a trifurcation, a three-way split. From this it was hypothesised that eutherian mammals in Africa, in the Eurasian–North American continent and in South America were separated almost contemporaneously by continental drift that disrupted the Gibraltar and Brazilian bridges. As the continents slipped apart, the loss of gene flow gave rise to Afrotheria, Boreoeutheria and Xenarthra. Be that as it may, the bottom line is that humans share multiple clonal markers (and therefore ancestors) with elephants and sloths.

Who could have anticipated that the study of clonal markers could illuminate cell lineages so removed from the study of lymphocyte or tumour cell clones? TEs have provided a detailed and compellingly supported outline of eutherian evolution. A summary of the studies described hitherto is given in Figure 2.19.

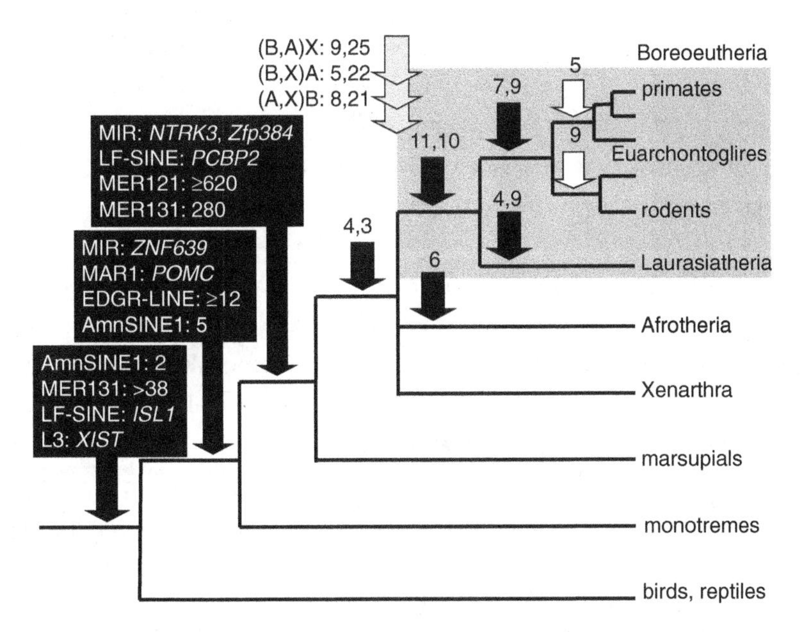

FIGURE 2.19. A PHYLOGENETIC TREE OF MAMMALS CONSTRUCTED FROM TE INSERTIONS
Numerals indicate the number of studied TEs inserted at each stage of phylogenetic development. Numbers of TEs from paired independent studies are shown. (B,A)X indicates that Boreoeutheria and Afrotheria are sister groups and Xenarthra branched out first [87].

2.4.4 *Mammals*

There remain two more orders that have not been incorporated into the TE-generated mammalian phylogeny. The most remotely related extant mammals are marsupials (such as the opossum) and monotremes (such as the platypus). Might TEs provide signposts regarding the monophylicity of all mammals? Aeons have elapsed since these lineages branched from our own, during which non-essential DNA will have decayed into oblivion. So if we share TEs with opossums or platypuses, we would expect to find them only in a form in which they are preserved as components of genes or regulatory modules.

We have in fact encountered such co-opted markers. Ty3/Gypsy retrotransposons have been exapted to support placental function (see Chapter 1). These TEs gave rise to the *PEG10* and *PEG11* genes. Each was inserted into a reproductive cell from which both eutherians and marsupials are descended. *PEG10* was transmogrified into a protein-coding gene in both marsupials and eutherians, but *PEG11* degenerated in the former group and persists as a protein-coding gene only in the latter.

MIR elements have been recruited as transcribed components of many genes. One MIR element has been assimilated into the *NTRK3* gene, and another into the *Zfp384* gene. We possess these elements in common with opossums (but not with platypuses). They thus date from a eutherian–marsupial ancestor. We share an element in the *ZNF639* gene with opossums and platypuses (but not with chickens) [88]. And a related CORE-SINE element is part of a highly conserved enhancer that drives expression of the *POMC* gene that encodes a set of hormones and the neurotransmitter β-endorphin. This TE also dates from an ancestor we share with platypuses. (A variety of other ancient elements also drive the *POMC* gene [89].) Indubitably, we share a common ancestor with platypuses. The monophylicity of all extant mammals is hereby established. These and further instances discussed below are included in the summary (Figure 2.19).

Subsequent evidence has only supported this conclusion. Scientists scrolling through genome databases have discovered particularly ancient TEs. Many contain sequences that are highly conserved between distantly related species. These TEs are strongly represented in *conserved non-coding elements.* They have survived and resisted changes in sequence – they are under purifying selection – because they have been domesticated to fulfil regulatory or structural roles. They are also signatures of very ancient evolutionary connections. Five of these fascinating TEs are introduced below.

- MER121 elements are present as 900 copies in the human genome. A substantial proportion (82%) of the individual elements are highly conserved in humans, rodents and dogs, and most (68%) are conserved also in marsupials [90].

- LF-SINEs were discovered as a class of TE in the genome of the coelacanth: a deep-sea lobe-finned fish. The acronym *LF-SINE* stands for *lobe-finned fishes* (or *living fossil*)-*short interspersed elements*. There may be 100,000 copies in the coelacanth genome, some of which retain target-site duplications. The human genome possesses 245 LF-SINE elements. Nearly all of these are shared with other eutherian and marsupial mammals, including one exonised into the *PCBP2* gene. At least some elements predate the divergence of amphibians from reptiles, birds and mammals (amniotes). One of these acts as an enhancer of the *ISL1* gene [91].
- EDGR-LINE sequences were discovered in the tuatara genome. The acronym stands for *endangered* LINE, reflecting the fact that the history of this class of TE might not have been known if the endangered tuatara had become extinct. Eighteen instances of this class of TE were subsequently found as conserved elements in mammalian genomes, at least 12 of these being common to humans and platypuses [92].
- Several hundred AmnSINE1 elements are scattered around the human genome. Humans share at least 130 of these with opossums and more than 40 with platypuses. A couple are found at orthologous locations in the chicken genome [93]. Two of these ancient AmnSINE1 sequences are present in highly conserved non-coding elements linked to the *FGF8* and *SATB2* genes, which are active in mammalian brain development. Perhaps AmnSINE1 exaptation, innovations in brain development and emergence of mammals are closely linked [94].
- MER131 elements were discovered only when the genome of the short-tailed opossum was sequenced. This genome is TE-rich, and analysis led to the discovery of 83 new families of TEs that are represented in amniotes (terrestrial tetrapods: mammals, birds and reptiles). Of 516 MER131 elements in the human genome, 280 are conserved in opossum. Of the 68 most-highly conserved elements, 38 elements were found to have orthologues in chicken [95].

Studies of the *XIST* gene identified a LINE-3 element that resides at orthologous positions in the genomes of mammals and of chickens. Ancient LINE-3 elements are another resource for identifying *really* distant relatives [96]. Indeed a large number of highly conserved elements are shared by us, other mammals and birds, and are comprised (at least in part) of ancient TEs. More than 100 families of TE are involved [97].

Sequencers have their eyes on the genomes of 10,000 vertebrate species. Analysis of the genome of the lizard *Anolis carolinensis* identified multiple families of ancient TE (recently active in lizards but long extinct in mammals) that are recognisable in our genome as conserved non-coding elements. Thirty of these elements are present in conserved non-coding elements common to all mammals. The time when such TEs actually entered the vertebrate germline may be considerably earlier than the last common ancestor of mammals. Eleven of these exapted elements are clonal markers of the mammal–bird–reptile (amniote) clade [98].

2.4.5 TE stories on other branches of the tree of life

This survey has necessarily been anthropocentric. We have considered TEs currently recognisable in the human genome, whether possessed only by one individual, or shared by all mammals, or by all amniotes. But TEs are ubiquitous. The genomes of all organisms seem to have their own communities of resident TEs. Such TEs are being used to disclose the phylogenetic development of other taxa.

The rodents comprise 40% of all mammalian species. TE analysis has elucidated the relationships between mice, guinea pigs and squirrels. Retrotransposons peculiar to lagomorphs have revealed how rabbits, hares and pikas are connected. Statistical approaches, such as the aligning of DNA sequences, had yielded hopelessly contradictory results. Both groups are monophyletic [99].

TEs have served to tease out an early Laurasiatherian history marked by extensive incomplete lineage sorting and by gene transfer between nascent, but still interbreeding, species. An initial burst of rapid speciation led to a network of relationships, not a bifurcating tree [100].

Until recently, whales were classified in an order of their own, Cetacea. A surprising result shattered this old understanding. A set of TEs was found to be shared by whales and hippos. This discovery was the first (and sufficient) indication that whales and hippos evolved from a common ancestor. Whales were promptly incorporated into the order Cetartiodactyla alongside even-toed ungulates (which includes hippos, pigs, giraffes, deer, sheep and cattle) [101]. Within the Cetartiodactyla, TEs have clarified whale and dolphin relationships. A retrotransposon insert that is a specific marker of baleen whales led to the loss of enamel-capped teeth [102]. Similar analyses in ruminants have demonstrated that giraffes have evolved from an ancestor shared with sheep and deer. There is no doubt that the giraffe evolved from short-necked forebears, even if the mechanism requires clarification [103].

The distribution of TEs has elucidated Afrotherian, and Xenarthran phylogeny [104]. Ancient TEs have demonstrated that the marsupials of Australia and South America have emerged from the same stock. Common marsupial origins evince an early history on the ancient southern landmass of Gondwanaland [105].

TE analysis has been applied productively to the phylogeny of galliform birds (chickens, quail and peafowl), to penguins [106], and to the passerine birds (loosely, perching and songbirds, which comprise most avian species), parrots and falcons. And TEs also show that all living birds – from hummingbirds to ostriches – are monophyletic [107]. The great majority of TE insertions in birds reflect phylogenetic relationships unambiguously. One study involving 66 insertions identified two sites that might have been independently targeted in different taxa (although in one of these the TEs were oriented differently), and two sites where exact deletion of a TE seems to have occurred [108].

TEs have provided a definitive classification of the cichlid fish of the African Rift Lakes, of which a large number of species have appeared with spectacular rapidity. One might expect that incomplete lineage sorting would be rife in this situation of explosive speciation – and indeed it is. Nevertheless, geneticists have learned to live with incomplete lineage sorting [109].

Anatomists have struggled doggedly for centuries, constructing taxonomies on the basis of the morphology of living and fossil species. In the last few decades, geneticists have discovered that organisms may be characterised by their own intricate barcodes – assemblages of TEs stitched into their genomes. The morphologists did a pretty good job, hampered as they were by the complexities of convergent evolution. However, of very recent years, they have been greatly assisted by the advent of TEs, which are close to being ideal phylogenetic markers.

2.5 EXAPTATION OF TES

Primate genomes have been invaded and expanded by millions of segments of DNA in the form of LTR, LINE-1, Alu and SVA elements. The genomes of mammals share the DNA footprints of more ancient invasions by LINE-2, LINE-3 and MIR elements, as well as a rich diversity of DNA transposons. To what extent has the accumulation of TE-derived DNA served as raw material to provide new functions to the host organisms? TEs that are harmful reduce the viability of their host organisms. TEs that perform no function would tend to degenerate and disappear with time. TEs remaining in our genomes are to some extent a selected group. Surviving TEs were once summarily dismissed as 'junk'. But that prejudicial assessment failed to take into consideration the extraordinary proclivity of organismic systems to co-opt unclaimed expanses of DNA.

TEs have been entrained to contribute to genetic function in many ways. Three are described below. Firstly, TEs have provided DNA sequences out of which new genes have been formed. Secondly, they have added variety to pre-existing genes by providing new exons. As noted, they have been *exonised*. And, thirdly, they have complexified genome function by donating segments of DNA that have been co-opted into regulatory circuitry.

These genome-constructing consequences of TE activity are the outcome of familiar genetic processes. Repeatedly, TEs have been transformed, under the constraining influence of natural selection, from randomly firing units of mutagenic DNA (that were of no use to the organism in which they were generated) into modules possessing genetic functionality (that have contributed to the viability of that organism's descendants).

2.5.1 *Raw material for new genes*

Alu elements have contributed new genes to our genome. A classical example is the *BC200* gene (later renamed *BCYRN1*). This arose from an Alu element that spliced itself into the genome of an ancestor of the apes and monkeys. This Alu element spawned a multitude of daughters (some 200 are scattered around the human genome), but itself became domesticated into a gene that currently functions in anthropoid primates. The *BCYRN1* gene generates an RNA transcript that is not translated into a protein (that is, the transcribed RNA is *non-coding*). This RNA molecule is expressed in the brain, particularly in neurons, and is part of a molecular complex that controls protein synthesis in dendrites. It may function in the control of neuronal plasticity, an integral part of learning and memory. This is a huge makeover for what was once a parasitic piece of DNA [110].

An Alu-derived gene family has been described. A primitive Alu element (of the FLAM_C sub-family) was transformed by a series of mutations into an ASR element (in an ancestor of apes and monkeys) and thence into a family of CAS elements (in an Old World primate ancestor). Subsequently, a CAS element underwent two small duplications to generate a first *snaR* gene in an ancestor of the African great apes. This progenitor was copied many times by being embedded in sequentially duplicated segments of DNA, generating families of *snaR* genes in African great apes. The duplicated units are marked by fragments of a gene that resided near the original FLAM_C element, by a LINE element and by other Alu elements. The *snaR* genes are transcribed into small RNA molecules that, like *BCYRN1*, do not encode proteins. They may regulate protein synthesis [111].

The *PMCHL1* and *PMCHL2* genes were assembled from several segments of genetic flotsam, one of which was an Alu element. These genes were created in a stepwise fashion that spanned several branch points of the primate family tree. Events included the copying of part of a precursor gene together with the insertion of an Alu element (in an Old World primate ancestor), the recruitment of downstream exons by mutations generating splice sites (in a hominoid ancestor), and gene duplication (in a human–chimp ancestor). These genes are active in the testis and fetal brain, and also encode non-coding RNA transcripts [112].

Finally, several Alu elements collaborated to form the *FLJ33706* gene, including one that was spliced into the primate germ-line in an ancestor of apes and monkeys, and that much later provided a protein-coding sequence. The gene came to acquire protein-coding capacity only on the lineage leading specifically to humans, when two critical enabling mutations occurred. First, a pre-existing TAG *stop* codon, which prevented production of a protein, mutated into an amino-acid-specifying TGG codon. Secondly, one base (a 'G') was deleted. This frameshift mutation created an *open reading frame*, a DNA sequence that specifies an amino acid chain of sufficient length as to generate a functional protein. The human-specific protein is expressed by neurons in the brain, and the gene is linked to a marker for nicotine addiction [113].

2.5.2 *Raw material for new exons*

Protein-coding genes generally consist of discontinuous segments of DNA that are destined to become parts of messenger RNA transcripts (called *exons*), separated by intervening sequences that are spliced out and discarded during the processing of transcripts (called *introns*). One gene may generate multiple messenger RNA species because various combinations of exons may be selected during RNA processing. The strategy by which one gene generates diverse messenger RNA species increases the complexity of genetic function.

In the human genome, TEs have been recruited repeatedly into genes as novel exons. TE sequences occur in 4.4% of all gene transcripts, and in 0.5% of all DNA sequences coding for proteins [114].

Table 2.2. *Examples of genes containing exonised TEs*

Gene	TE	Lineage into which TE was inserted	Lineage in which exonisation was completed	Ref.
leptin receptor	SVA	Human	Human	115
survivin	Alu	Hominoid	Hominoid	116
PKP2B-4	Alu	Hominoid	Hominoid	117
MTO1–3	Alu	Old World primate	Old World primate	117
*ADRA1A**	Alu	Old World primate	Old World primate	118
NARF	Alu	Simian	African great ape	119
RPE2–1	Alu	Simian	Great ape	117
TNFR2/ TNFRSF1B	Alu	Simian	Old World primate	120
c-Rel-2	Alu	Simian	Old World primate	117
SFTPB	Alu	Primate	Human?	121
ZRANB2	Plat_L3	Primate	Simian?	122

* MIR3 and L1MC5 elements common to simians were also exonised.

The mechanisms by which TEs have been exonised have been elucidated for several examples (Table 2.2).

The *survivin* gene provides an example. The splice sites required for incorporation of the Alu sequence into the messenger RNA product were present from the time of insertion. The *acceptor* splice site was present in the left-hand target-site duplication, and two possible *donor* splice sites were present in the Alu element itself, yielding two differently sized Alu-derived exons (Figure 2.20). This suggests that the Alu element could have been exonised immediately upon entry into the *survivin* gene [116]. In other cases, series of mutations were needed to incorporate TE sequences into exons.

A survey of 330 Alu-derived exons has shown that most are minor components of gene output. However, six cases were described in which Alu-derived exons are present in all transcripts arising from their respective genes. Initially, exonisation may occur inefficiently and be of minor consequence but, with time, natural selection may

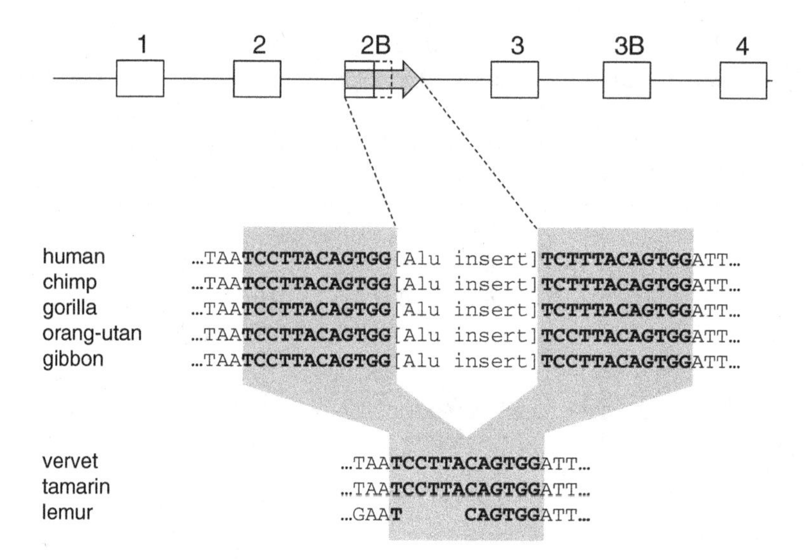

FIGURE 2.20. ALU ELEMENT EXONISATION IN THE *SURVIVIN* GENE [116]
Numbered boxes represent exons. The *dotted box* indicates an extension of exon 2B that arose from the inserted Alu (*arrow*). Not to scale.

lead to more efficient incorporation of advantageous TE-derived exons into mature transcripts [123].

2.5.3 *Raw material for new regulatory modules*

TEs have assumed a plethora of regulatory roles. Alu elements, for example, have been recruited to provide regulatory input into all levels of gene expression [124]. The genomic age has made life much more complicated for people who study genetic function. Not long ago, it was thought that the small proportion of the genome that was occupied by protein-coding genes (1.5%) was responsible for most of the RNA that was transcribed. Now it is known that nearly all of the genome is transcribed and so performs vital (albeit diverse and hitherto poorly understood) roles in specifying the biological properties of organisms [125].

DNA sites mediating control of genome functions may be defined as sequence motifs (such as the conserved non-coding elements described above) that are believed to act as enhancer and repressor sites. Alternatively, regulatory sites may be defined functionally by their ability to bind transcription factors. These groups are not mutually exclusive. TEs feature in each.

Analysis of the opossum genome (published in 2007) showed that a large number of *conserved* but *non-coding* DNA elements appeared in eutherian genomes. That is, they are absent from chicken or opossum, but present in human plus rodent or dog. At least 16% of all such conserved elements (numbering 33,760) were shown to be formed either wholly or in part by TEs representing multiple families (including ERVs, LINE elements, MIR elements and DNA transposons) [126]. Just 5 years later, a comparison of 29 eutherian genomes (published in 2012) identified 285,000 conserved non-coding elements that are derived from TEs. That is, at least 11% of all conserved non-coding elements identifiable in our genome are derived from TEs. Or to express the relation from the perspective of the TE population: 6% of all TEs have acquired stable functions as shown by the fact that they contain sequence motifs that resist change [127].

Experimentally, regulatory DNA sequences may be identified by their sensitivity to cleavage by endonucleases. A favourite approach is to look for *DNase 1 hypersensitive sites*. Several million of these have been found in diverse cell types, and 900,000 have been located in LTR, LINE, SINE and DNA elements. There may be extensive overlap between conserved non-coding elements and DNase 1 hypersensitive sites, but the involvement of TEs in each emphasises the vital, life-sustaining responsibilities assumed by our erstwhile parasites [128].

Many of the transcripts found in human tissues are initiated from specific sites within TEs. The proportion of such TE-initiated RNA molecules varies according to the tissue: from 3% in fibroblasts, 4% in liver, 5% in brain, to 16% in embryos. Individual TE-initiated transcripts tend to be expressed in particular tissues only [129].

The initiation of RNA transcription is under the control of proteins that bind to DNA at specific short DNA sequences. A catalogue of such *transcription factor binding sequences* in the human genome revealed that more than 4% were located, either wholly or in part, in TEs. These included representatives of multiple major classes of TEs: LTR, LINE-1, -2 and -3, Alu and MIR elements, and DNA transposons. Older elements tend to be over-represented in such functional motifs. They have survived in the genome because they have come to perform functions (transcription factor binding in this case) that natural selection has preserved [130].

However, Alu elements, which are primate-specific and young in evolutionary terms, have also contributed many short sequence motifs as binding sites for transcription factors [131]. In one instance, an Alu element situated near the gene for cathelicidin (an antibacterial protein) contains an internal nine-base duplication (CG**GGTTCA**A) that created a binding site (GGTTCA ... GGTTCA) for the vitamin D receptor. The novel vitamin-D-mediated control of cathelicidin arose potentially in an ancestor of apes and monkeys. It is testimony to our common ancestry with NWMs, and to the capacity of natural selection to transform randomly accreted DNA into life-sustaining regulatory information [132].

Particular TEs often harbour particular transcription factor binding sites. Selected studies are summarised in Table 2.3. Functional sites range in age from retinoic acid receptor-binding sites in Alu elements (most instances of which are common to Old World primates) to LINE-2 elements (many instances of which have been shown to be shared by humans and their Boreoeutherian best friend, the dog). Genome-disrupting jumping genes have been entrained, repeatedly through mammalian history, to the task of reconfiguring genome-orchestrating regulatory networks.

A striking example of TE-mediated genome reorganisation is found in the appearance of binding sites for the protein CTCF. Genomes contain thousands of CTCF-binding sites, and these act to demarcate functionally distinct regions of chromatin. These genome-organising motifs are up to 34 bases long, and may be either

Table 2.3. *TEs containing binding sites for transcription factors*

TE	Age of TE	Transcription factor(s)	Roles of regulated genes	Ref.
Alu	primate only	retinoic acid receptor	development, differentiation	133
		thyroid hormone receptor	energy metabolism	
		hepatocyte nuclear factor 4α receptor	development, liver function	
		oestrogen receptor	female reproduction	
		p53	tumour suppression	134
		MYC	proto-oncogene, cell division	135
		Oct4	maintenance of stem cells	
		NANOG	maintenance of stem cells	
		NFκB	inflammation, cell growth	136
		heat shock factor	responses to stresses	137
MER20	early Eutheria	progesterone-responsive	placental development	138
LINE-2	early mammal	REST (repressor)	neural development	139

lineage-specific (found only in particular groups such as rodents) or highly conserved in all mammals. In the former case, new CTCF-binding domains have been generated in genomes as parts of lineage-specific TEs. In the latter case, ancient CTCF-in-TE binding sites are embedded within a diversity of ancient TEs, and are common to Boreoeutheria. Some are common to eutherians and marsupials, and date from a human–opossum ancestor. A few are shared by mammals and birds, and thus arose in germ cells that were ancestral to humans and chickens [140].

2.6 THE EVOLUTIONARY SIGNIFICANCE OF TES

Multitudes of TEs that have accumulated in our genome have been recruited to perform a diversity of roles. But they have done more than that. They have been players in reorganising the genome. And they have done so by participating in just the kind of reactions that are intrinsic to this extraordinary molecule called DNA.

This chapter concludes with some provocative hypotheses regarding the role of TEs in evolution. Perhaps genomes tolerate these endogenous mutagens over the long term because there are times when some genomic destabilisation is advantageous. Perhaps stressful environmental conditions promote TE activity as a means of increasing the variation upon which natural selection can act to facilitate adaptive responses.

2.6.1 TEs, genomic reorganisation and speciation

TEs have facilitated the reorganisation of genomes. We have considered the propensity of ERVs to undergo non-allelic homologous recombination involving paired LTR sequences. Such events result in the deletion of all the internal sequence and of one LTR-equivalent (Chapter 1). TEs scattered through genomes also act as substrates for recombination. Two TEs of the same class may undergo a recombination event, resulting in the loss of the DNA that lies between them as well as of one TE-equivalent. One chimaeric TE is retained.

On short timescales, recombination events, leading to the loss of intervening sequences, cause genetic diseases [141]. But on evolutionary timescales, the same processes reorganise genome content, contributing to the transformation of the genome of progenitor species into those of descendant species. After the human lineage diverged from the human–chimp common ancestor, more than 70 recombination events occurred between LINE-1 elements. These trimmed the human genome by some 450,000 bases [142]. During the same time period, more than 490 recombination events occurred between Alu elements. These excised another 400,000 bases from the human genome [143]. It goes without saying that the same processes have taken place in the chimpanzee-specific lineage also. Some 660 Alu-mediated deletion events have excised 770,000 bases between the human–chimp ancestral genome and the chimp genome [144]. The occurrence of such genomic changes demonstrates that human–chimp ancestral sequences have been transformed into our own (and into those of chimps) by familiar mechanisms intrinsic to the nature of DNA. Humans and chimps are connected by a story.

The *SIGLEC* gene family exemplifies such molecular interconversions. *SIGLEC* genes encode proteins (Siglecs: sialic acid-binding Ig-like lectins) that are located on cell surfaces and that bind to sugar derivatives called sialic acids. Such Siglec–sialic acid interactions are molecular handshakes that contribute to recognition between cells involved in immunity. One of the family, *SIGLEC13*, along with five Alu elements, is present in the genomes of chimps and OWMs, but absent in humans. En route to the human genome, the entire gene was neatly excised by a recombination event between the left-handmost Alu element and the Alu element just to the right of the last exon (Figure 2.21). The truncated human locus is marked by a chimaeric Alu element, consisting of portions of each of the two elements that participated in the rearrangement [145].

Recombination events involving two Alu elements may also generate chromosomal duplications. Alu-mediated recombination events have led to reiterated duplication events that have generated a large family of microRNA genes. Individual members are arranged in series, in a head-to-tail, bumper-to-bumper tandem configuration. MicroRNAs are powerful regulators of gene expression. They act upon other RNA molecules, either preventing their translation into proteins, or orchestrating their degradation. Alu-mediated recombination may have exerted major effects in the control of cellular behaviour. MicroRNAs generated from this gene cluster may function by suppressing genome-wide Alu retrotranspositional activity [146].

Side-by-side TEs also participate in recombination reactions that result in DNA inversions (rearrangements in which a segment

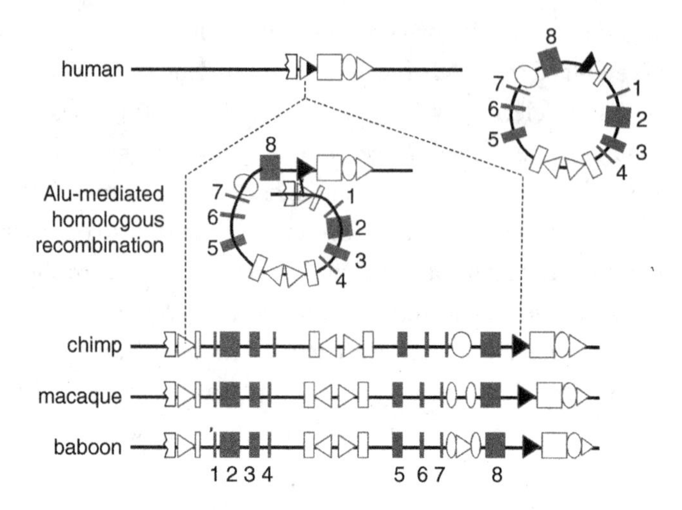

FIGURE 2.21. HUMAN-SPECIFIC DELETION OF THE *SIGLEC13* GENE BY HOMOLOGOUS RECOMBINATION BETWEEN ALU ELEMENTS [145]
The diagram depicts *SIGLEC13* exons (*numbered grey vertical lines* and *boxes*), Alu elements (*triangles*), LINE-1 elements (*white boxes*) and ERVs (*ovals*). The intermediate stage representing Alu-mediated homologous recombination, depicts the point at which two Alu elements underwent DNA breakage and exchange. The excised loop of DNA (*upper right*) was lost.

of DNA is flipped by 180 degrees with respect to the sequences on either side of it). Oppositely oriented Alu elements may align when DNA bends into a hairpin structure. This allows the TEs to undergo effective breakage-and-rejoining recombination. The result is an expanse of DNA that reads in the reverse orientation, bracketed by a pair of chimaeric TEs with non-matching target-site duplications and flanking sequences (Figure 2.22). Approximately 50 TE-mediated recombination inversions have occurred in humans and chimps since their common ancestor, and several genes have been disrupted. The familiar processes of TE-mediated deletion, duplication and inversion may have contributed significantly to speciation [147].

The genome contains many simple sequences that are present as repeated units. If the basic unit is composed of less than ten bases, such as the three-base sequence $[\mathrm{AAG}]_n$, the repeat is called a

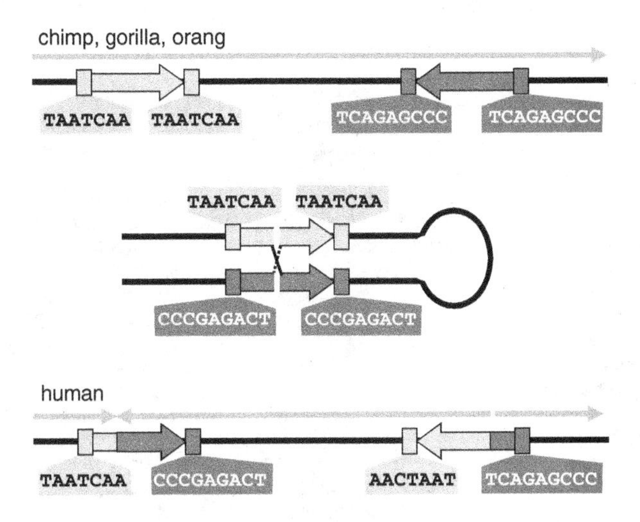

FIGURE 2.22. RECOMBINATION BETWEEN TES IN OPPOSITE ORIENTATIONS GENERATES CHROMOSOME INVERSIONS [147]

microsatellite; if the basic unit consists of ten or more bases, the repeat is called a *minisatellite*. These repeated sequences are often located within TEs such as LINE-1 and Alu elements. The polyA tails of Alu elements may be considered to be ready-made microsatellites, $[A]_n$, which can mutate readily into other microsatellite sequences [148].

Microsatellites tend to have high mutation rates. The number of repeat units changes readily. They underlie some 40 genetic diseases. The ability of TEs to introduce such sequences into genomes destabilises them. For example, an Alu element that appeared in a simian ancestor (and that has been identified in the genomes of each of 18 species of apes and monkeys) is the source of a complex microsatellite that is responsible for the development of the disease myotonic dystrophy type 2 (Figure 2.23). The microsatellite contains $[TG]_n$, $[TCTG]_n$ and $[CCTG]_n$ repeats. Expansion of the latter repeat, to hundreds or thousands of copies, perturbs gene function and leads to disease [149].

One practical use of microsatellite sequences is in the identification of criminals, human remains and children's fathers in

Alu insert

repeat sequence

human	...TATTACTGCCAG	[X ...AGCC] ACTACCAATTAT	TTCT...
chimp	...TATTACTGCCAG	[X ...AGCC] ACTACCAATTAT	TTCT...
bonobo	...TATTACTGCCAG	[X ...AGCC] ACTACCAATTAT	TTCT...
orang	...TATTACTGCCAG	[X ...	...
siamang	...TATTACTGCCAA	[X ...AGCC] ACTACCAGTTAT	TCT...
agile gibbon	...TATTACTGCCAA	[X ...AGCC] ACTACCAGTTAT	TTCT...
white-handed gibbon	...TATTACTGCCAA	[X ...AGCC] ACTACCAGTTAT	TTCT...
rhesus macaque	...	[X ...GGCC] ACTCGCCATTATTTGTTCT...	
bonnet macaque	...	[X ...GGCC] ACTCGCCATTATTTGTTCT...	
De Brazza's monkey	...	[X ...GGCC] ACTCGCCATTATTTGTTCT...	
patas monkey	...	[X ...GGCC] ACTCGCCATTATTTGTTCT...	
blue monkey	...	[X ...GGCC] ACTCGCCATTATTTGTTCT...	
hamadryas baboon	...	[X ...GGCC] ACTCGCCATTATTTGTTCT...	
mandrill	...	[X ...GGCC] ACTCGCCATTATTTGTTCT...	
silvered lutong	...	[X ...GGCC] ACTCGCCATTATTTGTTCT...	
hanuman langur	...	[X ...GGCC] ACTCGCCATTATTTGTTCT...	
marmoset	... ATTACTACCAATTA	[XI...AGTC] ACTACCAATTAT	TTCT...
owl monkey	... ATTACTACCAATTA	[XI...AGTC] ACTACCAATTAT	TTC...
tufted capuchin	... ATTACTACCAATTA	[XI...AGTC] ACTACCAATTATT	TTCT...
inferred ancestral sequence	... ATTACTACCAATTAT		TTCT...

FIGURE 2.23. AN ALU ELEMENT COMMON TO SIMIAN PRIMATES AND CONTAINING A MICROSATELLITE SEQUENCE (X) [149]
Also present is a small insertion (I) in NWMs.

paternity testing. All the lawyers in the world have never advanced a valid reason why DNA should not be used to establish relatedness between people [150].

2.6.2 *TEs and evolvability*

Genetic systems can be controlled *epigenetically*. That is, DNA or associated chromosomal proteins can be tagged by chemical signals (such as methyl or acetyl groups) that modulate gene activity. These tags are stable and heritable but are subject to regulation and are therefore reversible. The ability of TEs to transpose is subject to epigenetic control [151].

Evolutionary change is a compromise between two extremes. A low mutagenic burden is an advantage to the individual organism, because this would minimise the incidence of genetic diseases. For this reason, organisms possess regulatory mechanisms to suppress the insertional mutagenesis associated with TE activity. However, such genomic stability may be incompatible with adaptive change over the long term. A high mutational burden is disadvantageous to individuals, but promotes variation upon which natural selection can work, and hence promotes the development of adaptations. Under normal conditions, there may be an optimal level of TE activity that is compatible with both individual survival and lineage adaptability [152].

There is evidence that stress, acting through genetic or epigenetic mechanisms, increases the ability of TEs to replicate. Forms of such stress might include DNA-damaging agents, heat, oxidative stress, inflammation or viral infection [153]. Environmental agents are able to co-opt a variety of classical signalling pathways by which they may induce L1 retrotransposition and its concomitant insertional mutagenic burden. Many routes may connect the outside world and the hidden microworld of the genome [154].

TEs may therefore affect genome stability in a way that fluctuates according to environmentally imposed pressures. A well-adapted population of organisms may experience low levels of physiological stress and would have a low degree of transpositional activity. These conditions would engender a state of evolutionary stasis. On the other hand, a population of organisms confronted with novel or stressful environments might undergo increased rates of transposition, with resulting genetic instability, and a concomitant acceleration of evolutionary change. The result would be an increased frequency of harmful mutations, but also an elevated capacity for adaptation and speciation. Patterns of long-term stability interspersed by bursts of active speciation are well recognised and have been called *punctuated equilibrium* [155].

The environment–epigenetics–transposition connection might constitute a feedback mechanism according to which selective pressures elevate mutation rates and facilitate evolvability (Figure 2.24). Indeed, TEs may themselves modify epigenetic systems in the organisms that carry them. If these hypotheses are borne out, then the

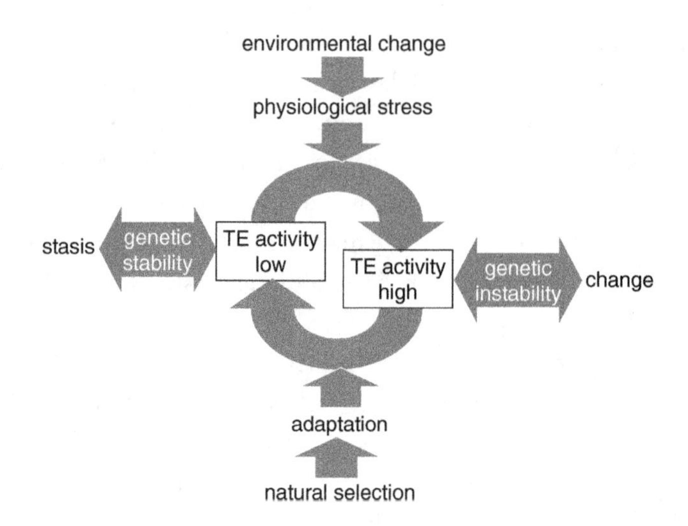

FIGURE 2.24. DO TES MEDIATE FEEDBACKS BETWEEN ENVIRONMENTAL STRESS AND GENETIC INSTABILITY? [155]

ubiquitous presence of TEs might be part of the essential novelty-generating requirement that makes life possible. The randomness of TE behaviour might be an essential part of the adaptability and fecundity of life.

Human evolution over the last few million years has been profound. Striking innovations in our biology include bipedality and the increase in brain size. It has been suggested that this rapid evolution may correlate with, and may have been driven by, a high rate of retrotransposition [156]. John Mattick has championed the paradigm that non-coding RNA has vital roles in the genetic regulation of complex organisms. He suggests that the waves of Alu insertions into primate genomes have provided a substrate contributing to the versatility of transcriptional and epigenetic regulation, necessary for interactions between the environment and epigenetic regulation of the genome, and thus contributing to the development of cognitive function [157].

It is generally stated that half of our genome is derived from ERVs and TEs. The application of more-sophisticated software that allows the identification of more degenerated (fragmented) TEs has raised this estimate to two-thirds of our genome [158]. TEs have expanded, modified and elaborated our ancestors' genomes at least as far back as genetic analysis can detect. Mattick's revolutionary theorising may be generalisable over biological history.

It may be that TEs are maintained in host genomes simply because they are efficient parasites, and not because they confer evolvability on lineages of organisms. Either way, they have had a major role in the evolution of genetic novelty and of new species [159]. For our purposes, they have provided an extraordinarily comprehensive (and to a cell biologist, elegantly satisfying) answer to the questions of the fact and the route of our phylogenetic history.

3 Pseudogenealogy

Geneticists have studied mutations intensively over the years. They recognise various categories and can anticipate their likely effects. Some mutations involve major reorganisations of the genome. Segments of DNA may be deleted, duplicated, inverted or exchanged between chromosomes. If these involve sufficiently large portions of chromosomes, they will be visible microscopically in mitotic chromosomes. At the other extreme of the size spectrum, mutations may change only one or a few bases. One base may be substituted for another. One or a few bases may drop out of the sequence (be deleted) or drop in (be inserted).

The genetic code is the information system according to which the sequence of bases in DNA specifies the sequence of amino acids in proteins. The code was spectacularly deciphered in the 1960s. A run of three bases in DNA (a base triplet, or *codon*) specifies the identity of an amino acid in a protein. Three particular codons have the function of specifying the end of the protein-coding section of a gene sequence. These are called the *stop* codons, and are `TAG`, `TGA` and `TAA`. Any base change mutation that transforms a codon specifying an amino acid into a codon specifying the *stop* command will terminate the protein. The resulting protein will lack all content between the *stop* mutation and the authentic *stop* codon. Such premature termination is typically disastrous for protein function.

The nature of the triplet code also means that the deletion or insertion of bases *that are not in multiples of three* will scramble genes by putting all the downstream triplets out of phase. Consider the sentences of three-letter words below, the second of which has a one-letter deletion:

The big fat dog ate the hot bun.
Teb igf atd oga tet heh otb un.

The loss of one letter reduces all subsequent text to gibberish. In precisely the same way, the deletion of one base (or of two, four, five, seven ... bases) is catastrophic for protein sequence and activity. Such insertions and deletions (or collectively, *indels*) that disrupt the triplet coding frame of genes are known as *frameshift mutations*. They are a familiar class of disease-causing mutations.

Other single-base changes may obliterate DNA motifs to which gene-regulating proteins bind. As a result, transcription from those genes may be suppressed. Mutations may destroy short DNA motifs that act as signals to direct splicing of messenger RNAs. These *splice-site* mutations will result in misassembled messenger RNA molecules with garbled information content. In short, there is an extensive catalogue of recognised mutation types, and an innumerable variety of possible mutations. It follows that the sharing of particular mutations by cells or organisms constitutes powerful evidence that those cells or organisms are related: that is, that they have inherited their particular, distinctive mutation from the progenitor in which the mutation occurred. This is the situation we have described as the *casanova phenomenon*.

Mutations indicate lines of descent. The logic of this statement may be illustrated from the history of Charles Dinarello, a scientist who was instrumental in the discovery of interleukin-1β (IL-1β), a cytokine that regulates inflammation. He submitted a manuscript, which contained the sequence of the IL-1β gene, to the prestigious scientific journal *Nature*. The editor of the journal sent the manuscript for review to scientists working for a biotech company. The biotech scientists told the editor that the manuscript was not worth publishing, and then proceeded to use Dinarello's sequence in a report of their own, which they submitted for publication (successfully) to the same journal, *Nature*. Dinarello had done the work. Others got the credit. Cunning indeed.

But then Dinarello discovered that his original IL-1β gene sequence contained seven errors – and that all of those errors were present in the sequence that the biotech scientists had claimed as their own. The presence of the same few errors in Dinarello's and the biotech company's gene sequences proved that one sequence had been copied from the other. The biotech scientists had appropriated Dinarello's data. The conclusion was unavoidable. The company did not even go to court, but agreed to pay out $21,000,000. If the original IL-1β sequence had been fully correct, the theft of Dinarello's data would not have been apparent [1]. Singular mistakes in manuscripts establish lines of descent.

A mutation that arose as a one-off event in a single cell, but is present in many descendants, is known as a *founder mutation*. In human populations these genetic markers can often be recognised as singular mutations embedded within a DNA sequence environment (the *haplotype*) that is relatively homogeneous compared with the same genetic locus found in the rest of the population. The shared mutation-plus-haplotype reflects the unique source of the mutation. Examples affecting humans are found in many people who suffer from iron-overload disease, sickle-cell anaemia and haemophilia [2].

A founder mutation underlies blue eye colour. The *OCA2* gene has a vital role in determining the colour of our eyes. It stretches over more than 345,000 bases of DNA, and specifies a protein that helps control the production of a brown pigment called melanin. Blue-eyed people first appeared in human history as a result of a base substitution mutation in a regulatory element that controls the activity of the *OCA2* gene (Figure 3.1). The equivalent DNA sequence is shown for nine other mammals. The central six-base `TAAATG` sequence highlights a protein-binding site that controls *OCA2* gene activity, and is invariant except for blue-eyed people, who have a `TAAGTG` sequence. Blue-eyed people from Denmark, Turkey or Jordan have the same mutation in the same haplotype. All are descended from the *one individual* in whom the A-to-G mutation occurred, approximately 10,000 years ago [3].

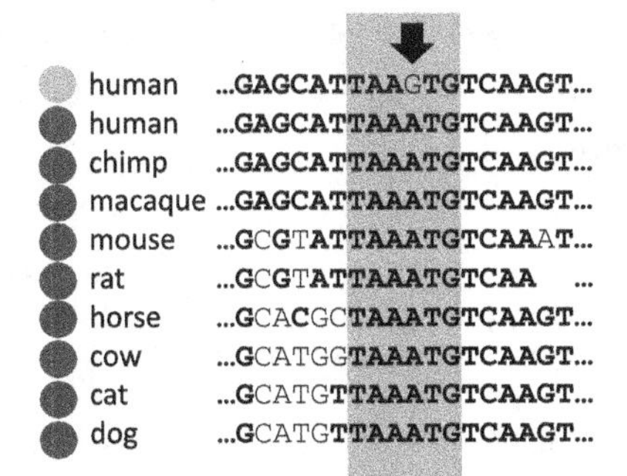

FIGURE 3.1. THE MUTATION RESPONSIBLE FOR BLUE EYES [3]
The protein-binding site is *shaded*. All bases that are the same as those from brown-eyed humans are in **bold**. The A-to-G mutation is *arrowed*. *Dark circles* represent brown (pigmented) eyes; the *light circle* represents blue eyes.

3.1 MUTATIONS AND THE MONOCLONAL ORIGINS OF CANCERS

Cancers arise when proto-oncogenes are mutated into deranged oncogenes, and when tumour suppressor genes are mutated into oblivion. Such mutations may occur spontaneously, or following, exposure to ionising radiation, ultraviolet radiation or chemicals, such as oxidants generated during long-term inflammatory conditions, hydrocarbons in cigarette smoke, fungal toxins in food or heavy metals in industrial air.

Regardless of the source, mutations arise randomly. A practically infinite number may affect the genes targeted in cancer – especially the tumour suppressor genes. If a particular mutation initiated oncogenic development, it would be present in every cell that was subsequently generated. Shared mutations demonstrate that cancers are monoclonal. This principle may be illustrated by the multiple cancers that may arise in the urinary systems of some individuals. These cancers may occur at the same time or in chronological sequence. Oncologists have asked whether these multiple tumours arise independently of each other (in which case each tumour might be expected to possess its own characteristic set of mutations) or whether they are all spawned by one progenitor tumour cell (in which case all tumours would share the mutations that give rise to the first malignant cell).

In the case of tumours in the bladder, a convenient tumour suppressor gene is at hand. This is the famous *TP53* gene that encodes the p53 protein. *TP53* is frequently mutated in bladder cancer, and sustains an extensive range of mutations in bladder cancers from different patients. Studies of *TP53* mutations have shown that, in individual patients with multiple tumours arising in lymph nodes, kidney, bladder or ureter, all of the tumours have the *same* mutation. This demonstrates that the multiple tumours are monoclonal, derived from the *one* delinquent cell in which the mutation arose [4]. Perhaps tumour-initiating cells spread locally through the plane of the epithelium, or over long distances via the hollow interior of the bladder and its associated ducting [5].

Studies of kidney cancers show how tumour development is branching (not linear), how all the cells that possess a particular mutation may be seen as descendants of the one founding cell in which that mutation occurred, how the succession of sub-clones may be delineated by a succession of new mutations, and how a single gene can be found to be independently mutated in different clones (Figure 3.2) [6].

Cell lineages in the colon have been studied intensively. The colon is lined by tiny glands called crypts. The cells populating each crypt are typically monoclonal, derived from one active stem cell that resides at the base. The study of unique mutations, appearing in mitochondrial DNA, has shown that single crypts may divide into paired crypts, and eventually into patches of crypts, all of which are tagged by the same mutation, and are therefore clonally related. Local spread of single clones of cells occurs in colon epithelial tissue that is morphologically normal [7].

The colon may degenerate into a chronic inflammatory condition called ulcerative colitis, in which long-term bombardment with damaging forms of oxygen wreaks havoc upon DNA. Abnormal

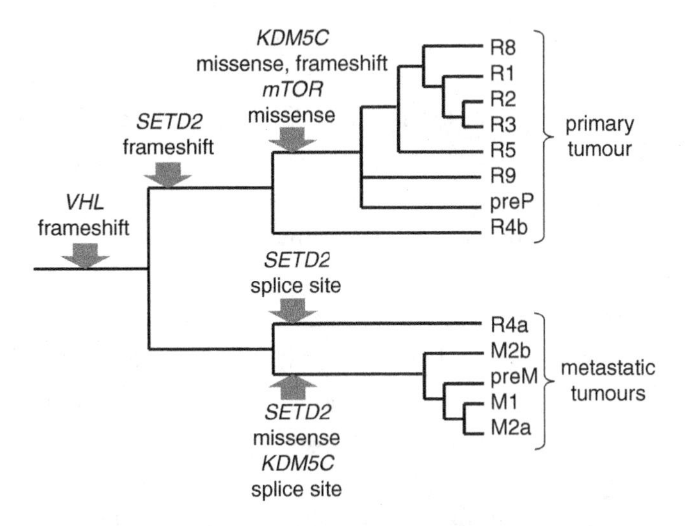

FIGURE 3.2. CLONAL ORIGIN OF A KIDNEY CANCER, AND DERIVATIVE SUB-CLONES [6]
Mutant genes that define clonal development are indicated.

crypts appear over extended areas of epithelium, and multiple tumours may emerge. Are these monoclonal or do they represent the growth of independently altered cells? Mutational analyses have shown that, in many cases, multiple altered crypts and tumours share mutations, indicating that one mutated stem cell had generated the expanding field of (monoclonal) progeny. In one case, a patient with ulcerative colitis developed two different sorts of cancer: a rare neuroendocrine carcinoma and several adenocarcinomas. Both types of cancer had the same *TP53* mutation, evidence that they were derived from the one altered stem cell. However, widely separated zones of the colon epithelium possess different mutations, indicating that clones of altered cells also arise independently [8].

Colorectal cancers arise from the expansion of one malignant clone, and progressively develop into multiple sub-clones, defined by the sequential appearance of new mutations. The appearance of new mutations distinguishes malignant cell populations from their benign precursors, and may allow discrimination between metastatic tumours and the primary tumours from which they were derived. Multiple metastases may share the same new mutations, indicating that the metastases were derived from one aggressive sub-clone [9].

The DNA sequencing revolution has allowed tracking of mutations to elucidate clonal development in cancers as diverse as B-cell chronic lymphocytic leukaemia, breast cancer, liver cancer and pancreatic cancer. The sequential appearance of new mutations shows that tumour progression involves a linear succession of one clone by another, as well as branching patterns and the production of clonal dead ends [10]. Whole-genome sequencing of multiple *individual* cells from single tumours indicates that progenitor clones may generate distinctive sub-clones that are defined by multiple idiosyncratic genetic markers. Cancer cell populations with transitional collections of genetic markers are not detectable. Tumours thus seem to evolve in a stepwise manner via transitory 'missing links' – intermediate sub-clones that are rapidly replaced by much more populous and enduring descendent sub-clones. Cancers may develop by *punctuated* clonal expansion [11].

Mutational analyses of tumours have confirmed and extended the postulate of the clonal evolution of cancer, proposed by P C Nowell in 1971. Clonal populations of cells in cancers can be identified on the basis of shared mutations, and their evolution generates the same sorts of patterns as are seen with populations of organisms over evolutionary time. The cancer genome is an archaeological record of the sequence of genetic changes that have accumulated as the cancer has developed. 'Oncogenealogies' and 'tumour suppressor genealogies' provide detailed histories of tumour evolution. Cancer cell biologists must think in a Darwinian fashion [12].

3.2 OLD SCARS ON DNA

Every nucleated cell in our bodies contains approximately 2 metres of DNA. A microlitre of blood, with 5,000 white blood cells suspended in it, contains 10 kilometres of DNA. Our genomes are composed of 46 DNA molecules, each of which is packaged in proteins to form a chromosome. DNA molecules vary in length from 1 to 8 centimetres. Such extremely delicate threads can be broken – say by a wayward high-energy photon – and the cell may die. The induction of DNA breaks creates a cellular emergency. Cells have elaborate means of recognising and repairing double-stranded breaks.

One of these repair mechanisms is called *non-homologous end-joining* (NHEJ). This process rejoins broken DNA ends, but inexactly. The stitched-up break might be held together by extraneous segments of DNA, usually copied from a site nearby on the same chromosome, but sometimes copied from anywhere else in the genome. The repair site may have lost some of the original base sequence, either when the injury happened or during trimming of the loose ends by exonucleases. And a few bases may have been inserted that were not copied from any DNA template (so-called *non-templated* bases) [13]. In a repair-or-die emergency, the DNA repair machinery cannot be too fastidious. Any one double-stranded DNA break spliced together by NHEJ generates a unique mutational fingerprint that would provide a perfect marker of monoclonality in the descendants of the damaged cell.

3.2.1 *Classical marks of NHEJ*

Random DNA breaks followed by repair by NHEJ can occur in any cell – including those of the germ-line. If that is the case, we will have inherited recognisable scars scattered throughout our genomes, representing repair events that occurred in the bodies of our forebears. Scientists have scrolled though genome databases to search for segments of DNA that bear the fingerprints of DNA repair patches. The key indications are that a piece of DNA has been copied from one site into another, concomitantly with losses and gains of bases at the recipient site.

Many such repair patches have been found. Some of these scars have been found *only* in humans, and one of these is polymorphic – that is, in some people the relevant DNA site is intact, while in others the sequence has been disrupted and a DNA bandage

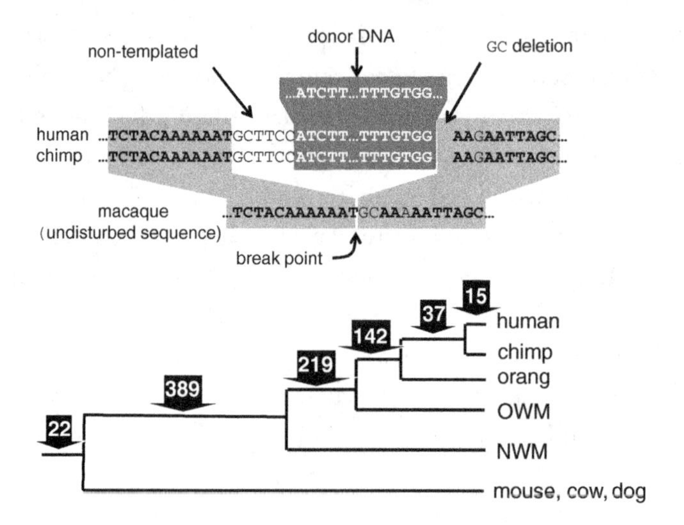

FIGURE 3.3. SCARS REPRESENTING DNA REPAIR BY NHEJ [14]
An example is shown (*upper diagram*). The length of DNA copied from elsewhere is indicated by *white lettering* starting ATCTT... on a *dark grey background*. Evolutionary stages at which NHEJ markers arose (*lower diagram*). *Numerals* indicate the number of repair sites identified.

has been interpolated. Remarkably, other repair jobs are *shared* by humans and chimps. One of these is shown in Figure 3.3 (upper diagram). The original, undisturbed sequence in the macaque genome is shown. However, in humans and chimps this sequence is disrupted, and an extensive length of foreign DNA (starting ATCTT ...) has been imported to join the ends. Six randomly inserted filler bases (GCTTCC) are present on the left-hand side of the break, and two bases (GC) have been deleted from the right-hand side. This example provides compelling evidence that humans and chimps share a common ancestor. Moreover, 36 other examples of shared repair patches linking humans and chimps to a common ancestor have also been found.

Hundreds of such DNA scars have been identified (Figure 3.3, lower diagram). Some are shared by humans, chimps and orangutans, but by no other species (142 in total). These scars arose in a great ape ancestor. Others are shared by apes and OWMs, but by no other species (219). Each of these dates from an ape–OWM ancestor. Many are common to all anthropoid primates (389). And some scars in our DNA are shared with mice, cows or dogs. These represent a relatively well-preserved selection of sites that have survived the vast tracts of time since the Euarchontoglires and Boreoeutheria ancestors lived [14].

The distribution of shared repair patches generates a phylogenetic tree that is congruent with that generated from ERV and retrotransposon insertions. NHEJ and retrotransposition are wholly independent processes, and so provide wholly independent signatures that illuminate our evolutionary past.

Our genomes possess a second class of small DNA duplication. Some 2,500 short lengths of DNA, 25–100 bases long, have been duplicated and inserted into new sites that are scattered around our genome. More than 90% of these duplications are shared with chimps. This general class of duplication is present in plants and animals, and may represent another (as yet undefined) mechanism of DNA break repair. It provides further evidence that humans and chimps share common ancestors [15].

3.2.2 *LINEs and Alus*

If a cell has a DNA break that urgently requires repair, any source of DNA that connects the loose ends could be co-opted. Cells in culture have been shown to acquire new LINE-1 inserts in a way that has nothing to do with the standard mechanism of retrotransposition. The LINE-1 endonuclease is not needed, and no target-site duplications bracket the insert. Rather, these anomalous LINE-1 sequences are associated with deleted or randomly inserted (non-templated) bases. It seems that cells, faced with the prospect of DNA break-induced death, readily use parasitic DNA to cobble repair patches together [16].

Scientists have searched the human genome for recently formed patches occupied by bits of LINE-1 sequence. (The criterion for being *recently formed* is that the LINE-1 elements have sequences that are very close to that of the standard type.) Twenty-one such LINE-1 elements were discovered, of which 14 instances were shared with chimps [17]. A similar search was conducted for Alu elements that manifest the properties of repair patches, and that are limited to great ape genomes. Thirteen such atypical Alu elements were found in the human genome. Four are specific to humans, one is shared by humans and chimps, and eight are shared by humans, chimps, gorillas and orang-utans. These numbers represent only a selection of the most-recent (least-diverged) retrotransposon-derived DNA patches in our genome [18].

DNA breaks are random and potentially catastrophic. NHEJ functions under urgency to rejoin the broken DNA ends. The molecular details of the NHEJ mechanism are yielding to stringent molecular analyses, which have revealed that the telltale messiness of the repair patch is inherent to the repair system. Random breaks are fixed by the desperate co-option of any available DNA, including bits of retrotransposon-derived sequence. Such patches are sufficient evidence of great ape monophylicity.

3.2.3 Numts

Other exotic sources of DNA may be appropriated to hold chromosome ends together. Medical geneticists investigating a family with an associated mental disorder discovered that the condition, tracking through successive generations, was carried by a particular chromosomal translocation (an exchange of portions of two different chromosomes). Translocations occur when *two* DNA breaks are mis-repaired by sticking the wrong ends together. When the junction of one of these translocations was sequenced, a length of mitochondrial DNA (41 bases long) was found at the breakpoint [19]. A fragment of the tiny chromosome located in a mitochondrion had been used to repair a chromosome break in the nucleus.

A second case implicating mitochondrial DNA as a molecular Band-aid also involved a mental retardation disease, lissencephaly. The DNA insertion was found in a child, but in neither parent, indicating that it had arisen afresh, probably in the germ cells of a parent. A length of mitochondrial DNA (130 bases from the *ATP8* and *ATP6* genes) had been spliced into exon 2 of the *PAFA1B1* gene. One base (a 'C') was lost at the insertion site, and two non-templated bases (TT) added – features typical of NHEJ [20].

Bits of mitochondrial DNA are found in the nuclear genomes of nearly all organisms, whether fungi, plants or animals. They are part of the DNA that characterises each species. It is likely that they were inserted into chromosomal DNA as fragments directly derived from mitochondrial DNA, and their chromosomal locations have the hallmarks of NHEJ products. These repair patches are called *nuclear sequences of mitochondrial origin*, abbreviated to *numts* (pronounced new-mites). They are also called *mitochondrial pseudogenes*, because they look like mitochondrial genes but are in fact non-coding derivatives thereof, spliced into nuclear genomic DNA [21].

The number of numts in the human genome varies according to the stringency of the criteria used to find them. Values have ranged from 211 to more than 1,200. Current compilations enumerate approximately 600. The smallest numt included in the catalogue is 31 bases long, and the longest is 14,904 bases long – almost the entire mitochondrial chromosome [22].

But when were they inserted? As before, this question requires that we know whether any one numt is present or absent in the genomes of each of a selection of primate species. Several studies have focused on particular numts. Some are found only in humans and are polymorphic. These arose so recently that they have been transmitted to only a fraction of the human population. Two large inserts that are found only in humans among extant species have been located in the genome of Neanderthals. The ages of these pseudogenes have been estimated at 620,000 and nearly 3 million years. At the other end of the temporal scale, numts have been shown to date from early in anthropoid (ape–monkey) history, as instances are shared by humans, OWMs and probably NWMs (Table 3.1) [23].

Table 3.1. *The distribution of numts in primate genomes [23]*

Mitochondrial gene from which the numt was derived	Species possessing the numt	Ancestor into which the numt inserted
NADH dehydrogenase subunit 5 (ND5)	Humans (polymorphic)	Human lineage
Displacement (D)-loop (control region)	Humans (polymorphic)	Human lineage
Multi-gene inserts (2)	Humans and Neanderthals	Human–Neanderthal
D-loop	All apes, humans to siamangs (a lesser ape)	Hominoid
16S rRNA	Humans and OWMs but not NWMs or prosimians	Old World primate
Cytochrome *b*	Humans, OWMs and probably NWMs	Anthropoid

Whole-genome surveys have indicated that most of the numts in our genome are shared also with chimps. Of 616 numts in the human genome, 502 (that is, over 80%) are present also in the chimp genome [24]. The presence of individual numts in both human and chimp genomes has abundantly demonstrated the monophylicity of humans and chimps. We and the chimps are indeed sister species, derived from common ancestors. But analysis of the presence and absence of individual numts in multiple primate genomes has also been initiated. Several partial studies have shown that humans, chimps and gorillas are monophyletic. So are the great apes (incorporating the orang-utans) and, further back in time, the apes (incorporating the lesser apes or gibbons). It appears that the majority of numts in our genome are shared with OWMs [25].

Are numts functional today? Quite possibly – but functionality is irrelevant to the issue of whether numts constitute markers of descent. It is the complex molecular *pathway* by which numts arose that makes them such compelling signatures of our shared ancestry with other primates. Each numt is a potent demonstration that the individuals and species possessing it are clonally related products of the reproductive cell in which the initiating repair event occurred. The expanding number of published genome sequences is providing scope for ever more in-depth numt analyses. These uniquely arising molecular Band-aids will continue to illuminate our phylogenetic pedigree.

3.2.4 Interstitial telomeric sequences

The ends of chromosomes are composed of specialised sequences called *telomeres.* Telomeres function as protective caps that maintain the integrity of chromosome structure, and so ensure continued genetic stability. If telomeres were not present, DNA repair enzymes would interpret the normal ends of chromosomes as being broken ends, and then 'repair' them by joining them together with other ends. But chromosomes joined end to end possess two centromeres, and cannot be separated normally when cells divide. Chromosome fusions scramble genomes, leading to cell death or cancer development. (Fused chromosomes require prompt silencing of one centromere if they are to behave appropriately.)

Telomeres consist of a distinctive base sequence. They are highly repeated units of the hexamer `TTAGGG`. It is notable that some telomeric repeats are present, not at the *ends* of chromosomes, but *within* them. These are called *interstitial telomeric repeats.* How did telomeric repeats get from the tips into the internal regions of chromosomes? Two pertinent stories must be told.

The first dates from 1982 when two pioneers of chromosome structure, Jorge Yunis and Om Prakash, spread the chromosome sets of the four great ape species on microscope slides, stained them with a dye and examined their structures using a microscope. The four sets of chromosomes looked strikingly similar – a graphic demonstration that humans are rightly classified among the great apes. But there was one radical difference between the human chromosome set and those of the other great apes. Human chromosome number

2 was shown to be a head-to-head (telomeric) fusion of two chromosomes that are separate in all the other great apes. Our chromosome set was derived from those of the other great apes by familiar genetic mechanisms, the most dramatic of which was a telomeric fusion [26]. More subtle microscopic differences between great ape chromosomes need not concern us here [27].

But if it is true that our chromosome 2 arose as a fusion of two smaller chromosomes, then some fossilised remnants of the ancestral telomeric repeats should be apparent at the point of fusion. The fossil fusion point was identified and cloned in 1991. It is somewhat degenerated, because highly repeated units of DNA are unstable. Nevertheless, the TTAGGG telomeric repeats are clearly visible (upper strand, left to right; lower strand right to left). The exact point of the fusion (*) is apparent – even though the event happened thousands of generations ago [28].

5′...TTAGGG TTAGGGG TTAGGG TTAG*CTAA CCCTAA CCCTAA...3′
3′...AATCCC AATCCCC AATCCC AATC*GATT GGGATT GGGATT...5′

Subsequent sequencing has confirmed this finding [29]. Different individuals vary in the sequence of this internal telomeric segment. But chromosome 2 has been sequenced in its entirety, and the vast tracts of sequence extending away from this head-to-head telomeric fossil correspond precisely with the two distinct progenitor chromosomes of the other great apes. We may conclude that our chromosome 2 is derived from an aberrant DNA repair event that stuck two chromosomes together. Humanity is descended from the reproductive cell in which that one-off mis-repair event occurred.

The second story relates an expedition into the genome to characterise other interstitial telomeric repeats. Some 50 short, well-conserved interstitial telomeric repeats, $(TTAGGG)_n$, are present in the human genome. Sequencing studies indicated that they arose as distinct insertion events, probably generated by the action of the enzyme *telomerase*, which has the function of adding TTAGGG units at the authentic telomeres. These inserts have the characteristics of

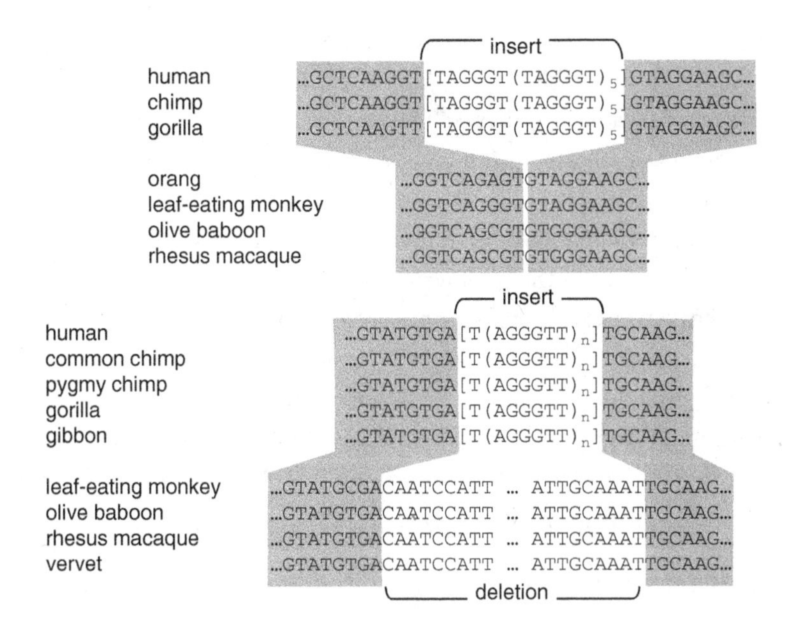

FIGURE 3.4. INTERSTITIAL TELOMERIC REPEATS SHARED BY THE AFRICAN GREAT APES (*UPPER DIAGRAM*) AND THE APES (*LOWER DIAGRAM*) [30] Flanking DNA is *shaded*. *Subscripts* indicate multiple copies of the repeat sequence.

emergency DNA repair patches that were recruited to hold double-stranded breaks together.

The presence or absence of ten of these telomeric repeats was ascertained in humans and other primate species. Each repeat was shown to have arisen at a particular stage of primate evolution. In some cases they precisely interrupt the pre-insertion sequence. The insert shown in Figure 3.4 (upper diagram) is not associated with any gains or losses of base sequence. It is a clean insertion, shared by humans, chimps and gorillas. The Asian great ape (orang-utan) and OWMs retain the uninterrupted pre-insertion site. This repeat unambiguously demonstrates African great ape monophylicity.

In other cases, the inserts are associated with the telltale random insertions and deletions that are the hallmarks of NHEJ. The example depicted in Figure 3.4 (lower diagram) shows a telomeric repeat that is present at precisely the same site in all ape species tested, including gibbons. In all species that possess this insertion, there is a deletion of 42 bases. The unique locus of the insertion and the size of the deletion provide a strikingly singular marker of hominoid monophylicity. This patch arose in a reproductive cell in a hominoid ancestor, establishing that we and the gibbons are descendants of the cell in which the telomeric insert arose.

This work can be summarised as follows. Three inserted telomeric repeats were shown to be specific to humans, two to the African great apes, two to all hominoids, two to Old World primates (apes and OWMs) and one to all simians. Such results provide cogent support to the fact of our evolution and to the validity of the now conventional anthropoid family tree [30].

3.3 PSEUDOGENES

About 5–10% of cancers have a strong hereditary component. The predisposition to developing cancers is transmitted by mutant, dysfunctional tumour suppressor genes that exist at low frequencies in the general population [31]. Databases of mutations for many such genes have been compiled, and they indicate that the responsible genes have been disabled by an enormous diversity of mutations. Sometimes mutations cluster in particular regions of genes called mutational *hotspots*.

For some genes featuring in hereditary cancer syndromes, such as the *RB1* and *MEN1* genes, and the *BRCA1* and *BRCA2* genes (implicated in breast and ovarian cancers), mutations are indiscriminately scattered over the gene sequence without evidence of recurring hotspots. *PTCH1* gene mutations (skin and brain cancers) are also widely distributed through the gene sequence with a small proportion (15%) found at ten favoured sites. But each one of these hypermutable sites generates multiple mutations. The *VHL* gene is inactivated in diverse cancers by a multitude of mutations – 87% of all codons have been targeted – although some parts of the gene are targeted more often than others [32].

The enormous diversity of mutations in familial cancer syndromes indicates that such mutations are powerful markers of family relationships (in kindreds in which cancer-predisposing mutants are transmitted) and of underlying relatedness (in populations through which founder mutations have diffused). In general, common mutations point to a shared genetic history. Even in the case of *TP53* gene mutations, which cluster in pronounced hotspots, mutations are potent markers of inheritance within families [33] and, in the case of non-inherited cancers, of tumour monoclonality (see Section 3.1).

Surprisingly, there are several classes of disabled genes that *all* of us share. Such derelict genes are an intrinsic feature of the genome that identifies us as human. These genes have lost the information content, once possessed by their progenitors, that is needed to make proteins. They are called *pseudogenes*, and are known to be mutants by precisely the same universally accepted criteria that identify disease-causing genes as mutants. They have been disabled by deletions and insertions, by *stop* mutations and frameshifts, and by the loss of regulatory and splice sites [34]. The recognition of such genetic aberrations is the everyday concern of geneticists, such as those who work in hospital laboratories. Pseudogenes may be classified into one of three sorts.

1. Some are degenerated forms of unique genes that, in a wide diversity of other organisms, retain their protein-specifying functionality. In the human genome, protein-coding capacity has been destroyed, and ancient functional roles extinguished. These relics comprise a minor proportion of the pseudogene complement in our genome, and are known as *unitary pseudogenes*.
2. Some are derived from parental genes as copies situated within large duplicated segments of DNA. In other words, they are included as parts of *segmental duplications* (see Chapter 4). In some cases, the whole gene has been duplicated, but one of the copies has decayed as it is surplus to requirements and has not been maintained by natural selection. In other cases, only part of the gene has been copied. In either situation, gene-like sequences exist that have lost the capacity of their parent genes to make proteins. These are known as *duplicated pseudogenes*.

3. Some are derived from unique parental genes via an RNA intermediate. This is suspiciously reminiscent of TEs, which generate multiple copies of themselves via RNA intermediates. Indeed, these pseudogenes arise because TE-encoded enzymes randomly select RNA transcripts of genes and copy them back into the genome. The RNA molecules so copied have been at least partially processed (for example, introns have been chopped out). The resulting inserts are known as *processed pseudogenes*. The inserts typically lack functional capacity, because they lack the normal assemblage of regulatory sequences (such as promoters and enhancers), and with time they decay into the genetic background.

There are approximately 20,000 pseudogenes in our genome (roughly as many pseudogenes as genes). They are evenly distributed between unitary and duplicated pseudogenes, on the one hand, and processed pseudogenes, on the other. Such gene relics may perform new functions, but can no longer perform the protein-coding functions that the parent gene once did, or that the parent gene from which they were copied still does. Pseudogenes are defined not by their *current* functionality but by the mechanisms by which their *original* function was lost.

Pseudogenes provide excellent markers for genealogical studies. They have this capacity because of the sequence of events by which they arose. Any current functionality is irrelevant. (The timber retrieved from a dilapidated old garage may be used to make a doghouse or a rustic fence, but its current use does not alter the fact that it once provided shelter for the car, and that it no longer does so. The timber may perform any number of new functions, but these do not hide the fact that the original function has been destroyed. Both the recycled timber from the garage and the 'recycled' base sequence in a pseudogene represent processes of transformation, sequences of discrete events, stories.) In this section, we consider the burgeoning field that we might designate as *pseudogenealogy*. We consider unitary and duplicated pseudogenes first, and then processed pseudogenes.

One example will illustrate the principles of pseudogenealogy. Ivermectin is a drug that kills parasites (worms and arthropods such as fleas, mites and lice) that infest farm animals. The drug kills the parasites by opening an ion channel in the membranes of their nerve cells. But Ivermectin is not toxic to farm animals, because the relevant nerve cells in mammals occur in the brain where they are protected by a cellular pump that keeps the Ivermectin out of the brain. The protective pump is called P-glycoprotein, and is encoded by the *MDR1* gene. (P-glycoprotein has been a preoccupation of cancer biologists, as it allows cancer cells to exclude anti-cancer drugs [35].)

In the early 1980s, Ivermectin was shown to be toxic for some collies. Drug administration may be followed by muscle tremors, breathing difficulties associated with fluid accumulation in the lungs, coma and even death. The basis of this toxicity is that vulnerable collies have a frameshift mutation in their *MDR1* gene – the gene is inactivated by a four-base deletion. It has become a pseudogene.

A breed of Australian sheepdogs that share common ancestry with collies also share Ivermectin sensitivity and the identical *MDR1* mutation. A survey of more than 90 dog breeds subsequently found the same mutation, on a shared haplotype, in seven other breeds. Some of these breeds were known to be related to collies (such as the Old English sheepdog), whereas other breeds (such as the longhaired whippet) had no documented pedigree relationship to collies. The unique mutation arose, and had been transmitted to the progenitors of multiple breeds, before formal records were started in 1873. The distribution of a unique pseudogene in pedigree dogs identified genealogical connections of which The Kennel Club was oblivious [36].

What of the other alternative – that the mutation had appeared independently in multiple breeds? Mutations in the *MDR1* gene are rare. Frameshifts are only one type of inactivating mutation. There are innumerable possible frameshift mutations. And there is the matter of the shared haplotype, which indicates that the surrounding sequence has one recent source. The chances of obtaining multiple independent mutations are exceedingly remote. In such a context, shared mutations indicate shared ancestry.

3.3.1 *Human-specific pseudogenes*

In the human gene pool, hundreds of genes currently consist of both protein-coding versions and non-protein-coding versions (or *alleles*). The latter have incapacitating mutations – *stop* codon, frameshift or splice-site – that preclude production of functional proteins. None of us contains the full set of active human genes. We are all mutants. We carry about 100 loss-of-function mutant alleles and about 20 genes that are fully inactivated. The inactivating mutations may have occurred thousands of years ago, and have been transmitted to a substantial proportion of people on planet Earth [37].

Some genes possessing non-functional alleles are involved in immunity. For example, several genes encoding chemokine receptors include variant alleles that have lost protein-coding capacity. (Chemokines are proteins that control the deployment of cells in the immune system.) The *GPR33* pseudogene (disrupted by a *stop* mutation) constitutes 98% of all the *GPR33* gene copies in the human population [38]. A *CCR5* pseudogene (scrambled by a 32-base deletion) is found only in Europeans, in whom it constitutes about 10% of all *CCR5* gene copies [39]. The *DARC* gene encodes the so-called 'Duffy antigen', expressed on red blood cells. A *DARC* allele with a mutant transcription factor binding site does not express the protein. This allele is particularly common in Africans [40].

Why should disabled alleles attain high frequencies in human populations? Perhaps the respective proteins have outlived their usefulness. They might have become liabilities by providing docking sites for pathogenic microbes. The *CCR5*-encoded protein currently enables HIV to infect cells, and the Duffy antigen is used by *Plasmodium vivax* (a malaria parasite) to infect red blood cells.

The *CASP12* pseudogene (disrupted by a *stop* mutation) is present at frequencies of 20–80% in different populations of sub-Saharan Africa, and of >99% outside Africa [41]. It has been hypothesised that the *CASP12* protein became a liability in the face of microbial challenges arising from animal domestication. This hypothesis has been challenged by the finding that DNA recovered from remains of Europeans who lived in the early days of domestication (5–12,000 years ago) already possessed only the pseudogene. Conversely, *CASP12* activity may have been *retained* in sub-Saharan Africa because it provides protection against pathogenic microbes that are native to that region [42]. Underlying selective pressures remain unknown – but gene disablement is a fact of life.

Functional forms of other genes have been *completely* lost from the human gene pool. In these cases, only a mutant version survives. Continuing the theme of immunity, pertinent examples are provided by the loss of functional *SIGLEC13* (Chapter 2) and *SIGLEC17* genes. The normal forms of these genes encode receptors that bind sialic acids: sugars that decorate proteins on the cell surface. Siglec–sialic acid interactions represent modes of communication between cells of the immune system. In humans a one-base deletion in *SIGLEC17* has caused a gene-destroying frameshift (Figure 3.5, upper diagram). Experimental work has suggested that pathogenic bacteria expressing sialic acids may adhere to Siglec proteins as a way of getting under the radar of immune defences. We are better off without proteins that can be so nefariously exploited [43].

But history has thrown up other surprises with sialic acids. The human body contains low concentrations, relative to other mammals, of a sugar called *N*-glycolylneuraminic acid (a sialic acid derivative). That is because we lack the enzyme that makes it. And we lack the enzyme because the *CMAH* gene that encodes the enzyme has been scrambled by the insertion of an Alu element. In other mammals, including chimps, the gene and the enzyme are functional, and the sugar is present abundantly on proteins expressed on cell surfaces [44].

It has been hypothesised that, initially, the loss of *N*-glycolylneuraminic acid was advantageous. The sugar acts as an anchor by which malaria parasites adhere to cells, and so our ancestors who ceased to make this particular sugar may have become resistant to infection (at least until the malaria parasite acquired the ability to adhere to other sugars).

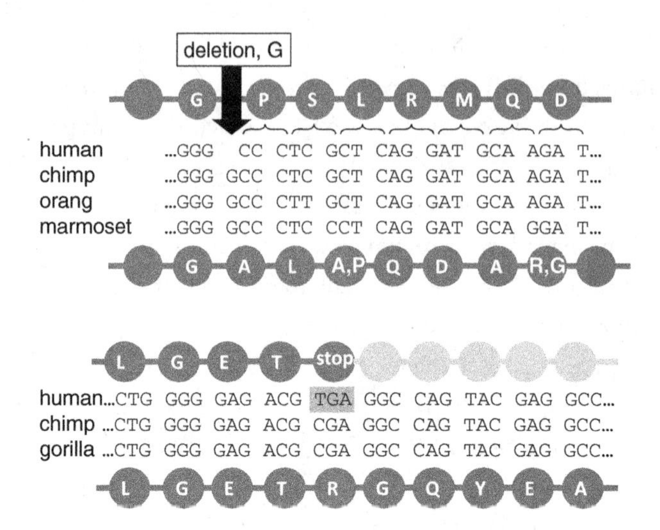

FIGURE 3.5. HUMAN-SPECIFIC PSEUDOGENES
A fragment of the *SIGLEC17* pseudogene (*upper diagram*) shows a single-base deletion [43]. *Brackets* indicate the new reading frame.

Lower diagram: the *ψhHaA* pseudogene shows a C-to-T mutation that creates a TGA *stop* codon (*shaded*) [46]. *Letters in circles* represent amino acids in encoded proteins.

But there may be a distinct downside to the loss of this sugar. In humans, dietary *N*-glycolylneuraminic acid (in red meat) is scavenged by the body and by resident bacteria, and chemically attached to cell surface proteins. Our immune systems seem to regard this exogenously acquired *N*-glycolylneuraminic acid as 'foreign'. Consequently we make antibodies against it. These antibodies in turn may cause chronic inflammation – a condition that promotes the development of cancer. Or those same antibodies may react with the sugar when it is present on endothelial cells (that line blood vessels), and cause blood vessel damage. This may promote atherosclerosis and heart disease. These same antibodies may also neutralise therapeutic proteins (such as immunoglobulins used as cancer treatments) if they have been produced under conditions where they become tagged with *N*-glycolylneuraminic acid [45].

Immune reactions occur out of sight, but the loss of other genes has had overt effects on our appearance. Humans have a distinctive distribution of hair, the basis of which may be multifactorial. One particular gene has been identified, however, that encodes a type of hair keratin (a fibrous protein) in chimps and gorillas but can no longer do so in humans. The human-specific pseudogene is known as *ψhHaA* (where the Greek letter ψ or psi indicates that it is a pseudogene). It has normal exon–intron boundaries, and is still transcribed into RNA. But it cannot generate a protein because a C-to-T mutation has changed a `CGA` codon (which encodes the amino acid arginine) into a `TGA` codon (which stipulates the *stop* command) (Figure 3.5, lower diagram). All humans possess this mutation and are descended from the progenitor cell in which the mutation occurred – some time before humans left Africa [46].

Interestingly, the Mari and Chuvash populations of the Volga–Ural region of Russia include many apparently unrelated families with an extreme form of hair loss. This condition arose from an Alu–Alu recombination event that deleted an exon of the *LIPH* gene. The mutation is part of a shared haplotype, indicating that everyone who carries the mutant allele (at least 100,000 people) inherited it from one founder [47].

The morphology of human faces is decidedly un-apelike. Muscles used for chewing are reduced in volume in humans relative to those of other primates. Muscles are composed of two sorts of fibres: slow twitch and fast twitch. Histological examination of normal human cheek muscle reveals that our slow twitch fibres are normal, but our fast twitch fibres are deformed. The basis for this resides in the derangement of a gene that encodes a muscle protein called *myosin heavy chain-16*. The *MYH16* gene has been passed on to all humans in a form that is essentially intact but for a single devastating mutation – the loss of two bases that has frameshifted the remaining gene sequence. A functional protein cannot be made (Figure 3.6) [48].

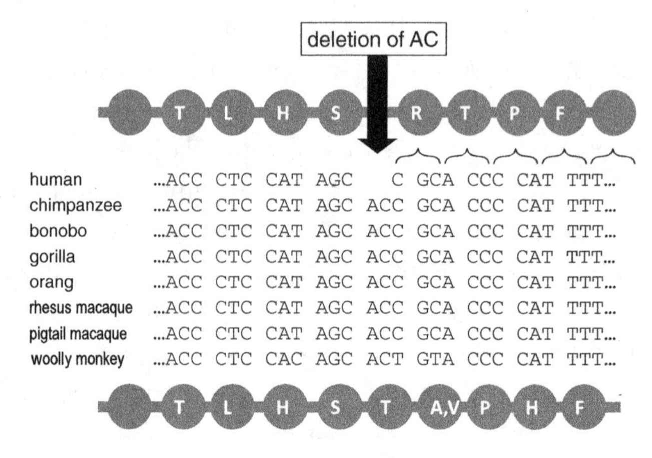

FIGURE 3.6. HUMAN-SPECIFIC *MYH16* PSEUDOGENE [48]
The gap in the human sequence represents the loss of two bases, AC.

What selective advantage may have accrued from the loss of this gene? It has been suggested that *MYH16* loss, with its concomitant decrease in muscle strength, was associated with changes towards a more modern skull shape. Perhaps it allowed the braincase to expand. The jury is still out, but what is certain is that a unique frameshift mutation in *MYH16* was a historical event that now distinguishes our *MYH16* gene from those of non-human primates. And the complex nature of the mutation suggests that all people on planet Earth are descended from the reproductive cell in which the mutation occurred.

In these cases we have inherited a gene in almost pristine form except for one devastating mutation that has occurred since our last common ancestor with chimps. (To return to our analogy: we no longer park the car in the old garage. However, the garage remains more-or-less intact, and we have not started to take the wood for any other use.) To be human is to be unable to make certain sugars, to have a defective pelt and to have malformed cheek muscles. About 40 genes have been inactivated specifically in human beings [49]. We have simply lost (and are still losing) some of the capacities shared by our primate relations.

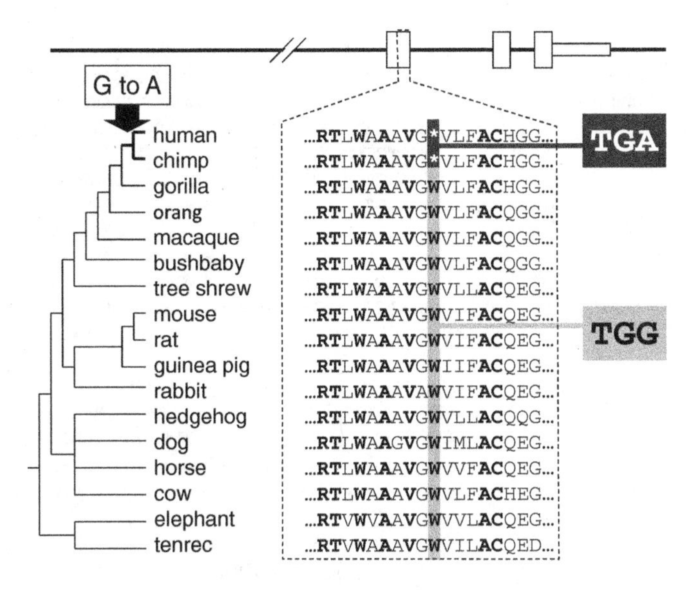

FIGURE 3.7. PROTEIN SEQUENCES SHOWING THE DISABLING MUTATION IN THE ACYLTRANSFERASE 3 ENZYME OF HUMANS AND CHIMPS [50]
Amino acids are identified by a standard *one-letter code*; the abnormal *stop* codon is indicated by an *asterisk*. In this and other figures, amino acids are shown in *bold* where they are the same in all species.

3.3.2 *Ape-specific pseudogenes*

But other derelict genes are shared with other primate species. For example, the *ACYL3* gene is an ancient gene, present in bacteria, fungi and animals. It encodes an enzyme called acyltransferase 3, which may have signalling or transporting activities. But the gene is dysfunctional in humans and chimps (and *only* in humans and chimps), and both species have the same function-destroying mutation (Figure 3.7). A single (G-to-A) base change has converted a `TGG` codon (which specifies the amino acid tryptophan, W) into a `TGA` *stop* codon [50]. The effects of this singular mutation on the biology of humans and chimps are not known. But what can be confidently inferred is that the mutation occurred in one reproductive cell, from which it was transmitted to the sister species humans and chimps.

11-base del.

human	...TTGATGTC AGCAACTCATGGA...
chimp	...TTGATGTC AGCAACTCATGGA...
gorilla	...TTGATGTC AGCAACTCATGGG...
crested gibbon	...TTGATGTCGGCAGATGCTCAGCAACTTATGGA...
rhesus macaque	...TTGATGTCAGCGGATGCTCAGCAACTCATGGA...
Goeldi's marmoset	...CTGATGTCAACAGATGCTCAGCAACTCATGGA...
spider monkey	...TTGATGTCAGCAGATGCTCAGCAACTCATGGA...
dog	...TTGATGTCAGCAGATGCCCAGCAGCTCATGGA...
Asiatic tapir	...TTGATGTCAGCAGATGCCCAGCAGCTCATGGA...
horse	...TTGATGTCAGCAGATGCCCAGCAGCTCATGGA...

55-base del.

human	...CTGACTGGAACC CATCATTGTAGA...
chimp	...CCGACTGGAACC CATCATTGTAGA...
gorilla	...CCGACTGGAACC CATCATTGTAGA...
orang	...CCGACTGGAACCTTGCC...CAGTCCCATCATTGTAGA...
squi. monkey	...CCGACTGGAACCTCGCC...CAGCCCCATCATTGTAGA...

FIGURE 3.8. PSEUDOGENES SHARED BY THE AFRICAN GREAT APES: *ABCC13, UPPER DIAGRAM* [51] AND GLUCOCEREBROSIDASE, *LOWER DIAGRAM* [52] In this and other figures, bases are shown in *bold* when they are identical in all species.

Other pseudogenes are shared not only by humans and chimps, but also by gorillas. We have noted that these species comprise the African great ape clade. The *ABCC13* gene is one of a large family of genes that encode membrane-spanning transporter proteins. But this gene contains an eleven-base deletion, at precisely the same site, in humans, chimps and gorillas, but in no other species (Figure 3.8, upper diagram). The part of the gene that is mangled in the African great apes is intact in other primates as well as in the dog, tapir and horse – distant Laurasiatherian relations [51].

A more sinister, potentially disruptive, pseudogene is responsible for some cases of the genetic condition known as Gaucher's disease. This is a disease caused by deficiency of the enzyme glucocerebrosidase, and it results in the accumulation of particular lipids (glucocerebrosides) in the body. The inexorable accumulation of this product may result in a greatly enlarged spleen and liver. In rare cases, the condition involves neural tissues, and may cause death. More than 80,000 people are affected worldwide.

One of the mutations that cause this disease arises from a complex chain of events. The glucocerebrosidase gene has been copied to generate a duplicated pseudogene that is located 16,000 bases away from the functional gene. The pseudogene contains a 55-base deletion – a frameshift mutation that puts downstream sequences out of frame. Blocks of DNA sequence may be transferred between the original, functional gene and the pseudogene by a process known as *gene conversion.* When sequences that include the deletion are copied from the pseudogene into the functional gene, the coding capacity of the latter is destroyed.

The unique mutation that makes the pseudogene such a menace is present only in humans, chimps and gorillas. The fourth great ape, the orang-utan, and a representative monkey lack this deletion (Figure 3.8, lower diagram). It is a randomly arising and unique aberration that establishes the monophylicity of the African great apes [52].

A third example of a pseudogene that establishes the reality of the human–chimp–gorilla clade must suffice. Fucosyltransferases are enzymes that add units of fucose (a sugar) to cell surface proteins. A cluster of genes encoding these enzymes is located on chromosome 19. One of these is a pseudogene, and it is shared by the African great apes. A variety of mutations have accumulated in this gene in the three species, but one is common to all: a C-to-T base change. This mutation converted a CAA codon (encoding the amino acid glutamine, Q) into a TAA *stop* codon. All other tested primates and the rabbit lack this mutation (Figure 3.9). It arose in an ancestor of African great apes after the orang-utan lineage had branched off [53].

Some pseudogenes are shared by *all* the great apes (including the orang-utan) but not any other species, demonstrating that the great apes too are monophyletic. A fascinating example is the gene that once encoded the enzyme urate oxidase, a gene of long standing that is widely present in living organisms, including bacteria. Urate oxidase is the first of a series of enzymes that breaks down

FIGURE 3.9. PROTEIN SEQUENCES SHOWING A DISABLING MUTATION IN THE α2-FUCOSYLTRANSFERASE ENZYME OF THE AFRICAN GREAT APES [53]

uric acid in the body, with a view to excreting the end-products. But humans lack this enzyme, and cannot degrade uric acid. As a result we have more uric acid in our blood than do other mammals. This high concentration of uric acid may predispose us to gout, a painful inflammatory condition that arises when sodium urate crystals are deposited in joints. Gout has afflicted humans throughout their history. Egyptian mummies show the signs of gout-associated arthritis. It has been considered a wealthy man's disease because it tends to occur in people who eat rich food that contains abundant amounts of the nutrients (purines) from which uric acid is derived.

We lack urate oxidase because we have inherited a mutant *UOX* gene. A single C-to-T base change has transformed a CGA codon (that encodes the amino acid arginine, R) into a TGA *stop* codon. A comparison of the surrounding DNA sequence is shown for ten primates (Figure 3.10). The sequence is highly conserved, but the *stop* mutation is shared by all the great apes (and only the great apes). The great apes are co-inheritors of this mutant gene that acquired its incapacitating lesion in a great ape ancestor. Interestingly, gibbons also lack the urate oxidase enzyme, due to independent mutations at other sites in the gene, shared by each of six species of gibbons [54].

If the uric acid degradation pathway no longer works, then genes encoding enzymes that function *after* urate oxidase in the

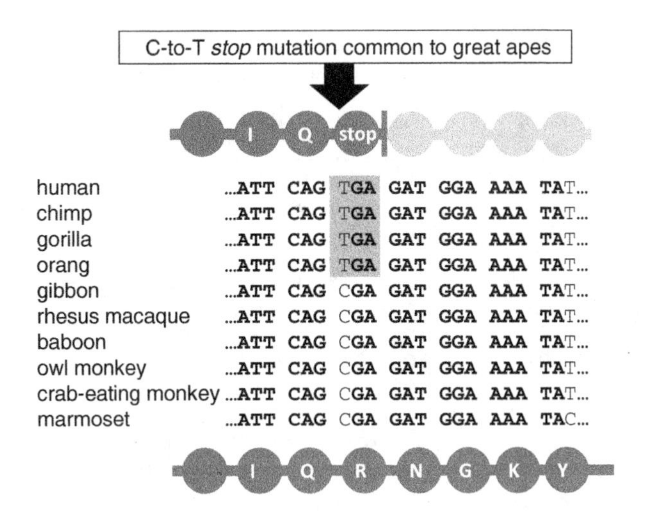

FIGURE 3.10. THE URATE OXIDASE PSEUDOGENE, SHARED BY THE GREAT APES [54]
The abnormal *stop* codon is *shaded.*

same pathway may have fallen into disuse and disrepair. A survey of such genes has confirmed this hypothesis. The gene encoding HIU hydrolase is indeed inactive in humans, chimps and gibbons. In this case, the three species show independent mutations, indicating that the inactivating events occurred relatively recently on the respective lineages, subsequent to the time when common ancestors lived [55].

We may speculate as to why such an ancient gene was lost, especially as its loss can have debilitating consequences. Uric acid is chemically related to caffeine, and it has been suggested that elevated uric acid concentrations may have stimulated brain activity. Do we live on a uric acid-induced high? Uric acid may act as an antioxidant. The presence of that extra uric acid may protect us against the effects of reactive forms of oxygen that damage tissues, cause cancer and accelerate ageing. We will return to this theme later.

Reproduction also has primate-specific features. The endozepine-like peptide (ELP) is highly expressed in the testes of many mammals, particularly in maturing male germ cells. It is

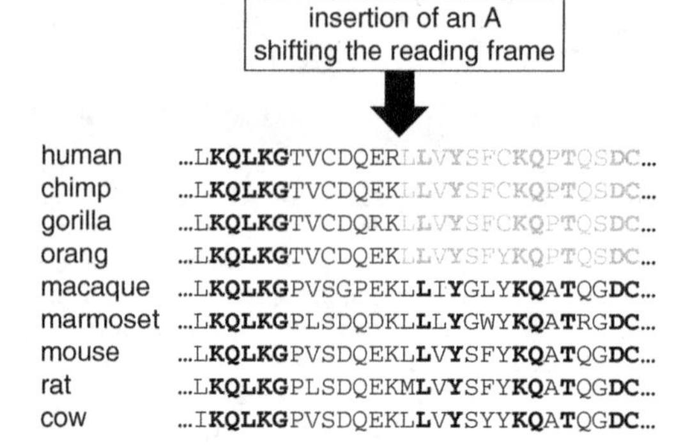

FIGURE 3.11. PROTEIN SEQUENCE SHOWING THE DISABLING MUTATION IN THE ENDOZEPINE-LIKE PEPTIDE OF THE GREAT APES [56]
Grey lettering indicates what the sequence would have been if there had been no insertion.

involved in transporting metabolic intermediates, and may have signalling roles inside cells. However, the protein cannot be found in human tissue. Human tissue does contain two types of messenger RNA molecules that are highly related to the mouse *ELP* transcript, but they cannot specify the production of ELP protein because they are incapacitated by frameshift mutations.

It appears that, in primates, an original *ELP* gene was duplicated, but each copy has been inherited by humans in a non-functional form. The *ELP1* gene is inactivated in all the great apes. It remains intact in representatives of OWMs, NWMs and non-primates. But in all those species in which the gene has been disabled, the inactivating mutation is identical: it is the insertion of one base (an 'A') at one particular site (following base 354). This singular frameshift mutation occurred in a germ cell, and all the great apes have received their scrambled copy of the *ELP1* gene from that cell (Figure 3.11).

Even in the monkey species with a normal gene sequence, the protein is not made. It appears that an unknown regulatory mutation has occurred, which precludes the production of the ELP protein. Perhaps the ELP protein became dispensable for male reproduction during the course of primate evolution. The authors of this study have speculated that losses in absolute male fertility were offset by the development of sociality [56].

3.3.3 *Simian-specific pseudogenes*

Loss-of-function mutations have affected enzymes involved in energy metabolism. The mitochondrion is an organelle that converts the energy locked up in food into ATP, the energy currency of cells. Mitochondrial energy conversion is driven by a flow of electrons along a series of proteins. The final step in the chain, the transfer of electrons to oxygen, generating water as end-product, is catalysed by an enzyme complex called cytochrome C oxidase. Most mammals have two genes encoding alternative subunits of this enzyme: *COX8H* (heart-type, expressed in muscles) and *COX8L* (liver-type, expressed in many tissues). Humans, other apes and OWMs lack the *COX8H*-derived protein. These species retain the gene, but it is deranged by a singular mutation, a deletion of 14 bases that removes the normal *stop* codon (present in intact sequences as TAA or TGA; Figure 3.12). Energy metabolism has become progressively more efficient as primates have evolved, possibly to support the activity of cells in the brain. Paradoxically, the loss of the *COX8H* gene in an ape–OWM ancestor may have been part of this streamlining process [57].

We have discussed how humans possess disabled versions of genes that would otherwise be responsible for the placement of certain sugars on the surfaces of cells (a sialic acid derivative in the case of *CMAH*, and fucose in the case of fucosyltransferase). We encounter this situation again in the case of a cell surface structure involving the sugar galactose. All mammals except apes and OWMs have a moiety on their cell surfaces consisting of two galactose units (the so-called αGal structure). We lack αGal because we lack the enzyme (α1,3-galactosyltransferase) necessary to make it. And we lack the enzyme because the respective gene (*GGTA1*) has accumulated many mutations. Two of these are shared by apes and OWMs (represented by the macaque) but not by NWMs (marmoset, howler monkey, capuchin), which retain the active gene. A frameshift mutation

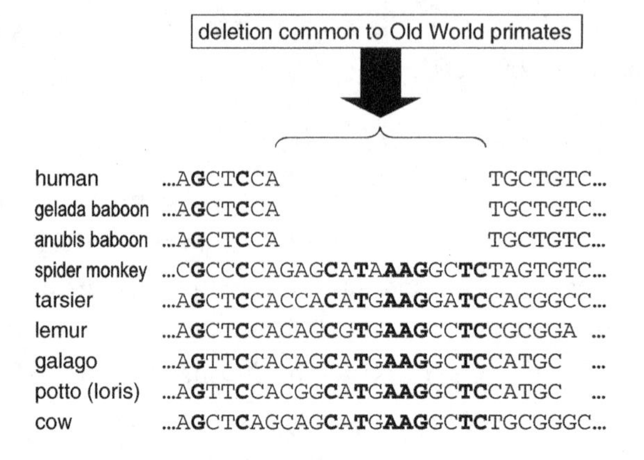

FIGURE 3.12. A DELETION IN THE *COX8H* PSEUDOGENE, SHARED BY OLD WORLD PRIMATES [57]
The bonobo and the gorilla share the same deletion (not shown).

involving the loss of a single base, G, and common only to Old World primates, has occurred (Figure 3.13). The mutation arose in a reproductive cell from which we and macaques are descended [58].

So why might the loss of the αGal structure have occurred? Organisms lacking the ability to synthesise αGal recognise the structure as immunologically 'foreign'. When αGal that is derived from gut bacteria enters the body, we Old World primates make antibodies against it. We have high concentrations of αGal-directed antibodies in our blood. Perhaps these antibodies confer protection against pathogens that possess the αGal structure. But in that case why have no other mammalian groups lost the *GGTA1* gene? These questions remain unanswered.

The loss of *GGTA1* coding capacity has implications for transplantation medicine. Someone who is the recipient of an αGal-expressing organ from a non-primate species will mount a vigorous antibody-mediated attack on that organ, resulting in the rapid destruction of the transplant. This has led to the selection of αGal-deficient breeds of pigs for use as tissue donors. One enterprising research team has sought to turn our natural anti-αGal antibodies to advantage. They have shown experimentally that treating wounds

a 1-base deletion common to Old World primates

human	...AAAAAGGAAGAGAGGAG ACCAAAGGAAGGAAAAT...
orang	...AAAAAGGAAGAGAGGAG ACCAAAGAAAGGAAAAT...
macaque	...AAGA GGAAGAGAGGAG ACCAAAGAAAGGAAGAT...
marmoset	...AAAAAGGAAGAGAGGAGGAACAAAGAAAGGAAGAC...
howler	...AAAAAGGAAAAGAAGAGGAACAAAGAAAGGAAGAC...
capuchin	...AAAAAGGAAGAGAGGAGGAACAAAGAAAGGAAGAT...
lemur	...AAAAAAGAAGAGAAAAGGAACAAAGGAAGGAAGAT...
loris	...AAAAAGGAAGAAAAAAGGAACAAAGAGAGAAAGAT...

FIGURE 3.13. A DELETION IN THE *GGTA1* PSEUDOGENE, SHARED BY APES AND OWMS [58]

with αGal, attached to tiny droplets of fat, recruits anti-αGal antibodies to the site of αGal application. A local inflammatory reaction is induced, and wound healing is accelerated [59].

The loss of αGal entails a further consequence of clinical importance. People who are bitten by certain species of tick generate anti-αGal antibodies of the IgE class – and these mediate allergic reactions. When such people eat meat or kidney from mammals that retain the capacity to make αGal, or when they are treated with therapeutic antibodies that are decorated with αGal, they may experience severe allergic (anaphylactic) reactions, characterised by hives, swelling of the tongue and wheezing, which may be life-threatening. People might even be sensitised by breathing dust containing αGal-tagged proteins from pets [60]. There are alarming dangers in opting out of the αGal club.

The TPC3 ion channel is one of a small family of intracellular channels that allows the release of calcium ions from membrane-enclosed stores in animal cells. It is believed to function in such diverse processes as fertilisation, insulin secretion and activation of cells in the immune system. One might imagine that this protein plays an indispensable role in human physiology – but no, the gene that encodes the TPC3 channel in other species has been well and truly pseudogenised in ours. The earliest gene-scrambling mutation,

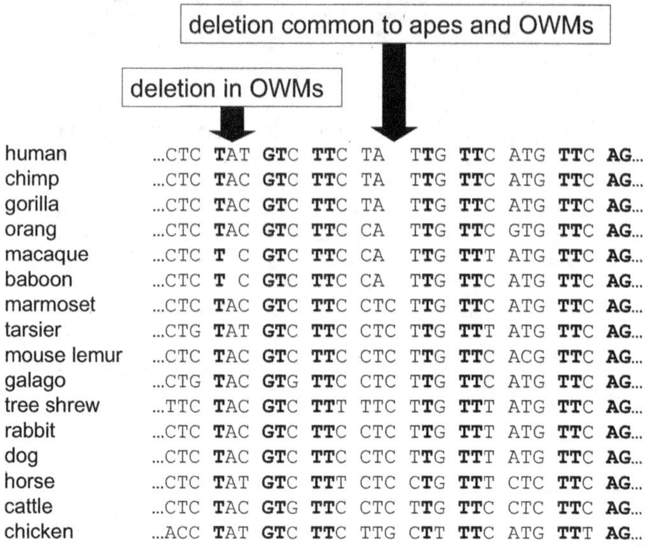

deletion common to apes and OWMs

deletion in OWMs

```
human        ...CTC TAT GTC TTC TA  TTG TTC ATG TTC AG...
chimp        ...CTC TAC GTC TTC TA  TTG TTC ATG TTC AG...
gorilla      ...CTC TAC GTC TTC TA  TTG TTC ATG TTC AG...
orang        ...CTC TAC GTC TTC CA  TTG TTC GTG TTC AG...
macaque      ...CTC T C GTC TTC CA  TTG TTT ATG TTC AG...
baboon       ...CTC T C GTC TTC CA  TTG TTC ATG TTC AG...
marmoset     ...CTC TAC GTC TTC CTC TTG TTC ATG TTC AG...
tarsier      ...CTG TAT GTC TTC CTC TTG TTT ATG TTC AG...
mouse lemur  ...CTC TAC GTC TTC CTC TTG TTC ACG TTC AG...
galago       ...CTG TAC GTG TTC CTC TTG TTC ATG TTC AG...
tree shrew   ...TTC TAC GTC TTT TTC TTG TTT ATG TTC AG...
rabbit       ...CTC TAC GTC TTC CTC TTG TTT ATG TTC AG...
dog          ...CTC TAC GTC TTC CTC TTG TTT ATG TTC AG...
horse        ...CTC TAT GTC TTT CTC CTG TTT CTC TTC AG...
cattle       ...CTC TAC GTG TTC CTC TTG TTC CTC TTC AG...
chicken      ...ACC TAT GTC TTC TTG CTT TTC ATG TTT AG...
```

FIGURE 3.14. THE DELETION IN THE *TPC3* PSEUDOGENE SHARED BY APES AND OWMS [61]

common to the widest variety of primate species, is a one-base deletion. This singular mutation is present in all apes and OWM species examined – but in no other species [61]. It occurred in a single cell, and has been inherited by all living apes and OWMs (Figure 3.14).

Other mutations are also found in the *TPC3* gene, including a one-base deletion in OWMs (Figure 3.14). Once a gene has been inactivated, there are no longer selective constraints on maintaining its integrity, and it steadily degenerates. An old pseudogene contains multiple mutations that vary in age, and cataloguing these in a panel of species may enable those species to be arranged in a full phylogeny. Every pseudogene has a story to tell.

Scurvy was a potentially lethal disease that affected sailors. In the mid-eighteenth century, James Lind, a British naval surgeon, performed an experiment on a group of sailors that showed that the consumption of fresh fruit could alleviate scurvy. The active ingredient is ascorbic acid (vitamin C). We require ascorbic acid in our diet because it participates in several biochemical reactions and also has antioxidant effects. But the consumption of ascorbic acid may not always be as beneficial as the health industry publicity claims: under some conditions, ascorbic acid may act as a potentially harmful oxidant.

Most mammals make their own ascorbic acid. It is a vitamin for us only because we cannot produce the enzyme (L-gulono-γ-lactone oxidase, or GULO) that is required for the final step in its synthesis. And we cannot make the enzyme because we have inherited a severely degenerated copy of the gene. Only 5 of the original 12 exons remain, the locus has been bombarded with retrotransposons, and those parts of the gene that are identifiable are riddled with mutations [62]. (To return to our garage analogy: once it fell into disuse, no effort was made to maintain it, and a process of unimpeded deterioration set in. The shelves collapsed, the windows fell off and the spouting rusted.)

The *GULO* pseudogene contains multiple indel and *stop* mutations. The oldest appears to be a *stop* mutation, shared by representatives of all simian primate groups – apes, OWMs and NWMs. A natural inference is that this mutation initiated the process of decay, and it arose in an ancestor of the simian primates [63]. Subsequently, exons 2 and 3 were lost from the genomes of apes and OWMs by a DNA deletion event that eliminated approximately 2,500 bases from the genome. A representative *stop* mutation, shared by apes and OWMs, is shown in Figure 3.15. A codon specifying the amino acid arginine (possibly CGA) has ended up as a gene-truncating TGA codon [64].

The vitamin C business owes its millions to a mutation that knocked out the *GULO* gene in a simian ancestor. But why should such an apparently vital gene turn out to be dispensable? We can only speculate – and take cognisance of the fact that relatively recent, and independent, losses of *GULO* gene coding capacity have occurred in bats [65]. Perhaps the mutation occurred in a small population, replete with dietary ascorbic acid, for which the surreptitious take-over of a defunct *GULO* allele was inconsequential [66]. Millions of

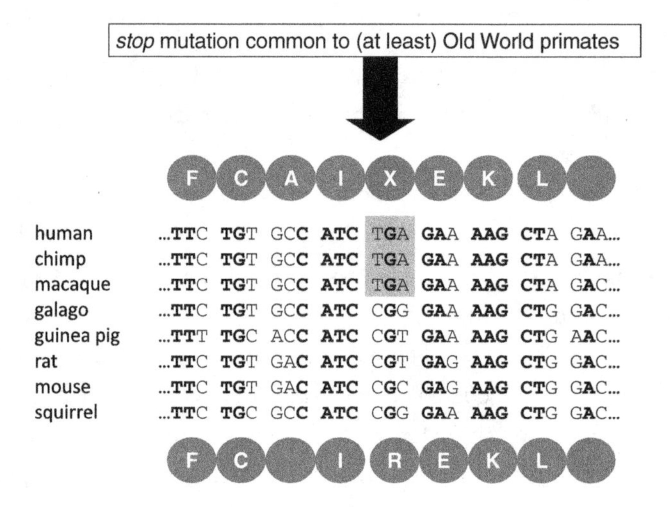

FIGURE 3.15. A *STOP* MUTATION IN THE *GULO* GENE, SHARED BY OLD WORLD PRIMATES [64]

years later, a time of nutritional deprivation followed. The source of dietary ascorbate dried up, antioxidant levels in the body declined, and an alternative antioxidant was needed. In this situation, the inactivation of the urate oxidase (*UOX*) gene, which led to elevated concentrations of uric acid in the body, may have provided an alternative means of coping with the oxidative burden.

A more recent hypothesis accounting for why we are *GULO–UOX* double knockouts has been suggested. Biochemical evidence has been adduced to argue that *reduced* ascorbate concentrations, and *elevated* uric acid concentrations, might have facilitated the transformation of dietary fructose (fruit sugar) into stored fat. This would have given our ancestors a survival advantage in times of dietary insufficiency. The problem is that the *GULO–UOX* double knockout genotype may be frankly maladaptive in contemporary people consuming Western diets. High levels of fructose intake may promote that condition of excessive fat storage known as obesity. And uric acid accumulations in excess of those experienced by our distant ancestors may predispose us to diseases of Western civilisation including high blood pressure and, as noted above, gout [67].

We may speculate with some plausibility on the loss of *UOX* and *GULO*. But the reason why the *arpAT* gene should have been lost is a mystery. This gene encodes a protein that (in a great many other organisms) transports amino acids into cells. But the human version of the gene is riddled with frameshift mutations, two of which we share with OWMs and NWMs. This indicates that pseudogenisation started in a simian ancestor, and that the gene has been accumulating mutations ever since [68]. Putting aside considerations of why such a venerable old gene should have been scrapped, the *arpAT* and *GULO* pseudogenes provide independent confirmation that all simians are derived from the same ancestor.

Long-standing pseudogenes may accumulate a multiplicity of mutations sufficient to provide a full genealogy in their own right. There are five members of the serum albumin protein family. They are made in the liver and act to carry a range of ligands in the blood. The most-recently discovered member of the gene family, *alpha-fetoprotein-related gene* (*ARG*), is present in the genomes of many non-primate animals, but is incapable of making a protein in any species of ape or monkey. In fact, multiple mutations are shared by humans, chimps, macaques (representing OWMs) and marmosets (an NWM) (see Figure 3.16). All simian species studied share one *stop*, three splice-site, three frameshift and two TE insertion mutations. In addition, apes and OWMs share mutations that are absent in NWMs, and apes share a splice-site mutation that is absent in OWMs and NWMs. This gene relic provides a 'one-stop shop' for an anthropoid phylogeny [69].

One would not expect to come across specifically marked pseudogenes that would enable us to peer much further back into our history. Protein-coding genes that have lost their coding capacity tend to decay into the genetic background. Specific sequence features fade from sight. But a gene encoding a form of cytochrome C provides an exception.

Cytochrome C is a protein that has two lives. It normally acts inside mitochondria to shuttle electrons around: a life-sustaining

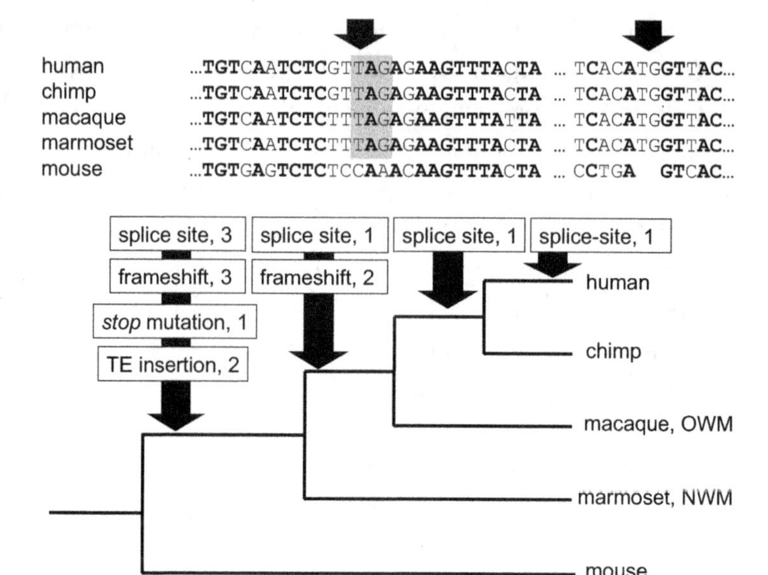

FIGURE 3.16. MUTATIONS IN THE *ARG* PSEUDOGENE [69]
The sequences (*upper diagram*) show a *stop* mutation and a two-base insertion that are shared by all simian primates. The family tree (*lower diagram*) indicates types and numbers of mutations at each ancestral stage (*arrows*).

role. But if it escapes into the cytoplasm, it acts to initiate cell suicide. Primates have one cytochrome C gene; other mammals have two. The difference resides in a testis-specific gene (*CYCT*) that is scrambled in primates, but which specifies the production of a functional protein in non-primates. The primate pseudogene has been fragmented by the intrusion of numerous ancient TEs, suggesting that it descended into decrepitude a long time ago. This is evinced by the finding that all tested primates, including a representative prosimian, have the same CGA to TGA *stop* mutation, clearly visible in a well-preserved segment of the pseudogene (Figure 3.17).

Conveniently for us, this part of the pseudogene has been magnificently conserved only because it overlaps an unrelated gene, which is read in an alternative reading frame, and upon which selective pressure has acted to retain the local sequence. From the

		W	G	L	F	G	X	K	T	G	K
ape	human	...TGG	GGC	CTC	TTT	GGC	TGA	AAA	ACA	GGA	AAA...
	chimp	...TGG	GGC	CTC	TTT	GGC	TGA	AAA	ACA	GGA	AAA...
	gorilla	...TGG	GGC	CTC	TTT	GGC	TGA	AAA	ACA	GGA	AAA...
	orang	...TGG	GGC	CTC	TTT	GGC	TGA	AAA	ACA	GGA	AAA...
	siamang	...TGG	GGC	CTC	TTT	GGC	TGA	AAA	ACA	GGA	CAA...
O	macaque1	...TGG	GGC	CTC	TTT	GAC	TGA	AAA	ACA	GGA	CAA...
	macaque2	...TGG	GGC	CTC	TTT	GAC	TGA	AAA	ACA	GGA	CAA...
	guereza	...TGG	GAC	CTC	TTT	GAC	TGA	AAA	ACG	GGA	CAA...
	mangabey	...TGG	GGC	CTC	TTT	GAC	TGA	AAA	ACA	GGA	CAA...
N	marmoset	...TGA	GAC	CTC	TTA	GGC	TGA	AAA	ACA	GGA	CAA...
	tamarin	...TGG	GGC	CTC	TTT	GGC	TGA	AAA	ACA	GGA	CAA...
PS	aye-aye	...TGG	GGC	CTC	TTT	GGC	TGA	AAA	ACA	GGA	GCA...
E	mouse	...TGG	GGC	CTC	TTT	GGT	CGA	AAG	ACA	GGA	CAA...
	rat	...TGG	GGC	CTT	TTT	GGC	CGA	AAG	ACT	GGA	CAA...
L	horse	...TGG	GGC	CTT	TTT	GGC	CGA	AAA	ACA	GGA	CAA...
		W	G	L	F	G	R	K	T	G	Q

FIGURE 3.17. THE *CYCT* PSEUDOGENE OF PRIMATES [70]
Species are apes, OWMs (O), NWMs (N), a prosimian (PS), other Euarchontoglires (E), and a Laurasiatherian (L). Shown are encoded amino acids (for apes, *top*; non-primates, *bottom*), the *stop* codon (*shaded*) and termination of the amino acid sequence (X).

perspective of genetic palaeontology, we have an immaculately conserved *stop* mutation transmitted from an ancestor of the primate clone. From the perspective of reproductive potency, we have a possible reason for the relative decline of primate fertility [70].

This survey of unitary and duplicated pseudogenes – protein-coding genes that have lost protein-coding capacity – outlines an almost complete genealogy of human beings through primate history. This pseudogenealogy is summarised in Figure 3.18. Exhaustive searches through multi-species genomic databases have generated a catalogue of 76 unitary pseudogenes that have accumulated in human DNA since the primate–rodent divergence [71]. The stage of primate phylogeny during which many of these appeared has been determined by observing the range of species that share unique inactivating mutations. The numbers of pseudogenes so identified (in addition to those discussed in the text) are indicated in Figure 3.18. Pseudogenes have been accumulating steadily throughout primate history. The same processes that produced pseudogenes in simian ancestors are still at work.

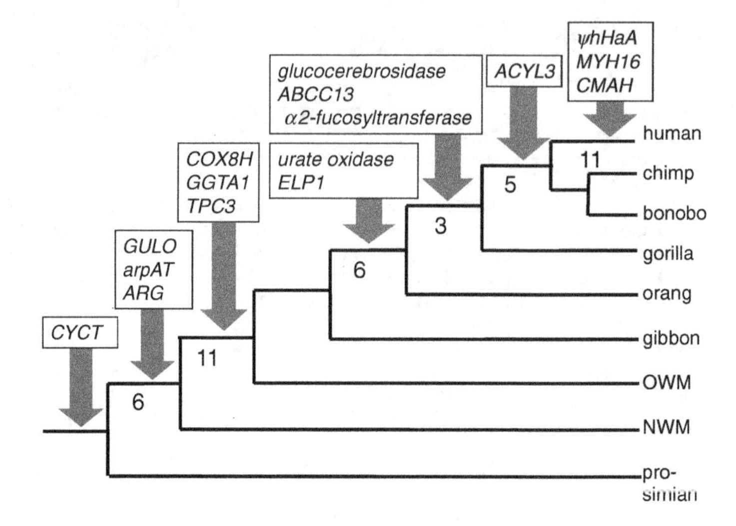

FIGURE 3.18. THE TIMES AT WHICH PSEUDOGENES APPEARED DURING PRIMATE PHYLOGENY

Genes identified *in boxes* are described in the text. *Numerals under lines* indicate the numbers of additional unitary pseudogenes [71].

3.3.4 Pseudogenes and sensory perception

One of the marvels of being a complex organism is the way in which we sense our environment. We have molecular machinery that interacts with a part of the electromagnetic spectrum, and that enables us to *see* – and in colour. We have cells with little hairs that are sensitive to oscillations in the density of molecules in the air, so that we can *hear*. And we possess proteins that can latch on to a vast variety of molecules that enter our noses and mouths, and that underlie the richness of chemosensory experience. These include our perceptions of taste (*gustation*) and smell (*olfaction*).

The proteins that bind such molecules are called *receptors* and are an essential first step in generating neural signals that are ultimately experienced as taste and smell. There must be an extensive repertoire of receptor molecules because of the sheer variety of molecules that we can identify. The easiest way to study the multiplicity of receptor molecules is to consider the genetics of chemosensation.

Most mammals possess three systems of chemosensation but, surprisingly, these have been reduced – to various extents – en route to humans [72].

First, there is the *vomeronasal* system. The vomeronasal organ (VNO) is present in the nasal cavity. It contains the neurons that respond to *pheromones* – chemical signals that communicate between members of the same species and regulate behaviours such as aggression, territoriality, mating and nursing. It is uncertain whether humans respond to pheromones. There is one telling argument against the proposition: almost the entire set-up appears to be derelict in humans and closely related primates. The VNO itself is absent and is replaced by a small pit in the nasal cavity. The part of the brain to which (in other organisms) the VNO nerves send signals is also absent [73].

In a functional VNO, the binding of pheromones to their receptors induces a flux of calcium ions through the nerve cell membrane. This generates the nerve impulse that transmits the binding signal to the brain. The protein channel through which ions flow is called the *transient receptor potential cation channel-2* (TRPC2), but its gene is thoroughly incapacitated in humans, and indeed in all apes and OWMs. In humans there are two frameshift mutations in the second exon. The second of these corrects the reading frame, leaving the 60 intervening amino acids scrambled. There are also four *stop* mutations.

Comparative genome studies have shown that the earliest two mutations occurred in an ancestor of apes and OWMs. Both were *stop* mutations. In the case of the *TRPC2* gene, the phylogeny has been obscured by reversion mutations, in which each of the *stop* mutations appears to have been further mutated into another codon. One reversion occurred on the lineage leading to great apes; the other on the lineage leading to orang-utans (Figure 3.19). Reversions of single-base change mutations are expected at low frequencies.

Three more mutations occurred in the African great ape lineage, and others in the lineages leading to humans, orang-utans,

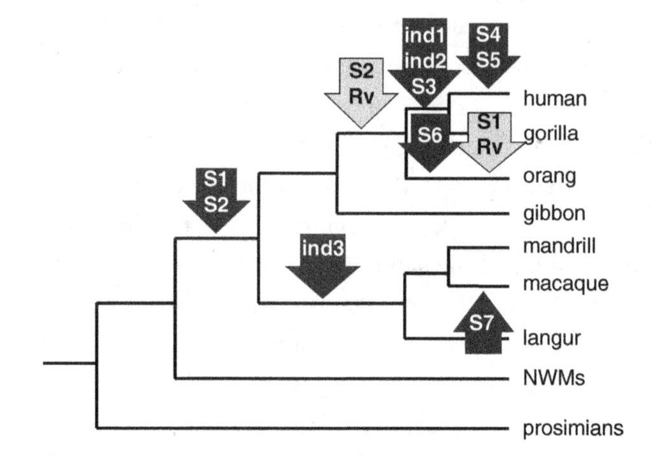

FIGURE 3.19. MUTATIONS IN THE *TRPC2* GENE OF APES AND OWMS [74]
Mutations indicated are *stop* mutations (S), indels (insertions or deletions; ind) and reversions (Rv).

OWMs and macaques. Overall, the *TRPC2* gene is non-functional in all ape and OWM species tested, but retains protein-coding functionality in NWMs and prosimians [74].

The adoption of an aquatic lifestyle is also associated with the loss of *TRPC2* gene protein-coding functionality. This gene is active in the Californian sea lion, but has accumulated inactivating mutations in the river otter and harbour seal (semi-aquatic carnivores), and in the dolphin (a toothed whale) and fin whale (a baleen whale), which share three inactivating mutations. These mutations establish that dolphins and fin whales (whose marine environment would render pheromone signalling ineffective) are derived from a whale progenitor [75].

Loss of the TRPC2 channel would be expected to inactivate VNO signalling. What then would happen to the dedicated pheromone receptors? If receptor binding to pheromones cannot generate a calcium flux, and so cannot generate a neural signal, the receptors would provide no advantage to the organisms that possess them. Selective pressure to retain the respective genes would be lost. The genes would degenerate. In primates lacking TRPC2 function, the pheromone receptor genes, which are classified into two unrelated families, V1R and V2R, have decayed almost completely into decrepitude.

- The human genome contains 120 VIR genes, and 5 retain the capacity to encode a functional receptor protein. One of these genes shows signs of positive selection, indicating current functionality. However, two of these five genes retain coding capacity in only a proportion of the human population. They have been caught in the act of undergoing pseudogenisation. All of these five genes have degenerated to pseudogenes in at least two of the other great apes [76]. Mammals differ hugely in their V1R repertoire. The platypus, mouse and mouse lemur have large numbers (at least 160 genes). At the other extreme, dolphins and whales, some bats, and OWMs and apes have depleted V1R repertoires [77].
- The human genome contains 20 V2R genes, and all are pseudogenes (as they are also in chimps and macaques). Pheromone signalling associated with the V2R receptor set has been repeatedly lost in mammals. Such a loss may characterise species with pronounced sexual dimorphism. When animals can see potential mates (or competitors) on the basis of body size and shape, fur markings, horns, tusks or behaviours, pheromone sensitivity may become redundant. On the other hand, opossums, rats and mice retain approximately 100 functional V2R genes [78].

Pheromone signalling has diminished progressively during primate evolution. To be human is to have inherited the complex genetic infrastructure of an elaborate sensory system, albeit in an almost wholly non-functional state. Humans do not develop social relationships by sniffing for pheromones. We use other sensory modalities such as sight. We can even talk to conspecifics.

The second type of chemosensory system is *olfaction*. Olfactory receptors are expressed by nerve cells of the main olfactory epithelium that lines the nasal cavities. In the human genome, the genes that encode olfactory receptors comprise the most populous of all the gene families, consisting of some 850 members. Half of these have acquired mutations that prevent the production of functional receptor proteins.

The number of functional genes is approximately the same in representatives of all primate groups (300–400). The proportion of pseudogenes is similar in apes and OWMs but appears to be lower in NWMs (in which <40% of the olfactory receptor genes are pseudogenes). Net gene inactivation has occurred in both the human and chimp lineages, as it has in all major primate lineages [79].

A sample of olfactory receptor pseudogenes has been analysed for mutations in different primate species. The presence of shared mutations indicates the stage of primate evolution at which pseudogenisation occurred. Five pseudogenes shared the same inactivating mutation in all apes and OWMs (although one subsequently reverted in the orang-utan lineage). These mutations arose in an ape–OWM ancestor. Other mutations appeared in great ape ancestors, in African great ape ancestors and in the human–chimp ancestor (Figure 3.20). A survey of olfactory receptor pseudogenes is itself sufficient to provide independent confirmation of the shape of primate phylogeny [80]. A more exhaustive search of 400 pseudogenes in humans showed that two-thirds share an inactivating mutation with their chimpanzee counterpart [81].

Four hundred human olfactory receptor genes retain coding capacity. These 'active' genes are characterised by a high frequency of variant forms, many of which would encode proteins with reduced activity. In fact, two-thirds of the 'active' olfactory receptor genes possess alleles that are non-coding pseudogenes. The degeneration of the olfactory receptor gene family, apparent through later primate evolution, is ongoing. No other gene family currently shows signs of relaxed purifying selection. Perhaps the benefits of stereoscopic vision (which entails the loss of a snout), the advent of bipedal gait (which lifts our noses above the ground) or the development of big brains (which process sensory information more efficiently) accounts for the decline in our olfactory receptor gene set [82].

Mammals vary hugely in their complement of olfactory receptor genes. Opossums, armadillos, elephants, horses, cattle and rodents are well endowed, with over 1,000 genes. Dogs are not far

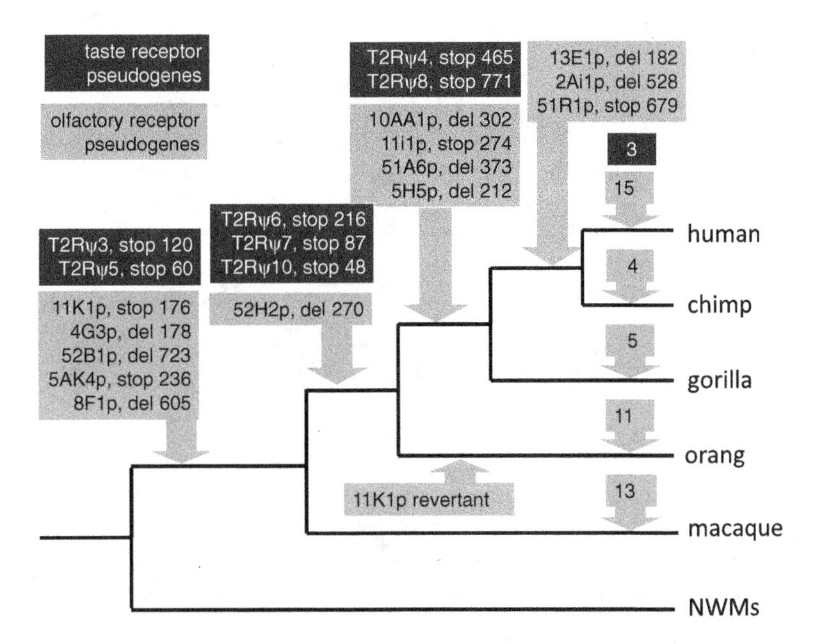

FIGURE 3.20. PRIMATE PHYLOGENY BASED ON OLFACTORY (*LIGHT BOXES*) [80] AND TASTE (*DARK BOXES*) [86] RECEPTOR PSEUDOGENES
Mutations are identified by the gene symbol (e.g. 11K1p), the type of mutation (e.g. *stop*) and the position of the mutation (e.g. codon 176).

behind. Primates have a modest repertoire of several hundred. At the other extreme, the aquatic mammals (dolphins and whales), along with their nearest living relative (the hippopotamus), have very few. The facile expansion and contraction of olfactory gene repertoires depends on the ecological niche occupied by each species [83].

The diminished role of olfaction in primates is attested also by the loss of a signalling molecule that is involved in a subset of olfactory nerve responses. The protein, known as guanylyl cyclase D (GC-D), is usually located in cell membranes. When the part *outside* the cell engages with its particular stimulant, the part *inside* the cell generates a messenger molecule called cyclic GMP, which opens ion channels to generate nerve impulses. (GC-D thus acts analogously to TRPC2.) GC-D is functional in non-primate mammals and in prosimians (lorises and lemurs). Protein-coding function, however,

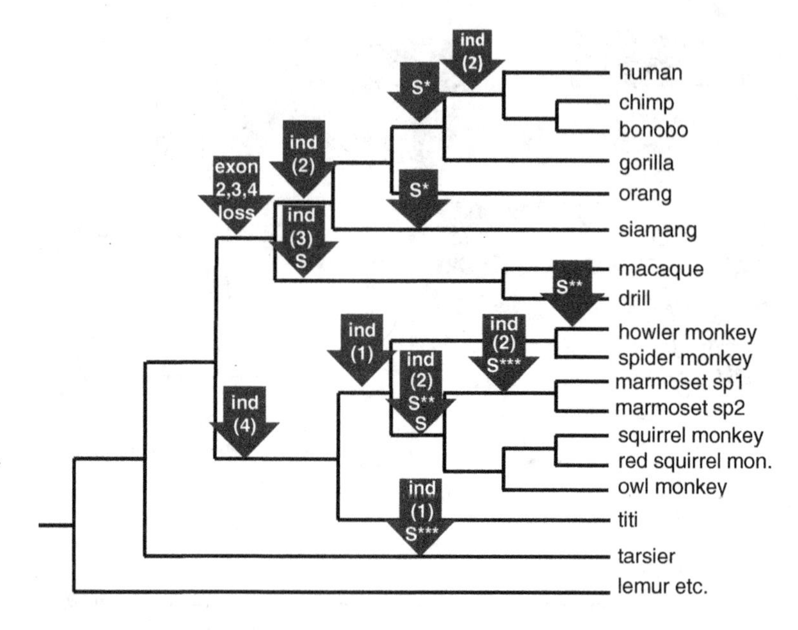

FIGURE 3.21. A PRIMATE PHYLOGENY BASED ON MUTATIONS IN THE *GC-D* GENE [84]
Three pairs of *stop* mutations, generated independently in two lineages, are indicated by *asterisks.* Mutations are *stop* mutations (S) and indels (ind). *Numerals in brackets* indicate the number of indels.

has been lost independently in the ape–OWM group, in NWMs and in tarsiers.

The accumulation of inactivating mutations in this gene provides an independent primate phylogeny (Figure 3.21). For example, two indels are common to all the apes, and four are common to all the NWMs. There is also a selection of *stop* mutations, although some of these single-base change mutations have occurred independently in different lineages [84].

A third category of chemosensation is that of *taste.* Our sense of taste arises from receptors expressed by nerve cells clustered in taste buds on the surface of the tongue and palate. People vary in the way they respond to certain bitter substances. The bitter-tasting test compound phenylthiocarbamide is classically used to categorise us into tasters and non-tasters. The reason for this variation is that one of our bitter taste receptors exists in alternative forms, which we share with Neanderthals. The same situation exists, but has arisen independently, in chimps [85]. Our vegetables may not taste the same to us as they do to other members of the family.

We have a relatively small family of taste receptors, 33 or 34 in all. Neither our ancestors nor we could afford to lose our capacity for certain tastes. Our sense of sweetness motivates us to value energy-giving (sugar-rich) food. Our sense of bitterness motivates us to avoid plant material rich in potentially toxic alkaloids. Only 10 of approximately 30 bitter taste receptor genes are pseudogenes. Judging by the presence of shared mutations, some of these lost their coding capacity in an ancestor shared with OWMs, others in an ancestor of the great apes, and others in an ancestor of African great apes (Figure 3.20) [86].

Cats are the consummate carnivores. They have a finely honed appetite for meat but are indifferent to sugars. Cats lack a sweet tooth because their taste buds lack the Tas1R2 sweet taste receptor. The gene has been inactivated in cats. Two mutations (a 247-base deletion in exon 3 and a deletion in exon 4) exist also in cheetahs and tigers. This demonstrates that domestic cats, cheetahs and tigers are descendants of the progenitor cat species – the primeval felion – in which the mutations occurred. The existence of this pseudogene with its unique mutations confirms that a warm fluffy cat is devoid of sweetness, a ferocious killer writ small. Indeed, pseudogenisation of sweet taste receptor genes is widespread in carnivorous mammals [87], including blood-eating vampire bats [88].

A similar story is illustrated by the giant panda – a bear and therefore classified as a carnivore. However, pandas have reformed their ways and now dine only on vegetables, of which bamboo is the staple. The sensory receptor that responds to meat is the Tas1R1 *umami* receptor (from the Japanese word denoting 'savoury'). We humans stimulate this receptor by pouring monosodium glutamate on our takeaways. The vegan panda, however, does not seek meat, and the gene for its Tas1R1 receptor contains two frameshift mutations. Surprisingly, some carnivorous mammals have also lost their umami receptors. These include various pinnipeds (seals and sea lions) and the bottlenose dolphin, which swallow their prey whole. Taste receptors do not serve much function if you don't bother to chew your food [89]. We might conclude that if you don't use it, you lose it.

3.3.5 Pseudogenes from further afield

Pseudogenealogy is in its infancy. As genome data accumulates, pseudogenes shared by humans and non-primate mammals will undoubtedly provide markers of more remote relationships. Analysis will be hampered by the fact that, as we have noted with TEs, pseudogenes tend to disappear into the genetic background with time. (To return to our analogy: whether or not the old garage has any function, most of its timbers ultimately decay into a barely recognisable wood heap.) Shared inactivating mutations that created pseudogenes before the origin of the primates will be obliterated by superimposed mutations. But such derelict genes are present. A fossil exon (with parts of the flanking introns) of an MHC class I gene is found in both humans and pigs, for example, and probably arose in an ancestor of primates and artiodactyls [90].

The vitellogenin genes function in egg-laying animals. Vitellogenin is a protein that transports nutrients from the liver to the egg yolk. Birds lay eggs and have three vitellogenin (*VIT*) genes. Monotremes (such as the platypus) are mammals that lay eggs, albeit with a diminished nutrient content relative to birds. They retain one functional *VIT* gene and one identifiable *VIT* pseudogene. Placental mammals do not lay eggs, and lack vitellogenin genes. However, fragments of each of the three *VIT* genes (present in birds) are identifiable in our DNA. They are riddled with mutations such as *stops* and indels. For example, a fragment from the third exon of *VIT1* in humans, dogs and armadillos shares two deletions of one base each. These are naturally interpreted as mutations that arose in a eutherian ancestor (Figure 3.22, upper diagram) [91].

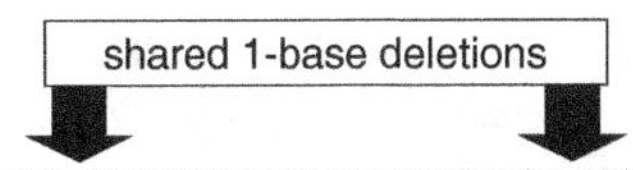

human	...TATGAAAGTACA TTTCTAGTGTATTTCCACA ACAGT...
dog	...TAAGAAAGGATA TTTCTGATGGACGGCCACA ATCAT...
armadillo	...TATGGAAGTATA TTTCCAGTGGACTGCTACA ACAGT...
chicken	...TATGAAAGCATACTTTTCAGTGGTATTCCAGAGAAGGA...

enamelin exon 4, crocodile (above) and chicken

...ATGTTCCTCCAGTTCCTGTGCCTCTTTGGCATGTCTATAGCAGTGCC...
...AAAAAACTCCTCTTCCTGTGCCTCTGTGCAATGTCCTGGGCAGTGCT...

amelogenin exon 2, crocodile (above) and chicken

...GTACTATATTTCGAGAAAGATGGAGGGCTG...
...ATACTGTAATTATAGAACGATGGAGGACAG...

FIGURE 3.22. FRAGMENTS OF ANCIENT PSEUDOGENES
Sequences from the *VIT1* pseudogene and the chicken gene (*upper diagram*) [91]. Bases common to at least three of the species are in *bold*. Sequences from the enamelin and amelogenin genes (*lower diagrams*) [97, 98]. Shared bases are in *bold*; *start* codons are *shaded*.

Also of interest are genes that encode the proteins needed to make tooth enamel, the hardest material made by any vertebrate. Mammals with peculiar dietary preferences either lack teeth, or lack teeth crowned with enamel. Baleen whales spectacularly lack teeth. They strain krill and small fish from the sea using a filter of baleen (chemically akin to hair, and composed largely of a fibrous protein called keratin). Pangolins and anteaters are insect-eaters and also lack teeth. Other mammals retain teeth but they are not crowned with enamel (Table 3.2) [92].

If the standard scheme of phylogenetic development is true, we would expect that these mammals evolved from progenitors (now extinct) that possessed teeth with enamel. We would hypothesise that enamel-less species would retain in their genomes degenerated copies of the enamelin (*ENAM*) gene.

And indeed, representatives of every group that lacks enamel do retain the *ENAM* gene in their genomes, but in every case its

Table 3.2. *Mammals lacking enamelised teeth*

Super-order	Group	Teeth	Diet
Laurasiatheria	Baleen whales	Absent	Krill, small fish
	Sperm whales	Lack enamel	Cephalopods, fish, crustaceans
	Pangolins	Absent	Insects
Afrotheria	Aardvarks	Lack enamel	Insects
Xenarthra	Anteaters	Absent	Insects
	Sloths	Lack enamel	Vegetable material
	Armadillos (most)	Lack enamel	Insects and grubs

coding function has been destroyed by *stop* and frameshift mutations. Indeed in species representing the taxa listed in Table 3.2, 125 different frameshift mutations in *ENAM* genes have been identified. Of these mutations, 123 were distributed among the diverse species in a way that supported the derivation of wholly consistent phylogenetic relationships. Of the two exceptions, one showed probable incomplete lineage sorting in related whale species. The other shared mutation, a one-base frameshift, probably arose independently in two anteater species. Overall, these frameshift mutations are powerful markers of evolutionary relationships [93].

We have already mentioned the loss of another tooth-forming gene in baleen whales (Chapter 2). This gene encodes enamelysin, a proteolytic enzyme that processes the structural proteins (enamelin, ameloblastin, amelogenin) when enamel is formed. All living baleen whales lack enamelysin. The protein-coding function of the gene was lost as a result of a TE insertion. This establishes that all baleen whales are descended from the ancestor in which this unique inactivation event occurred. Moreover, inactivation of enamelysin rendered the other enamel-making genes redundant, and started the processes by which they lost protein-coding function [94].

The ultimate toothless wonders are birds. But some fossil birds (such as *Archaeopteryx*) had teeth. Some show reduced patterns of dentition [95]. Moreover, a chicken mutant has been described that undergoes early steps in tooth formation. Chickens seem to retain the remnants of a developmental pathway that once culminated in tooth formation [96].

The chicken genome has been explored for four genes involved in tooth formation. Degenerated remnants of the *ENAM* gene have been tracked down, with each of exons 4–10 being recognisable [97]. Multiple fragments corresponding to the gene for amelogenin are also present in the chicken genome. Comparisons of parts of *ENAM* exon 4 and amelogenin exon 2 with the corresponding parts of the crocodile genes are shown in Figure 3.22 (lower diagrams). Parts of the ameloblastin and dentin sialophosphoprotein genes are also discernable. Particular indels are shared by multiple species of birds [98]. Clearly, birds have ancestors that had teeth.

It is interesting to note that genes active in tooth formation may have been modified to adopt new protein-coding functions. An example relates to the origins of proteins found in milk, a product peculiar to mammals. Caseins are major milk proteins that transport calcium phosphates, essential for the growth of bones and teeth in young mammals. Evidence suggests that the casein gene family was derived from the *odontogenic ameloblast-associated* (*ODAM*) gene by a series of DNA duplications that involved parts of exons (making them longer), whole exons and entire genes. The casein genes arose before mammals appeared, and originally may have encoded proteins that transported calcium in eggs or that controlled mineralisation of teeth [99]. The formation of new genes is the topic of Chapter 4.

3.4 PROCESSED PSEUDOGENES

Approximately half of the pseudogenes located in our DNA have arisen from the unscheduled activities of retrotransposons. LINE-1 elements encode enzymes required to 'copy and paste' themselves into new sites. But sometimes these enzymes act promiscuously. They may disengage from the LINE-1 transcript that produced them and attach themselves to bystander RNA molecules, which they reverse-transcribe into DNA copies. This process generates copies of genes that are inserted into genomic DNA at randomly chosen positions [100]. Such sites typically lack the regulatory sequences required to orchestrate gene transcription. These processed pseudogenes would thus lack the capacity to generate RNA copies and make their respective proteins. They have been described as being 'dead on arrival'.

This random process is ongoing. Some genetic diseases of humans arise because fragments of RNA are commandeered by the enzymatic machinery of LINE-1 elements and spliced back into genomes. Tragically, the pseudogenic fragments may disrupt functionally important genes, destroying protein-coding capacity and precipitating clinical abnormality. In a case of Duchenne muscular dystrophy, a fragment of non-coding RNA (from chromosome 11) was retrotransposed into exon 67 of the *dystrophin* gene (on the X chromosome). The mutation was not present in the constitutional DNA of either parent, and occurred either in one of the parents' germ cells or in the affected child [101].

The human genome contains several thousand such randomly generated processed pseudogenes. Estimates range from 3,600 to over 13,000. The variation arises because different researchers have arbitrarily chosen different minimum sizes as criteria of what they would accept as a pseudogene. Some inserts are full-length copies of messenger RNAs, whereas others are fragments. A gene will contribute processed pseudogenic copies to the germ-line only if that gene is expressed in germ cells, which alone transmit genetic novelties to future generations. The more highly a gene is expressed, and the more stable its messenger RNA molecule, the more copies tend to be found in genomes [102].

Some processed pseudogenes retain base sequences that are extremely similar to those of their parent genes. Such copies were generated relatively recently. Others have accumulated many mutations relative to their source gene. These are likely to be of relatively greater age. Clearly, processed pseudogenes have been accumulating in mammalian genomes over a long history. This is hardly surprising, since they have been generated by TEs, which themselves have colonised genomes over a long history. It follows that if any of these unique markers reside in the genomes of multiple species, those species must have been derived from the progenitor in which the pseudogene arose.

Consider the *NANOG* gene, which is active in early embryos and in stem cells. It is expressed in the germ-line and is a master regulator of gene expression. In the human genome, it has spawned ten processed pseudogene copies, and nine of these are present also in the chimp genome. That is, they were already present in the genomes of human–chimp ancestors. One of these, *NANOGP4*, has accumulated a number of classical gene-disrupting mutations. The human version has four *stop* mutations (three of which are shared with chimps) and three deletions (two of which are shared with chimps). This gene embodies multiple markers of monophylicity [103]. The human- (and Neanderthal-)specific pseudogene, *NANOGP8*, is normally silent, but is expressed in the dysregulated environment of tumour cells. It has been suggested that *NANOGP8* acts as an oncogene and is responsible for the increased propensity of humans (relative to other primates) to develop cancers [104].

A fragment of the *ATM* gene exists as a processed pseudogene in the human genome. It encompasses exon 30 and parts of both flanking introns, including part of an intronic Alu element. It is bracketed by target-site duplications – as expected of a LINE-1 endonuclease-generated product. The same *ATM* pseudogene and target-site duplications are present in chimps and gorillas, whereas the orang-utan genome retains the uninterrupted target site (Figure 3.23, upper diagram) [105]. The African great apes are descended from the progenitor in which this unique pseudogene arose.

Other such retrotransposed genes, such as the *RNASEH1* pseudogene, are present in all apes, but in no other species. The

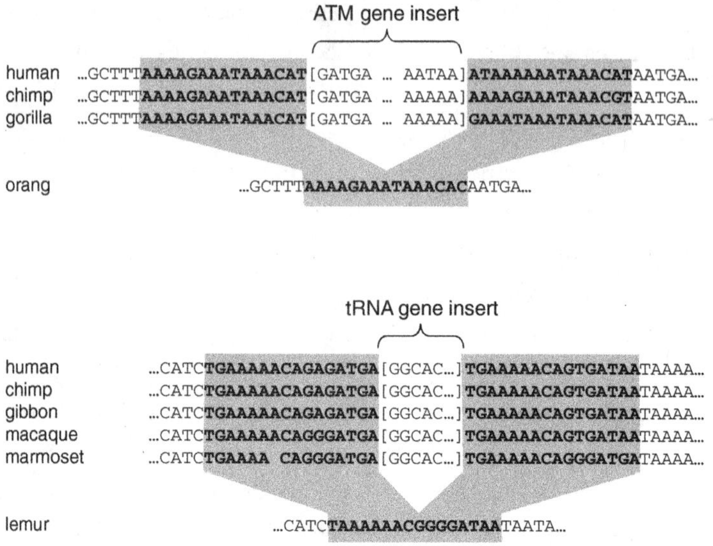

ATM gene insert

human ...GCTTTAAAAGAAATAAACAT[GATGA ... AATAA]ATAAAAAATAAACATAATGA...
chimp ...GCTTTAAAAGAAATAAACAT[GATGA ... AAAAA]AAAAGAAATAAACGTAATGA...
gorilla ...GCTTTAAAAGAAATAAACAT[GATGA ... AAAAA]GAAATAAATAAACATAATGA...

orang ...GCTTTAAAAGAAATAAACACAATGA...

tRNA gene insert

human ...CATCTGAAAAACAGAGATGA[GGCAC...]TGAAAAACAGTGATAATAAAA...
chimp ...CATCTGAAAAACAGAGATGA[GGCAC...]TGAAAAACAGTGATAATAAAA...
gibbon ...CATCTGAAAAACAGAGATGA[GGCAC...]TGAAAAACAGTGATAATAAAA...
macaque ...CATCTGAAAAACAGGGATGA[GGCAC...]TGAAAAACAGTGATAATAAAA...
marmoset ...CATCTGAAAA CAGGGATGA[GGCAC...]TGAAAAACAGGGATGATAAAA...

lemur ...CATCTAAAAAACGGGGATAATAATA...

FIGURE 3.23. INSERTION SITES OF PROCESSED PSEUDOGENES DERIVED FROM AN *ATM* GENE TRANSCRIPT, *UPPER DIAGRAM* [105], AND FROM A TRANSFER RNA MOLECULE, *LOWER DIAGRAM* [109]

Per4 pseudogene is present in humans and OWMs. The *SHMT-ps1* pseudogene is shared by all simian primates. Small-scale studies of the species distribution of processed pseudogenes are congruent with independently established patterns of primate evolution. One study with six such relics has provided additional evidence that the great apes, the Old World primates and the simians each comprise a monophyletic group [106]. Processed pseudogenes may acquire new functions. Keratin-19 and β-tubulin processed pseudogenes, each of which arose in an ape–OWM ancestor, are loci in which new micro-RNA genes have emerged [107].

Systematic studies using available human, chimp, orang-utan and macaque genomes have determined the order of appearance of processed pseudogenes, derived from protein-coding genes, in the primate germ-line. Forty-eight processed pseudogenes were found in humans only. Ninety-four were present in the genomes of humans and chimps, but in no other species. And 337 were found in the genomes of each of the three great ape species surveyed, but not in macaques. Such processed pseudogenes are only a small proportion of those described in the human genome, and so there must be a considerable number common to apes and OWMs [108].

Transfer RNAs and hY RNAs are small RNA molecules without protein-coding functions. They function alongside messenger RNAs in the complex process of protein synthesis. Approximately 800 retrotransposed pseudogenes arising from transfer RNA genes have been identified in the human genome. A selection of these have been investigated for their presence in other primate species. Of the nine investigated, one was restricted to humans, one to humans and chimps, five to humans, chimps and gorillas, one to all apes and one to all anthropoid primates (Figure 3.23, lower diagram) [109]. The four hY RNA genes in our genome have been copied and pasted into 966 pseudogenes, and 95% of these are shared with chimps [110].

The application of processed pseudogenes to phylogenetic studies (what one might call 'processed pseudogenealogy') owes its persuasiveness to the nature of the reverse transcriptase-dependent mechanism. Multiple events involve probabilistic outcomes. Each such pseudogene is a singular marker of those kindred species that are descended from the germ cell in which the pseudogene arose. And living systems opportunistically transform genetic flotsam into function.

3.5 RARE MUTATIONS THAT CONSERVE PROTEIN-CODING FUNCTION

We have outlined 'pseudogenealogical' relationships based on genes that have sustained either random devastating mutations or that have been generated by the haphazard process of reverse transcription. But pseudogenes tend to decay into the genetic background and ultimately disappear from sight. They may have a limited capacity to define distant family connections in deep time. Not all indel mutations, however, destroy genes. Indels may spare gene function when they involve the insertion or deletion of bases *in multiples of three*, because amino acids are gained or lost in such a way as to preserve the rest of the reading frame. The encoded protein may tolerate this change.

Such mutations are complex and rare, and (because they are associated with the retention of gene function) do not lead to decay and obliteration of the gene. The stability of genes that contain such mutations has the consequence that such function-sparing mutations may endure over huge evolutionary timescales. They have provided striking answers to tricky questions of distant phylogenetic relationships.

Shared multiple-of-three indels have shown that the closest relatives of primates are flying lemurs, followed by tree shrews. These three groups comprise the taxon known as Euarchonta (the 'real ancestors'), which is the sister group of Glires (rodents and rabbits) [111]. Other indels demonstrate that Euarchonta and Glires are the descendants of a Euarchontoglires (primate–rodent) ancestor. For example, the SCA1 protein has an 18-amino-acid deletion (reflecting a 54-base deletion in the gene) in multiple mammalian species (Figure 3.24). This extensive but function-sparing deletion occurred uniquely in a single germ cell that gave rise to the diversity of organisms classified (by independent criteria) as Euarchontoglires [112]. These relationships are congruent with those determined from the pattern of TE inserts.

Other indels have contributed to the identification of the Afrotheria as a super-order, and showed that Afrotheria and Xenarthra are the issue of a common ancestor [113]. The relationships between the three basal mammalian groups have been uncertain. These groups are the monotremes (platypus and spiny anteater), marsupials (such as the opossum) and eutherians. Multiple-of-three indels in two genes have demonstrated that the latter two groups share indels that are not shared by monotremes, birds or reptiles. We conclude that opossums and people share an ancestor that lived on a separate lineage to that of the monotremes – exactly what we have concluded from the wholly independent study of ancient TEs [114].

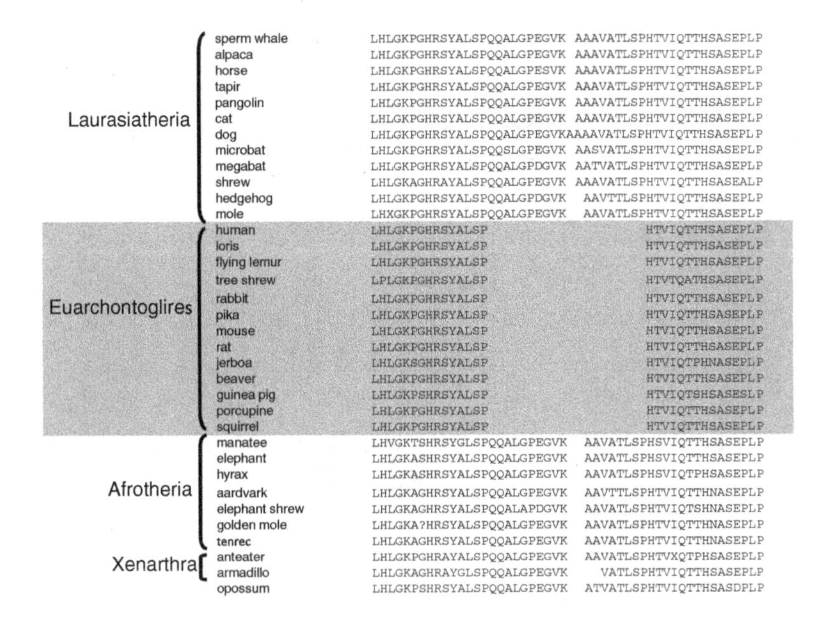

FIGURE 3.24. AN 18-AMINO-ACID DELETION IN THE SCA1 PROTEIN OF EUARCHONTOGLIRES [112]

3.6 CONCLUSIONS

Are there any alternative interpretations of the remarkable data presented in this chapter? It may be suggested that the apparently shared mutations are in reality independent mutations that have occurred at *hotspots* – regions of the genome that are susceptible to recurring mutations. In response to this, we may make several comments.

Firstly, many classes of mutations do not lend themselves to hotspot considerations. Many DNA breaks arise from radiation, and radiation is not fussy about where it interacts with genome sequences. High-energy photons (or the destructive hydroxyl radicals they generate) shatter DNA where they will. And yet the scars of ancient NHEJ events provide a full and consistent phylogeny of primates.

Other classes of mutations that are complex, unrepeatable and not amenable to hotspot regularity are being used to work out phylogenies independently of those discussed in this chapter. The distributions of ERVs and TEs yield the same phylogenetic relationships as do pseudogenes. Other mutations include indels scattered through intergenic DNA [115], inversions [116] and duplications of huge tracts of DNA (Chapter 4). These complex mutations generate phylogenetic trees congruent with those constructed by the approaches described here. A colossal study comparing some 35,000 bases from the genomes of each of 191 largely primate species has used the presence of indels to generate a primate taxonomy wholly consistent with what has been described hitherto [117].

Some of these complex mutations may have had significant impacts on the human form. The human genome has sustained more than 500 deletions of sequences that are highly conserved in chimps and other mammals. Most (but not all) of these deletions are features also of the Neanderthal genome, indicating that we share most of our history with Neanderthals since branching out from the chimp lineage. Deleted DNA segments include regulatory sequences that (in other mammals) direct the production of sensory whiskers and penile spines. The loss of penile spines, tiny barbs on the skin of the penis, has been taken to reflect the importance of pair bonding in human reproduction, and of increased paternal input in raising young. Another deletion has inhibited the action of a gene, *GADD45G*, that acts to limit the proliferation of neurons in the brain. This deletion may have contributed to increased brain size [118].

Secondly, the pattern of evolutionary relationships derived from the distribution of mutations is too consistent to be attributed to hotspot arguments. We have encountered occasional revertant or independently arising mutations in the case of one-base changes. The reversion of these most simple of mutations is to be expected at a low frequency. But the data presented in this chapter reveal a strikingly consistent order of human relatedness to chimp, gorilla, orang, gibbon, OWM, NWM and beyond. Hotspot mutations are still mutations, and mutations cannot produce the consistency disclosed in pseudogenealogical studies. Rather, hotspot mutations would present patchy distributions in multi-species comparisons. Take the

GULO gene, for example. A mutational hotspot could not account for the same *stop* mutation in each of every simian primate species tested, but in none of the non-simian species. Exhaustive analyses of tumour suppressor genes, mutant in human cancers [119], show that many genes do not show hotspots. And even those that do (such as the *TP53* gene) provide brilliant markers of cell or family lineages.

Thirdly, many widely disseminated mutations (such as those conferring sensitivity to Ivermectin in dogs, to those responsible for blue eye colour in humans) are located within defined haplotypes, extended DNA segments of shared variants. Each founder mutation establishes that pseudogenisation arises from a unique mutation, and then spreads by inheritance through the population.

One final example will close this chapter. The *IRGM* gene is involved in protection against infections, but was destroyed early in primate evolution by the insertion of an Alu element into its left-hand end. Lemurs (which are prosimians) lack the Alu element and retain the *IRGM* gene in active form. But the disruptive Alu element is present in every one of 4 NWMs, 11 OWMs and 5 apes (Figure 3.25). Following the Alu insertion, the gene could not retain its protein-coding function, and it started to accumulate *stop* mutations. (When the old garage fell into disuse, no-one bothered to maintain it. The walls started to cave in, and eventually the windows fell out.)

In each of the four NWMs, the *IRGM* pseudogene contains the same two *stop* mutations (TAA and TAG). These are not hotspot mutations because they are absent from any OWM or ape gene. Rather, both of these mutations occurred in a NWM ancestor. They have been inherited by all the species descended from that ancestor.

A frameshift mutation (deletion of a G) is present in each of the 11 OWMs. It cannot be a hotspot mutation because it is absent from any NWM or ape gene. It arose in the non-protein-coding gene of an OWM ancestor, and was transmitted to all OWM species that currently possess it.

But something curious occurred in an ancestor of the apes. An ERV9 inserted itself into the left-hand end of the pseudogene,

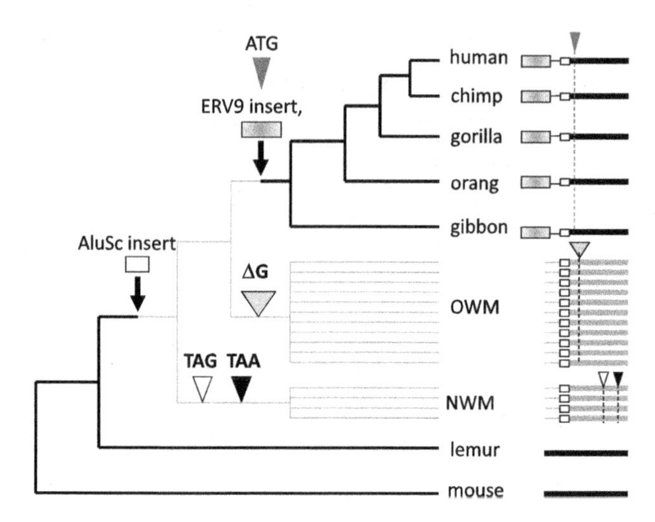

FIGURE 3.25. DEATH AND REBIRTH OF THE *IRGM* GENE [120]
On the cladogram, taxa possessing *IRGM* activity are indicated by *black lines*; taxa possessing the pseudogene by *grey lines*; the timing of mutations is indicated. The corresponding mutations are shown on the gene diagrams on the right. The Alu element is represented by a *small rectangle*; the ERV9 element by the *larger shaded rectangle*.

providing a brand new transcription start site. In addition, a substitution mutation generated an ATG codon, which signals the start of a protein-coding reading frame. These two mutations rejuvenated the gene, and it regained protein-coding capacity in an ape ancestor millions of years after it was destroyed in a simian (ape–monkey) ancestor [120].

To conclude: our genome contains thousands of disabled genes. They no longer possess the information content required to specify the production of proteins – information content that is still embodied in the progenitor genes from which they were derived. These pseudogenes have been disrupted by an extensive variety of mutations. When multiple species share one of these mutations, it is only because they have inherited it from the reproductive cell in which the unique mutation occurred. The burgeoning scientific field of pseudogenealogy establishes the concept of common descent in a way that would have been inconceivable before the DNA sequencing revolution. Humans and other apes have common ancestry. So too do humans and monkeys, and (progressively further back in history) humans and primates, mice, eutherian mammals and opossums.

Pseudogenes may, of course, acquire novel functions [121]. Biological systems may opportunistically co-opt unclaimed DNA resources. But it is the molecular *mechanisms* of mutagenesis (reflecting the characteristic chemistries of nucleic acids and the enzymes that process them), deciphered in basic genetics laboratories, that establish such mutations as powerful markers of descent. The progressive changes in base sequence provide the *history* of a pseudogene, and this *history* defines the evolutionary relationships of those species that share the pseudogene. Current functionality is irrelevant to the value of pseudogenes as evolutionary markers. The timber from the old garage may be put to innumerable and imaginative uses, but these can never alter the fact that the wood was scavenged from the old *garage*.

ERVs, TEs and pseudogenes were once dismissed as 'junk' DNA. Many are now known to possess diverse regulatory functions [122]. There is an irony here. Certain people have argued for years that mutations are overwhelmingly destructive (or *disabling*) and could never provide novel functions underlying evolutionary development. But to acknowledge (as such critics must) that ERVs, TEs and pseudogenes are the loci of myriad genetic innovations is to acknowledge the centrality of *enabling* mutational events in generating these functions. The experimentally demonstrated functionality of erstwhile 'junk' evinces the potency of natural selection. We examine enabling mutations in Chapter 4.

4 The origins of new genes

Progressively accumulating mutations have mapped out the route of human evolution (Chapter 3). Genes have been disabled at a measurable rate. Genes or fragments thereof have been 'copied and pasted' haphazardly around the genome. All this may leave us with the impression that the genome that we have inherited is heading towards decrepitude. But if humanity is the terminus of such a messy process of genomic shuffling, how is it that the species shows capacities such as manual dexterity and mental versatility that are unprecedented in the biosphere? The advent of these new powers implies the elaboration of genetic complexity.

ERVs and TEs have added vast amounts of raw material to primate genomes. Some of these units of genetic flotsam have been transformed into new protein-coding genes. Thousands of these have been exapted as regulatory sequences that control gene activity. Our genomes have been rewired by their semi-autonomous denizens. But could new genes also arise through cumulative mutational changes? A coding sequence would need to arise – and that seems unlikely enough. Such a nascent gene would require concomitant regulatory sequences that would assemble the transcription factors needed to transcribe it – in the appropriate cell type, and at the right time. But that coding sequence would also need to encode a protein with functional domains, able to connect with established proteins and integrate into elaborate networks.

At this stage, one old myth – that the development of new functionality must arise from mutations in established genes – must be laid to rest. This hypothesis for new gene function has long been discarded. Rather, the genome revolution has shown that large chunks of the genome, thousands of bases in length, are frequently duplicated. Any genes that are included within such duplications are superfluous to requirements, and may be subject to mutational modification with impunity. If the gene is scrambled, no loss is incurred – the gene was a spare. But if a new function appears, allowing a duplicated gene to confer a reproductive advantage on its carrier, then that gene will be preserved by natural selection. The result would be diversification and elaboration of gene function.

In fact, this situation is encountered all the time in the vastly compressed timescale of cancer development. In this context, the reproductive success of tumour cells is increased by genetic changes that enable those cells to escape from normal restraints. Tumour cells randomly but recurrently generate new genes that drive tumour growth. Duplications of genes and the generation of novel genes in cancer is surprisingly facile and will be described as a prelude to considering evolutionary development.

4.1 NEW GENES IN CANCER

We have two sets of chromosomes, and two copies of each gene (apart from genes on the X and Y chromosomes in males). But when cancers develop, they may generate scores of copies of particular genes. This well-recognised phenomenon is called *gene amplification*. The increases in gene copy number drive the abnormal proliferation that is a hallmark of tumour cells. Amplification of the *EGFR* gene is found in several types of carcinoma. This gene encodes a receptor protein that enables cells to proliferate in response to proteins of the epidermal growth factor (EGF) family. *EGFR* gene amplification leads to increased numbers of EGF receptors on cell surfaces, and an elevated tendency of the cells to proliferate in response to the usual concentrations of EGF.

The *EGFR* gene is amplified on a large segment of DNA called an *amplicon*. The actual extent of the amplicon varies between cancers (Figure 4.1). The only thing that matters is that the amplified

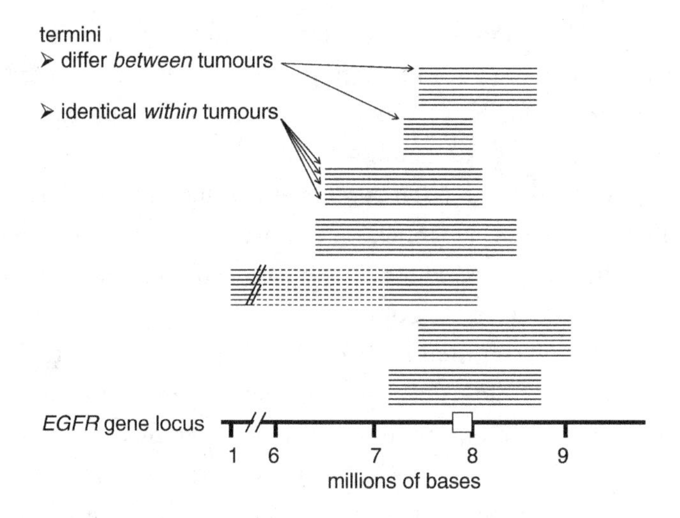

FIGURE 4.1. AMPLICONS IN THE *EGFR* GENE IN BRAIN TUMOURS [1]
The *lower line* represents the DNA locus containing the *EGFR* gene (*box*). The *groups of lines* above this represent amplicons found in seven tumours. In any one tumour, all the amplified lengths of DNA terminate at the same points. All amplicons were derived from one original.

segment includes the *EGFR* gene. The left-hand and right-hand ends of the amplicon differ for every tumour. This lack of consistency indicates that the events giving rise to the amplicon are random. But within *any one* tumour, all copies of the amplicon in every cell have precisely the same left-hand and right-hand ends. This indicates that every amplicon is derived from one original mutational event, and that every cell possessing this amplicon is derived from the cell in which the amplicon was generated. Such duplications demonstrate monoclonality [1].

Tumour cells are past masters at rearranging their genomes to assemble new genes. The classic example occurs in chronic myelogenous leukaemia (CML), which is characterised by an exchange of genetic material, or *translocation*, between chromosomes 9 and 22. This translocation joins most of a proto-oncogene (*ABL*) from chromosome 9 to part of a gene (*BCR*) on chromosome 22. The product is a fusion gene, *BCR-ABL*, which directs the production of a chimaeric protein with a novel suite of properties. This alien protein fits into and perturbs pre-existing protein-signalling networks and so orchestrates the abnormal behaviour of CML cells.

The breakpoints on chromosomes 9 and 22 are different in every case of CML. This means that there is no predetermined trigger. The chromosome breaks occur at random. But in the tumour of each patient with CML, all the millions of leukaemia cells have the *same* breakpoints. This means that every cancer cell is a descendant of the one cell in which that unique chromosome translocation arose. The possession of a unique translocation in the cells that populate a tumour is accepted as proof of monoclonality. An instance of this relates to a woman who suffered from CML during pregnancy. She delivered a healthy baby, but within a year the baby developed a tumour. Molecular analysis demonstrated that the baby's tumour possessed a *BCR-ABL* fusion oncogene, and that the *BCR* and *ABL* genes were joined to each other at precisely the same point as in the mother's tumour. This showed that, during pregnancy, cells from the mother's cancer had crossed the placental barrier into the baby: the baby's tumour was a monoclonal derivative of the mother's [2].

A colossal volume of research has demonstrated that particular chromosomal arrangements such as translocations establish the clonal status of tumours. This is summarised in a classic textbook:

> Evidence for the monoclonal origin of human tumours is provided by the observation that a unique identifying feature (a clonal marker) can be found in all of the constituent cells. Clonal markers include chromosomal rearrangements such as the Philadelphia chromosome in chronic myelogenous leukaemia, [and] uniquely rearranged immunoglobulins or T-cell receptors expressed by B-cell lymphomas or multiple myelomas and T-cell lymphomas respectively. [3]

Chimaeric genes feature frequently in cancer. Following the near meltdown at the Chernobyl nuclear power station in 1986, there has been an increased frequency of childhood thyroid cancer in the Ukraine. This disease arises from rearrangements of proto-oncogenes such as the *RET* gene with a number of partner genes, which form chimaeric transcriptional units and, in some cases, chimaeric proteins. The well-studied *RAF* proto-oncogene family participates with partner genes in the formation of chimaeras in a variety of cancers. *TMPRSS2-ERG* fusion genes are generated in prostate carcinomas, and the *EML4-ALK* fusion gene in lung carcinomas. These are randomly arising but recurring gene fusion events [4].

The presence of such novelties demonstrates the ease with which new genes are generated and (given the appropriate selection pressure) the efficiency with which they can modify cell behaviour. They show how uniquely arising chimaeric genes can define the clonal origin of tumours. We should not be surprised if analogous processes fabricate novel transcriptional units in the germ-line – that is, in cells which give rise to sperm and eggs, and that transmit their genes to future generations.

4.2 COPY NUMBER VARIANTS

Humans vary in appearance and physiology. Some of this familiar variation reflects subtle differences in our DNA – such as occasional differences in single bases (*single nucleotide polymorphisms*), which make each person's genome 0.1% different from any other person's. One of the surprises of the genome era is that each one of us has multiple copies of certain very large expanses of DNA, and that the number of such copies differs between individuals. Early results showed that any two people may differ by at least ten such variably repeated segments of DNA. These reflect both deletions and duplications, may be hundreds of thousands of bases in extent, and often include genes. The human genome is much more dynamic than had been assumed hitherto [5].

Five years after these first reports of copy number variations, higher resolution analyses have identified some 11,700 variable loci (involving DNA segments of >440 bases) in the human genome. Any two people differ at approximately 1,100 of these. Almost a thousand variable loci involving segments >50,000 bases long have been found. Such copy number variations affect gene structure, gene transcripts and protein sequences. Changes to genomic structure associated with copy number variations even generate fusion genes [6].

Any two of us differ in the numbers of copies of many genes. Perhaps 14% of recognised genes are present at variable number in the human population. This variability famously pertains to sensory receptor genes. Any two of us may differ in the number of copies of each of ten different olfactory receptor genes (out of a total of 385 functional genes) and in a few taste receptor genes too. We also vary in the number of opsin genes that feature in red–green colour discrimination. We each sense our world differently from others. Dozens of genes that exhibit variation in copy number in the human population appear to be subject to positive selection [7].

The completed Human Genome Project is significantly incomplete. Sequencing technologies that were used in this project cannot provide an inventory of those genes that encode ribosomal RNA (rRNA) molecules. These molecules are components of ribosomes – biochemical workbenches upon which proteins are assembled. A cell may contain 10 million ribosomes, and multiple copies of the rRNA genes are required to meet demand. Each genome set possesses one array of *5S rRNA* genes and five arrays of *45S rRNA* genes. Each array of rRNA genes contains multiple gene units, and these are so similar in sequence that individual rRNA repeat units cannot be distinguished on the basis of their sequence.

The copy number of rRNA genes varies widely in the human population. In different genome sets, the number of *5S rRNA* genes varies from 35 to 175 copies. The gene content of the *45S rRNA* gene arrays can vary a hundredfold. Each person has a unique constellation of rRNA gene arrays. Family studies have shown that the number of genes per cluster may change radically as a result of chromosome recombination events: in every generation, approximately 10% of the gene arrays are altered relative to those possessed by the parents. Not surprisingly, copy number changes in rRNA genes also occur when cancers develop, and they have been identified in approximately half of lung and colorectal carcinomas. Cancers simply exaggerate the plasticity that is already inherent in the genome [8]. To be human (or presumably any other mammal) is to differ from most other members of one's species in rRNA gene copy number.

The processes that generate and maintain copy number variation provide extensive resources of raw material upon which natural selection can work. This is demonstrated by the relationship between the copy number of the gene encoding salivary amylase, *AMY1*, and the amount of starch in traditional diets. Amylases are enzymes that digest starch, and salivary amylase comprises approximately half of the protein in saliva.

People possess different copy numbers of the salivary amylase gene. Some populations eat large amounts of starch, such as those comprising the agriculturally based societies of Europe and Japan. Others are much less dependent on dietary starch. These include forest hunter-gatherers (such as the Mbuti), pastoralists (such as the East African Datog) and pastoralist-fishermen (the Siberian Yakut). Individuals belonging to the European and Japanese populations maintain higher numbers of *AMY1* gene copies, and have more amylase in their saliva, than do individuals belonging to the African and Siberian groups. People with more salivary amylase digest starch more efficiently: when they start chewing a mouthful of starch they experience a more rapid decline in viscosity (stickiness) than those with less salivary amylase. Furthermore, *AMY1* gene copy number affects our chances of developing metabolic disorders such as diabetes. People with more salivary amylase (and accelerated digestion of starch) generate a correspondingly more rapid release of insulin into the circulation. This in turn appears to suppress maximum glucose concentrations in the blood after a meal – and reduce the chances of developing insulin resistance and diabetes [9]. When it comes to processing food, you are (in terms of copy number) what your ancestors ate.

4.3 SEGMENTAL DUPLICATIONS

Copy number variation is alive and well in human populations. But our genomes also contain a large number of copied segments of DNA that are the shared property of *all* members of the human species. Genome science has revealed that 5% of our genome is composed of large duplicated segments of DNA or *segmental duplications* (SDs). Each such duplication that is a common genomic character of our species must have occurred in an ancestor who was common to us all.

Segmental duplications have been defined (for practical purposes) as copies of large expanses of DNA (>1,000 bases long) that show a high degree of similarity between copies (>90% of the bases are the same). Copies may be found on the same chromosome as the parent sequence, or inserted into other chromosomes. They may be crammed together in bumper-to-bumper configurations adjacent to centromeres and telomeres, or distributed along the chromosome arms.

Segmental duplications may arise by several mechanisms. Firstly, during cell division that produces germ cells, *non-allelic homologous recombination* changes copy number. This process accounts for the facile changes in rRNA gene copy number that occur between generations. Secondly, errors may occur during DNA synthesis, leading to multiple copying of the same sequence. Thirdly, blocks of DNA may be duplicated or deleted when DNA breakage is followed by error-prone repair effected by *non-homologous end-joining* (NHEJ; see Chapter 3). However SDs may have arisen, they are of immense importance. They destabilise genomes. They feature in genetic diseases. They duplicate genes. They demonstrate common descent [10].

4.3.1 Some early pointers

Humans and other primate species possess particular segmental duplications *in common*. Cytogenetic approaches have demonstrated that a fragment of chromosome 1 (approximately 100,000 bases long) has been copied and pasted into the Y chromosome of humans, bonobos and chimpanzees. This duplication is not found in gorillas, orang-utans, gibbons or OWMs. Humans and the two chimpanzee species are the clonal issue of the cell in which the duplication arose [11].

Duplicate segments 24,000 bases long are present on chromosome 17. They flank a 1.5 million-base region that contains the *peripheral myelin protein-22* (*PMP22*) gene. The duplicated segments engender genetic instability, leading to extra copies of the *PMP22* gene, or its total loss. Either outcome results in neurological disease. The duplication is present only in humans and the two chimpanzee species, and shares the same boundaries, establishing its singular origins [12].

The *CD8β1* gene encodes a protein that contributes to immune surveillance of pathogens. Part of this gene, the first seven of the nine exons, has been duplicated as a truncated pseudogene (*CD8β2*) that is present in humans, chimps and gorillas, but not in orang-utans. One terminus of the duplicated segment occurs in a large intron. The breakpoint is precisely the same in each species possessing the pseudogene. This unique piece of genetic happenstance occurred in a reproductive cell ancestral to humans, chimps and gorillas [13].

A segment of DNA 76,000 bases long has been copied from chromosome 9 to chromosome 22. The presence of the same expanse of DNA on chromosomes 9 and 22 may induce these chromosomes to pair up in an illegitimate way, promoting the translocations that generate the *BCR-ABL* fusion gene of CML. The paired blocks of DNA are found in the genomes of humans, chimps and gorillas, but not of orang-utans or monkeys, indicating that the duplication event occurred in an ancestor of the African great apes [14].

Another cancer connection concerns the *CHEK2* tumour suppressor gene. Its encoded protein helps to orchestrate responses to DNA damage. The gene is found on chromosome 22, but part of it (20,000 bases long) has been copied to chromosome 16. This event occurred in an ancestor of the great apes, as humans, chimps, gorillas and orang-utans all possess this duplicated unit (or *duplicon*). Further duplication to numerous chromosomes followed in the African great apes. In humans, five of the duplications are interrupted by a particular LINE-1 element, and in two of these cases, a corresponding chimp duplicon exists. These retrotransposon-tagged duplications were all copied from one original in which the LINE-1 element appeared, and two of the duplications were already in place in a human–chimp ancestor [15].

These SD instances extend the principle that has been established (Chapters 1–3) – but also involve another kind of mutational event. Duplications of huge tracts of DNA arise as an inevitable outcome of the biochemical nature of DNA and of the enzymatic systems that maintain it. The presence of a particular, randomly arising duplication in multiple cells or in multiple species indicates that those cells or species are monoclonal, descendants of the cell in which the unique structure arose.

4.3.2 Systematic studies of SDs

The sequencing of the human and chimp genomes provided the opportunity of ascertaining, in a systematic and unbiased way, the proportion of duplications that are shared by both species. Of all the *bases* present in SDs in the human genome, approximately two-thirds are shared with chimps, and were already present in the genome belonging to ancestors of humans and chimps. Some of these SDs are concentrated near telomeres, and these tend to be younger than SDs found elsewhere. Accordingly, a lower proportion of telomeric SDs (50%) is shared by humans and chimps [16].

SDs that are *shared* by the two species tend to be clustered together with SDs that are *unshared* (or *species-specific*) and that exhibit copy number variation within a species. This indicates that pre-existing SDs may facilitate the ongoing generation of new duplication events [17]. SDs arising in human history represent the same genetic phenomena as SDs shared by multiple species.

Systematic analyses involving high-quality sequence data have allowed scientists to compare multiple SDs at single-base resolution,

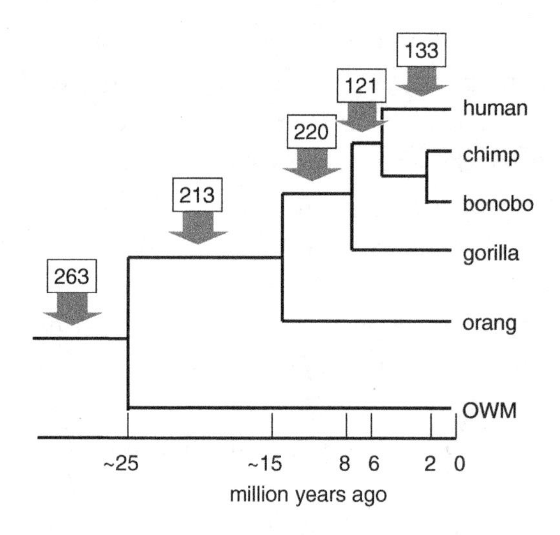

FIGURE 4.2. TIMING THE ORIGINS OF SEGMENTAL DUPLICATIONS [19] *Numerals* indicate the numbers of duplication events that occurred at each time interval in primate evolution.

and to locate the points at which SDs and the surrounding DNA sequences are joined. This ultimately rigorous type of analysis confirmed that many SDs are shared by humans and chimps; others by humans, chimps and gorillas; and others by humans, other apes and OWMs such as macaques and baboons [18]. This work has timed the origins of hundreds of SDs, and the results generate a primate family tree that is identical to that generated by other genomic markers (Figure 4.2) [19].

One class of SDs stands out because the reiterated units have a remarkable compound structure. They have a modified ERV at each end and a large core region between the two ERVs. The smallest of these duplicated units are about 30,000 bases long. The two ERVs and the central core region (derived from chromosome 19) are each about 10,000 bases long. These duplicated units have been nicknamed Xiao (Chinese for 'small'). With time, the core region accreted more blocks of DNA from various parts of the genome. Xiao expanded into DA ('large') duplication units. Full-length DA units span 300,000 bases.

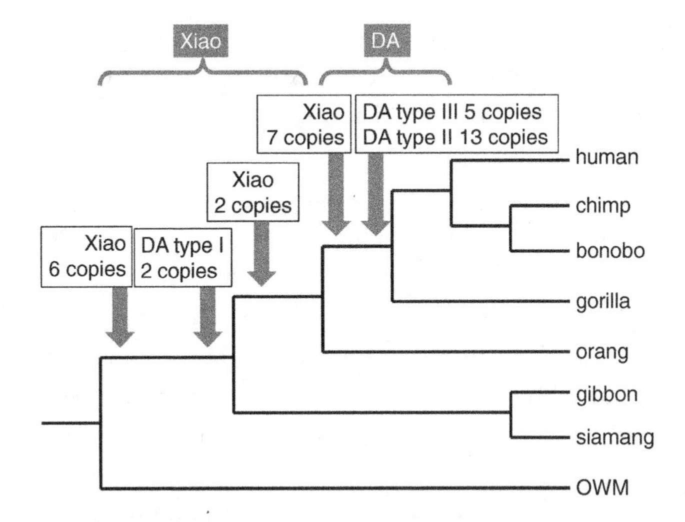

FIGURE 4.3. STEPWISE ADDITION OF SELECTED XIAO AND DA DUPLICONS TO PRIMATE GENOMES [20]
Each *box* indicates the number of new copies of Xiao or DA units acquired, and when they arose. *Brackets* indicate when most Xiao and DA units were duplicated.

The human genome contains 45 copies of Xiao and 20 copies of DA units, occupying about 5 million bases. Analyses of multiple primate genomes have shown that the Xiao units were first assembled in an ancestor of all apes, and that DA units increased in a stepwise manner during subsequent hominoid history. There has been no more amplifying activity since the human–chimp ancestral lineage. Some of the duplicated units seem to be bracketed by target-site duplications, implying that the duplicons proliferated by way of a copy-and-paste mechanism involving an RNA intermediate. An RNA intermediate hundreds of thousands of bases long is startling but credible. (The primary transcripts of the titin and dystrophin genes are respectively 280,000 and 2,400,000 bases long.) The mechanism of amplification remains speculative, but the hominoid phylogeny they evince is compelling (Figure 4.3) [20].

More ancient SDs (say those arising in pre-primate history) will have accumulated a heavy load of mutations and will be harder to identify. Nevertheless they are there. A class of large ancient duplications exists in *gene deserts*. These regions possess very few genes, and are composed of duplicated segments (two to six copies) that are at least a million bases long. Humans and other primates share them with mouse and dog (but not chicken). Such duplications date from ancestors shared by primates and other mammalian orders [21].

4.4 NEW GENES

Expanses of duplicated DNA are situated around our genomes. Any duplicon shared with another species establishes common ancestry. But, significantly, gene birth has occurred repeatedly in such SDs, by a variety of mechanisms and at particular stages of primate evolution. The documentation is so extensive that it cannot be reviewed in detail here [22]. We will outline a selection of histories that include molecular mechanisms of gene birth – mechanisms that connect all the species possessing a new gene to the reproductive cell in which each gene-forming event occurred.

The rate of change in gene copy number through primate evolution has been estimated by comparing the ability of genomic DNA samples from human and other primate species to compete for binding to arrays of DNA molecules representing the full human gene set (25,000 genes). Copy number gains and losses have occurred on every lineage, with gains outnumbering losses, particularly on the route leading specifically to humans. Copy number increases involving up to 100 genes were observed on the human lineage after it branched off from the human–chimp ancestor. Other copy number changes occurred on the lineage leading to humans and chimps (27); to humans, chimps and gorillas (80); to the great apes (124); and to the apes (105) [23].

Some gene families appear to be particularly dynamic. These include the highly amplified *Morpheus* [24] and *LLRC37* [25] families (of unknown functions) and the *NBPF* genes that were first encountered at the site of a DNA rearrangement in a cancer (and hence named ***n**euro**b**lastoma **b**reak**p**oint **f**amily*). *NBPF* genes encode a protein domain (DUF1220). The number of DNA motifs encoding DUF1220 domains has increased greatly on the human lineage. These domains are implicated in brain development and cognitive capacity [26]. Rapidly expanding families have been particularly well documented in hominoids, tend to show signs of positive selection and rapid evolutionary divergence, and may result in novel chimaeric transcription units.

A cunning twist on the gene duplication theme has been demonstrated for a gene that encodes a receptor for sialic acids. The *SIGLEC11* gene was duplicated in an ancestor of humans and chimps. In the lineage leading specifically to humans, the duplicate copy was pseudogenised, and part of the pseudogene sequence subsequently pasted back into the parent *SIGLEC11* gene. Such one-way sequence transfer is a well-known process known as *gene conversion*. A subsequent gene conversion went the other way, generating a revivified allele of the pseudogene. The duplication–divergence–conversion process generated two novel genes existing only in humans. The novel *SIGLEC11* gene encodes a protein with novel sialic acid-binding properties, and is expressed by microglial cells in the brain. It is also active in the ovary, where its product may contribute to the regulation of reproduction and, when abnormally expressed, to a uniquely human disease (polycystic ovarian syndrome) [27].

4.4.1 Reproduction

Sexual differentiation arises from the inheritance of specialised sex chromosomes. In mammals, females have the XX and males the XY chromosome pair. In birds, males have the ZZ and females the ZW chromosome pair. The X and Y chromosomes are very different from each other, as are the Z and W chromosomes. It is believed that the highly distinctive X and Y chromosomes have evolved from a standard non-sex chromosome (and that the Z and W chromosomes have arisen likewise). This is a fascinating notion – but is it true?

This hypothesis has been validated by the study of several genes on the avian Z chromosome, each of which has a relation on the W chromosome. One might hypothesise that such genes are related because they are descended from a single precursor gene that existed in the ancestral chromosome. In the case of three gene pairs, uniquely arising mutations (indels and retrotransposon insertions) are shared by a gene on the Z chromosome and its sibling gene on the W chromosome. Such shared mutations establish that the two genes evolved from a single precursor that was marked by the indel or retrotransposon. This in turn establishes that an avian Z-like chromosome evolved into the Z and W sex chromosome pair. By analogy, it is likely that a mammalian X-like chromosome evolved into the X and Y sex chromosomes [28].

Related, paired genes have arisen from a single progenitor as a result of sex chromosome differentiation. Such gene pairs are said to be *gametologues*. But evolving sex chromosomes have also been the locations of other types of gene traffic. The mammalian X and the chicken Z chromosomes have evolved from different progenitors, but each of them has gained a large number of genes, which are particularly involved in male (testis)-specific functions. Many such genes have been acquired though extensive tandem duplications. *La difference* has arisen, at least in part, from repeated rounds of segmental duplications [29].

A particular class of duplicon is abundant on the sex chromosomes. These are duplicated segments that are inverted with respect to each other, and in which the duplicates are separated by an unrelated spacer sequence. Broadly, they appear as

original sequence: A-B-C-D-E
duplicated sequence: A-B-C-D-E-spacer-E-D-C-B-A

Such inverted repeats are called *palindromes* when the same sequence is present on the complementary DNA strands of each duplicon (and read in the opposite direction).

The male-determining Y chromosome has a high density of palindromic duplicates. The human *male-specific region* of this chromosome contains eight such palindromes. These are colossal

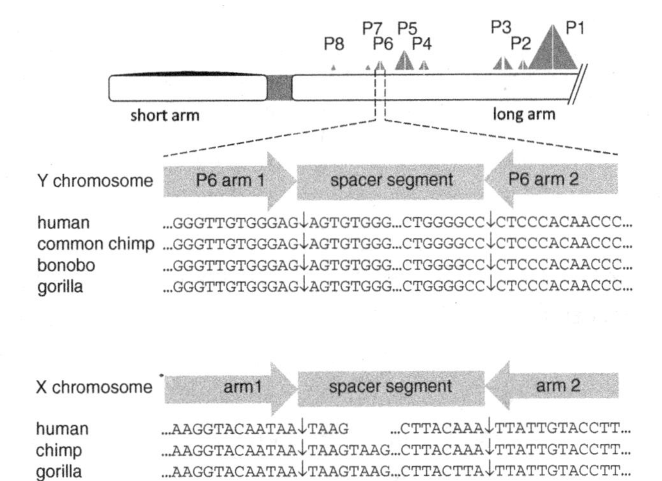

FIGURE 4.4. PALINDROMIC REPEATS
Repeats on the Y chromosome (*upper diagram*; P1– P8) indicated by *back-to-back triangles*, representing their symmetry. Boundaries (↓) between arms and spacer are shown for P6 [30]. A repeat on the X chromosome (*lower diagram*) [32]. Palindromic repeats may be recognised by reading the sequence of each arm in opposite directions, and substituting each base in one arm by its complement in the other.

structures. The arm lengths vary in length from 9,000 to 1,450,000 bases, and the spacers from 2,000 to 170,000 bases. Most of these palindromes exist also in other great ape species. In six cases, the boundaries between the arms and spacers have been identified in DNA from humans and at least one of the other species: common chimps (five of the palindromes), bonobos (four) and gorillas (two). In all cases where arm–spacer junctions have been identified, they occur at the *identical* point (↓) of the sequence in *all* the species that possess the respective palindrome. The spacer–arm boundaries of palindrome P6 are shown to illustrate this (Figure 4.4) [30]

There is only one way of interpreting the remarkable fact that these huge copied structures are shared by multiple African great ape species. Each DNA duplication event was a unique happening that occurred in one germ-line cell, and the multiple species that share the product received it by inheritance. Further evidence that these palindromes arose in duplication events arises from the presence of particular retrotransposed inserts in both arms of a palindrome. Several of the palindromes (P1, P3, P4, P5) contain paired ERV-K14C inserts. The ERV, its flanking sequence and its insertion point are the same in both arms [31].

Inverted repeats are distributed throughout the genome, with a high concentration also on the X chromosome. For some of these, the junctions between the left and right copies and the central spacers have been sequenced for several great ape species, and are identical in human, chimp and gorilla (four of eight inverted repeats tested) and in human and gorilla (two additional cases). An instance of such a repeat structure on the X chromosome is shown (Figure 4.4, lower diagram). All species possessing each inverted repeat are descended from the one reproductive cell in which that repeat arose [32].

Many genes lie within the compass of the inverted repeats (Figure 4.5). For example, on the Y chromosome, *DAZ* genes have been duplicated in palindromes P1 and P2; and *CDY* genes and pseudogenes in P1, P3 and P5. These duplicated genes have undergone sequence changes that now differentiate them from their parents [33]. Three phases have been discerned in the history of the *DAZ* genes.

1. In an ancestor of apes and OWMs, the *DAZL* gene on chromosome 3 was copied and translocated into the Y chromosome to form the first *DAZ* gene.
2. In a human–chimp ancestor, the *DAZ* gene was duplicated to generate an inverted head-to-head pair.
3. In species-specific lineages, the whole structure was reduplicated to form two pairs.

Ongoing non-allelic homologous recombination (NAHR) has led to further changes in gene number. Some men have two *DAZ* genes; most have four; some have six. Moreover, *DAZ* copy number is related to sperm count [34].

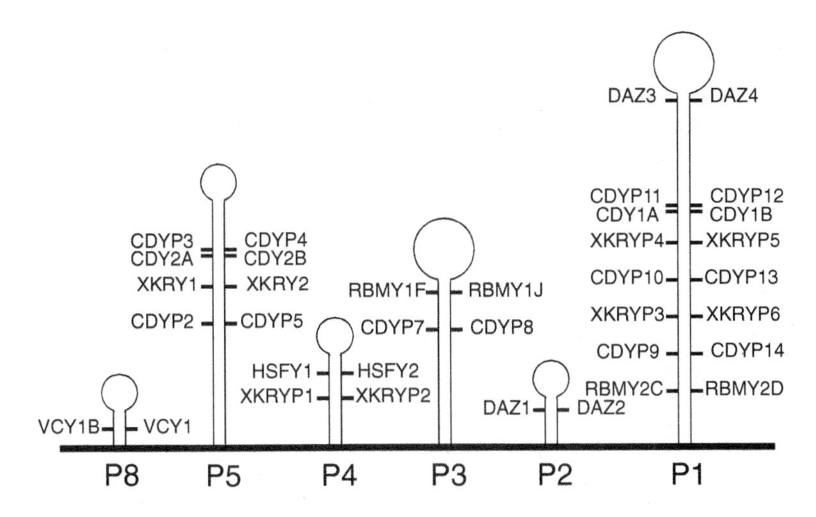

FIGURE 4.5. DUPLICATED GENES PRESENT IN THE PALINDROMIC REPEATS, Y CHROMOSOME [33]
The *thick horizontal line* represents part of the Y chromosome. The six hairpin-like structures represent palindromic repeats. *Stems* represent duplicated arms and *loops* represent spacers. Duplicated genes are indicated. Not to scale.

Some gene families on sex chromosomes have been studied in detail. The *SPANX* genes are located on the X chromosome. They are active largely in the testis, and their proteins are found in sperm cells. (*SPANX* is an acronym for *sperm protein associated with nucleus on the X chromosome.*) These genes may also have a connection with female fertility, as women with X chromosomal abnormalities that delete *SPANX* genes suffer from premature ovarian failure [35].

In non-primate mammals such as dogs and rodents, a single *SPANX-N* gene is found, and this is located on the long arm of the X chromosome (Figure 4.6). Primates, in contrast, possess multiple *SPANX* genes. The gene family has expanded in a stepwise manner through primate history, and the rate has accelerated especially in the great apes. The original *SPANX-N* gene underwent two duplications in an ancestor of the apes and OWMs (which share three *SPANX-N* genes), and another duplication in an ancestor of the great apes (which share four *SPANX-N* genes). Further duplications occurred

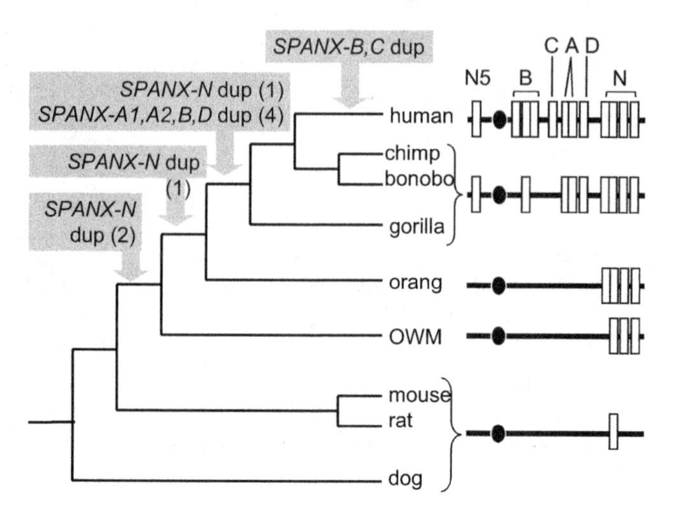

FIGURE 4.6. EXPANSION OF THE *SPANX* GENE FAMILY THROUGH PRIMATE EVOLUTION [36]
Gene content is depicted to the right of the phylogenetic tree: X chromosome (*thick horizontal line*), centromere (*oval*), *SPANX* genes (*open boxes*), *SPANX- N* genes (N), *SPANX-N5* (N5), and *SPANX*-B, C, A1, A2 and D (B, C, A and D). Not to scale.

in the lineage giving rise to the African great apes. These species share an additional five genes. These are *SPANX-N5* on the short arm of the X chromosome, and *SPANX-A1*, *-A2*, *-B* and *-D* on the long arm of the X chromosome. Finally, humans have acquired a further gene (*SPANX-C*) and up to 14 copies of *SPANX-B*. The flanking and intronic sequences of the *SPANX-A* to *-D* genes are very similar to each other, indicating that they are located on segmental duplications approximately 20,000 bases long.

The primate *SPANX* genes possess one intron. An ERV sequence is present within this intron. This is full length in *SPANX-N* genes, but has been reduced to a solitary long terminal repeat (LTR) in the five new members (*SPANX-A1*, *-A2*, *-B*, *-C* and *-D*). As the *SPANX* gene family expanded, the two LTRs of the ERV underwent a recombination event (see Chapter 1). The presence of this ERV and its derivative solo LTR is a graphic demonstration that all the family members have been generated as copies of the first ERV-containing

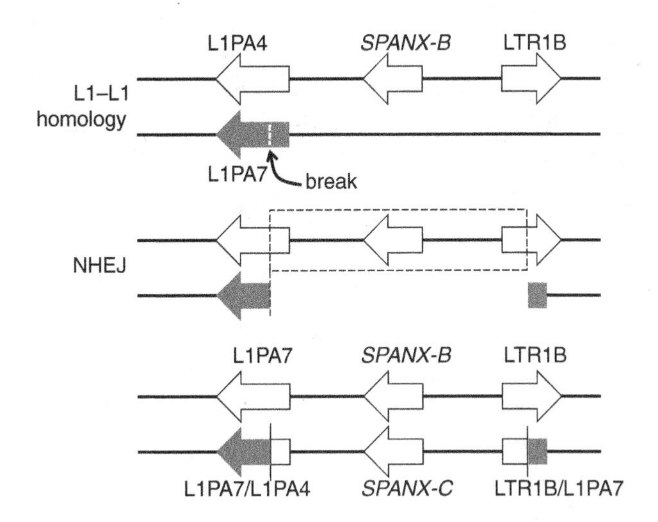

FIGURE 4.7. MECHANISM BY WHICH *SPANX-B* GENERATED *SPANX-C* [37]
A DNA break in the L1PA7 element (*shaded arrow*) led to NHEJ repair, initiated within a related L1PA4 element upstream of the *SPANX-B* gene. The *SPANX-B* gene and its flanking sequences (*dotted box*) were copied and pasted into the break.

member. A single *SPANX-N* gene with a full-length ERV is the progenitor of all the *SPANX-A* to *-D* genes with solo LTRs [36].

The mechanism by which the *SPANX-C* gene arose has been elucidated. A DNA break occurred in an L1PA7 element that was located some considerable distance upstream of *SPANX-B*. A segment of DNA (containing *SPANX-B*) was imported to join the broken ends, and this was selected on the basis of close similarity between the broken L1PA7 element and an L1PA4 element upstream of *SPANX-B*. The reorganised locus has the features of a repair job performed by non-homologous end-joining (NHEJ). The product contains a chimaeric LINE-1 element, the brand new *SPANX-C* gene and a chimaeric LTR1B/L1PA7 element (Figure 4.7) [37].

There is evidence that the diversification of this gene family represents the most-recent phase of an ongoing process of gene birth that reaches back into deep time. The *SPANX* genes and members of two other gene families found on the sex chromosomes have similar upstream (promoter) sequences. This suggests that the three gene families have originated from a common ancestral gene. Not only families of species, but also families of genes, proliferate and diversify from distant progenitors [38].

SPANX gene coding sequences have evolved rapidly, suggesting that they have been progressively modified under the influence of positive selection. This rapid evolution of proteins acting in reproduction has been linked with mechanisms of speciation. A downside of these genes is that they seem to be abnormally expressed in cancers. There are a whole class of rapidly evolving and populous gene families that are active in the testis and that are abnormally expressed in cancer cells. These encode the cancer/testis antigens, including the *BAGE* [39], *MAGE* [40] and *GAGE* [41] genes. They have been regarded as potential targets for anti-cancer therapies.

Other genes active in reproduction are not located on the sex chromosomes. The *TRE2* gene is a young gene, present on the short arm of chromosome 17. Like many young genes, it is expressed mainly in the testis, in which it functions to organise the cytoskeleton. *TRE2* is a chimaeric gene derived from two progenitor genes, *TBC1D3* and *USP32*, which reside on the long arm of the same chromosome. Copied portions of the parental genes came together on adjacent SDs. Exons 1–14 of the *TRE2* gene are derived from the *TBC1D3* parent, and exons 15–29 from the *USP32* parent. The *TRE2* gene is found only in the genomes of humans, chimps, gorillas and orang-utans, in which the joining point between the two parental segments is identical. Clearly, the chimaeric gene arose once [42].

Similarly, a side-by-side duplication of a DNA segment 100,000 bases long is located on chromosome 1, and is found in the genomes of apes, OWMs and NWMs (but not prosimians). The duplication event can be approximately dated by the presence of six ancient Alu elements that are present in both duplicons. They are representative of old FLAM, AluJ and AluS sub-families. They were already present in the original DNA segment before it was copied.

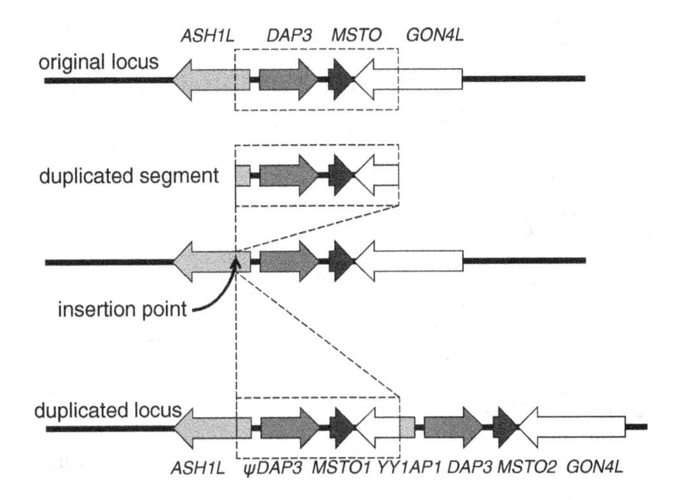

FIGURE 4.8. FORMATION OF A PSEUDOGENE, A NEW GENE COPY AND A CHIMAERIC GENE FOLLOWING A DUPLICATION EVENT [43]

New genes have arisen out of this copying event. The duplicated segment of DNA contained two complete genes and parts of two others (Figure 4.8). Of the two internally located genes, a new copy of *MSTO* was generated, but the extra copy of *DAP3* was subsequently pseudogenised (now identified as *ψDAP3*). The genes that were broken as a result of the duplication formed a new, chimaeric gene, the protein of which (YY1AP1) is a mediator of gene transcription. The new, simian-specific fusion gene is active in many tissues, including male reproductive tissue. Perhaps it contributed to speciation [43].

Hormones control development and metabolic activity. Growth hormone is produced in the pituitary gland and has many effects including the stimulation of growth, muscle mass and anabolic (synthetic) metabolism. The growth hormone gene locus contains one *GH* gene in most mammals, including prosimian primates, but has been subject to multiple rounds of duplication in simian primates. The locus in humans is present on chromosome 17, and contains a cluster of five genes, of which the four new members are all expressed in the placenta (and include the placental lactogens). The same gene set is present in chimps, indicating that these species share the same

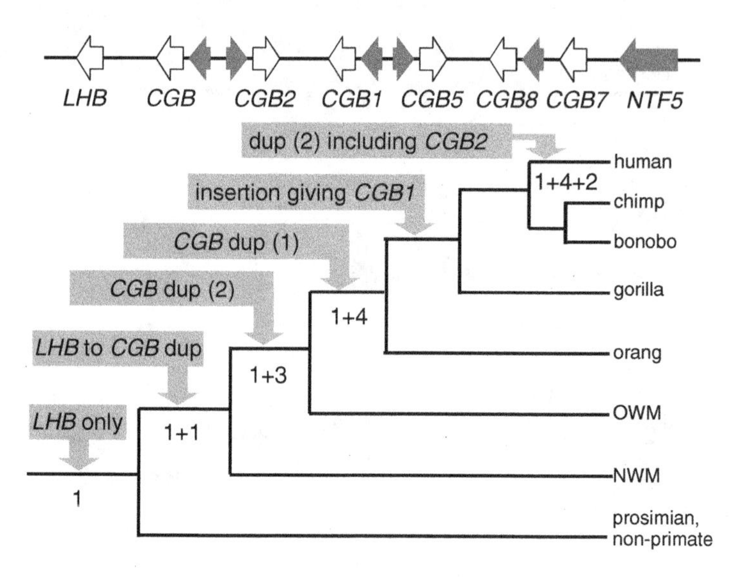

FIGURE 4.9. EXPANSION OF THE *LHB/CGB* GENE FAMILY BY DUPLICATION [46]
Top: genes represented are *LHB/CGB* (*open arrows*), *NTF5* (*long shaded arrow*) and *NTF5* pseudogenes (*short shaded arrows*). Not to scale.
Lower diagram: Stages in the generation of the *LHB/CGB* family.

history of gene multiplication [44]. Gorillas possess an additional placental lactogen gene as a result of a recent duplication peculiar to that species [45].

The luteinising hormone/chorionic gonadotropin-B (*LHB/CGB*) gene family is integral to reproduction. These genes encode hormones that control processes such as ovulation and the implantation of the embryo. *LHB* is the only member of the family present in prosimians, and is the founding member. It was duplicated to generate the first *CGB* gene in an ancestor of the simian primates, so that NWMs have one *LHB* and one *CGB* gene. New members were added sequentially through simian history (Figure 4.9). The human genome possesses one *LHB* gene and six *CGB* genes that are straddled along a stretch of chromosome 19. The individual members are highly similar in base sequence, both in coding regions and in the surrounding tracts of DNA [46].

Two of these genes, *CGB1* and *CGB2*, stand out because they have a length of foreign DNA inserted into their promoters. This

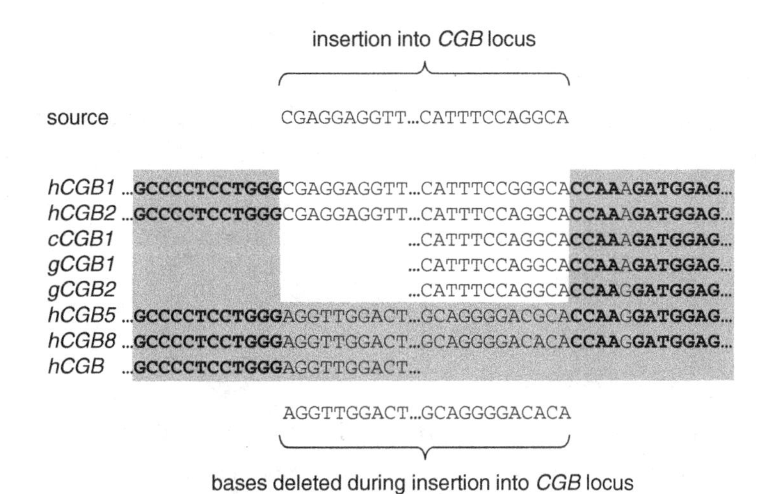

FIGURE 4.10. SEQUENCES AT THE INSERTION BOUNDARIES OF *CGB1* AND *CGB2* [47]
DNA sequences (*from top*) are of the source of the insert (located distantly on chromosome 19), the insertions in the *CGB1* and *CGB2* loci of humans (*h*), chimps (*c*) and gorillas (*g*), and the original undisturbed sequence, represented by *CGB5*, *CGB8* and *CGB*. Complete sequences for chimp and gorilla inserts are not available. Invariant bases adjacent to the insertion are in *bold*.

intruding piece of DNA is about 730 bases long, and was copied from elsewhere on chromosome 19. It is found in precisely the same site upstream (to the left) of both genes (Figure 4.10). This establishes that the two genes are duplicates. In fact, *CGB2* is a copy of *CGB1*, and both are derivatives of a *CGB* gene lacking the insertion.

Both chimps and gorillas possess the *CGB1* gene with its insertion (as well as gene copies thereof) in common with humans. Orang-utans, however, do not possess any such insert-tagged *CGB* genes. The insertion and initial duplication events are markers of monoclonality, showing that the birth of these genes occurred in an ancestor of the African great apes. The inserted length of DNA has altered the promoters of the *CGB1* and *CGB2* genes, so that they are expressed in the testis, performing novel functions in male reproduction.

There is a further baroque twist to this story. The sequences inserted into the *CGB1* and *CGB2* genes themselves contain genes that encode small RNA molecules, members of a gene family that has been amplified into over 20 copies on chromosomes 19, 2 and 3. These small non-coding RNA genes (or *snaR* genes) are expressed in several tissues but most abundantly in the testis. We have encountered them before because they are ultimately derived from an Alu element [48].

4.4.2 *Hydrolytic enzymes*

A classic case of gene duplication giving rise to a multi-gene family is provided by the amylase (*AMY*) gene consortium, the products of which are required to digest dietary starch. (We have already encountered one family member, *AMY1*, because it demonstrates striking variability in copy number in the human population; see Section 4.2.) NWMs possess one *AMY* gene, and this represents the ancestral arrangement. Hominoids possess three genes. The stepwise duplication process can be followed because of a series of insertion and deletion events that occurred in regulatory (promoter) sequences to the left of the gene. The whole process involved the following steps (Figure 4.11).

- A γ-actin pseudogene was inserted by retrotransposition into the promoter of the progenitor *AMY* gene. The insertion event occurred in an ape–OWM ancestor, and the derived gene conformation is retained in apes and OWMs as the *AMY2B* gene.
- The gene was subsequently duplicated, and in the new copy, the γ-actin pseudogene underwent a deletion. These events also occurred in an ape–OWM ancestor. OWMs retain a copy of this derivative.
- A retrovirus was inserted into the truncated γ-actin sequence (in an ape ancestor). This gave rise to the *AMY1* gene present in apes, and is singular because the gene is transcribed from a start site located in the γ-actin pseudogene (so generating a new exon).
- The *AMY1* gene was duplicated, and the ERV underwent a homologous recombination event between LTRs to produce a single LTR remnant. This locus is retained in apes as the *AMY2A* gene.

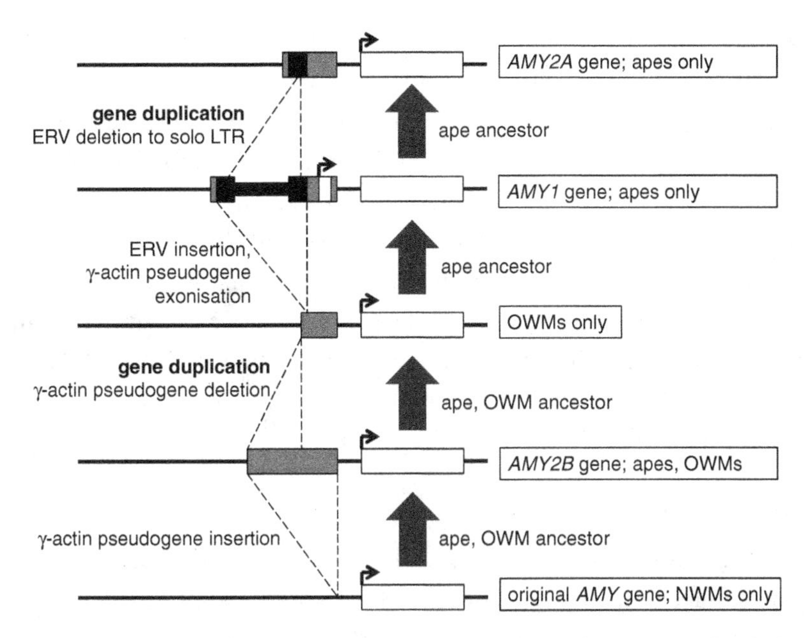

FIGURE 4.11. EXPANSION OF AN AMYLASE GENE FAMILY BY DUPLICATION [49] *AMY* exons (*white boxes*), γ-actin pseudogene (*grey boxes*) and ERV sequences (*black boxes*). The transcription start site is indicated by the *right-angled arrow*.

- There have been subsequent duplications involving the *AMY1* gene (together with its upstream ERV) specifically on the human lineage (Section 4.2). The *AMY1*-containing gene locus in the human population contains from one to more than five copies [49].

This remarkable history anticipates others pertaining to multi-gene families encoding hydrolytic enzymes. *Lysozymes* are enzymes present in tears and milk. They degrade peptidoglycan components of bacterial cell walls and induce bacterial cell rupture. Mammals possess a family of related proteins, and these perform diverse functions. The respective genes have proliferated by duplication events since pre-mammalian times. Humans have two new members of the gene family, each one on a recently generated duplicon [50]. *Ribonucleases* are encoded by a multi-gene family in primates. The various members of this family have diversified to perform roles in digestion, protection against microbes, and control of blood vessel development [51]. The proteolytic *pepsinogens* [52] and *kallikreins* [53] are encoded by gene families that have expanded and contracted during the course of hominoid history. Four of the 15-member kallikrein gene family are tagged by a particular TE (of the LTR40A class) that is common to Boreoeutherian species – evidence that these genes, and these species, are derived from single progenitors. A duplication of the kallikrein-encoding gene *KLK2*, generating *KLK3*, occurred in an ape–OWM ancestor. These genes are marked by a signature LINE-2 element. Their products, known as *prostate-specific antigen*, promote sperm motility.

4.4.3 Neural systems

Humans are unique (as far as we know) in their capacity to engage in abstract thought. Our cognitive capacities arise from the way in which our nerve cells interconnect and network. We have considered genetic innovations that have occurred during mammalian history and that may have contributed to the development of neural function. The *BCYRN1* non-coding gene has been generated, expression of the *GADD45G* gene has been lost in parts of the forebrain, the *FLJ33706* gene has been assembled and *DUF1220* domain-encoding genes have proliferated. In general, the expression of phylogenetically new (primate-specific) genes is not apparent in the *adult* human brain – but is apparent in the *fetal* human brain. Young genes are active especially in the developing neocortex, the part of the brain that is closely associated with cognitive functions, and include many that encode transcription factors [54].

NBPF genes, encoding DUF1220 domains in proteins, are expressed in neocortical and hippocampal neurons. The number of encoded DUF1220 domains has increased during primate evolution, and strikingly so in the human-specific lineage. The order in which new DUF1220-coding sequences arose can be inferred from the presence of certain TEs within them. A LINE-1 element is present in an intron of 16 of the genes, establishing that all of these genes are copies of the progenitor in which the LINE-1 inserted itself [55]. This LINE-1 element was subsequently fragmented by the insertion of a LIPA4 element, a marker of a more restricted (and recent) subset of *NBPF* genes. An ERV of uncertain function [56] and a duplicon of 4,700 bases (encoding a unit of three DUF1220 domains) each arose in an African great ape ancestor. Reiterated rounds of duplication of this unit were responsible for generating 140 new DUF1220-coding segments in the human lineage [57].

The number of copies of gene segments encoding DUF1220 domains correlates with head circumference (in individuals with abnormal brain development) and with the volume of grey matter (in people showing no pathology) [58]. These proteins are candidate regulators of brain size.

The *SRGAP2A* gene functions in brain development. This gene has been duplicated twice in the lineage leading specifically to humans – or rather, half the gene has been duplicated twice, as both derivative copies are truncated. Both termini of the initially duplicated segment lie in Alu elements, and these termini are preserved in the second duplicated segment, indicating that the first duplicon is entirely included within the second (Figure 4.12). At least one of the derivative genes, *SRGAP2C*, produces a protein that is present in neurons [59].

The original SRGAP2A protein appears to drive the maturation of neurons in the cortex of the brain, and to promote the formation of spines – the microscopic knobs protruding from neuronal membranes, which allow for intercellular communication. The new, human-specific SRGAP2C protein forms a complex with the parental SRGAP2A protein and antagonises its activity. This interaction may allow developing neurons to migrate for a longer time before settling down. An extended period of cell migration may be needed to establish appropriate cell placements in the context of the increased thickness of the human brain cortex. The interaction also delays the development of spines and increases their density. This may contribute to the increased complexity of neuronal communication as

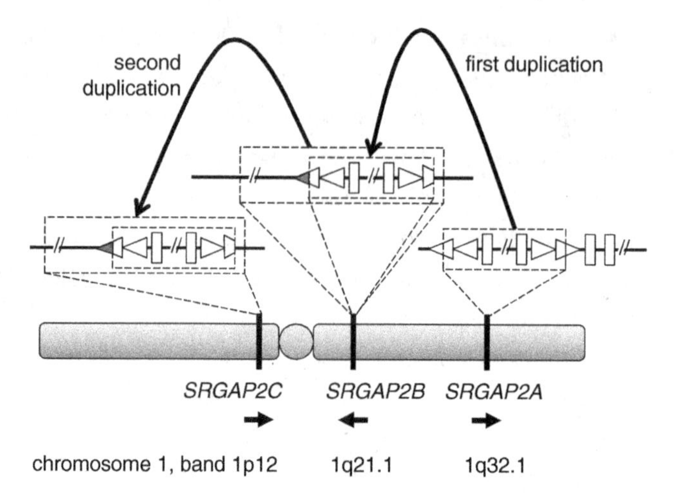

FIGURE 4.12. BIRTH OF *SRGAP2* GENES BY DUPLICATION [59]
The location of *SRGAP* genes on chromosome 1, the duplication events that generated *SRGAP2B* and *SRGAP2C*, Alu elements (*triangles*) and exons (*rectangles*) are shown. Not to scale.

seen in the human brain, which underlies activities such as thinking, learning and memory [60].

Segmental duplications have resulted in the elaboration of sensory function. Animals sense colour because nerve cells (the cones) in their retinas express two or more proteins (*opsins*) that contain a light-absorbing molecule (or *chromophore*), retinaldehyde. The wavelength of light to which the chromophore is maximally sensitive is affected by the particular opsin to which the chromophore is bound. In general, mammals (including prosimians and NWMs) possess two-colour (dichromatic) vision. Their genomes contain two genes for light-sensitive opsins, of which one is on the X chromosome. Apes and OWMs enjoy the wonders of three-colour (trichromatic) vision. They possess three opsin genes, of which two are located on the X chromosome [61].

The two opsin genes on the X chromosome are very similar in sequence. They are situated close to each other in a tandem left-to-right configuration. The long-wave (LW; red) opsin gene is followed

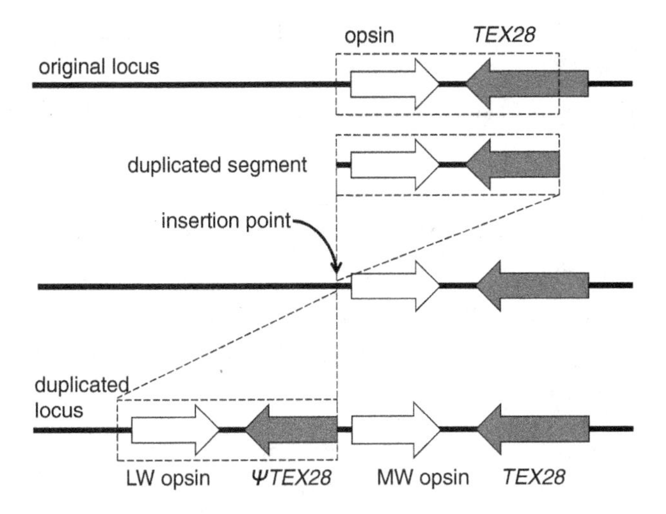

FIGURE 4.13. FORMATION OF A NEW OPSIN GENE BY DUPLICATION [62]
The original arrangement (retained in NWMs) is shown *above*. The duplicated segment (*dashed box*) and site of insertion are shown *below*. One of two possible insertion sites is shown, but the final product is the same. The derived arrangement present in apes and OWMs is depicted *at the bottom*.

by a truncated pseudogene called *ψTEX28*, and the medium-wave (MW; green) opsin gene by a full-length *TEX28* gene (Figure 4.13). This two-gene arrangement has come about because of a segmental duplication.

The original arrangement is preserved in NWMs, and consists of an opsin gene and its *TEX28* neighbour. A segment of 35,000 bases, encompassing the opsin gene, and part of *TEX28*, was duplicated to give the LW opsin–*ψTEX28*–MW opsin–*TEX28* array. This gene array is present in apes and OWMs and thus arose in an ape–OWM ancestor [62].

The left-hand end of the duplicated segment fitted seamlessly into the pre-existing DNA sequence, but the right-hand end generated a discontinuous junction, where the broken *ψTEX28* gene was joined to sequences to the left of the MW opsin gene. The actual breakpoint can be identified by comparing the unperturbed sequences to the left of the progenitor opsin gene of NWMs, and of the human LW opsin gene, with the derived sequence to the left of the MW opsin gene.

The blocks of sequence to the left of the NWM and human LW opsin genes are very similar throughout their lengths (Figure 4.14, see the three sequences in upper box). A comparison of these sequences with those to the left of the MW opsin genes in humans, other apes and OWMs (Figure 4.14, lower box) reveals a sharp discontinuity in the latter case. The abrupt break in the sequence occurs at the identical point in all four species (... TTACA↓**GGTTT** ...). This junction point leads to two conclusions.

1. Trichromatic vision arose through a segmental duplication that generated a new opsin gene. One terminus of the duplication is defined at molecular (single-base) resolution, because the junction point represents the exact location at which the copied segment (now containing the LW opsin gene) juxtaposes the interrupted original sequence (now containing the MW opsin gene). We may infer with confidence that increased complexity arises through familiar mutational events.
2. Old World primates (apes and OWMs) are a monophyletic taxon. There was one duplication event, as demonstrated by the unique junction point shared by all species. Humans, chimps and OWMs share this precise junction point because they inherited it from the ancestor in which the singular event occurred.

The opsin locus on the X chromosome remains subject to rearrangements. Segmental duplications and deletions still occur. Some males with normal vision have a LW opsin–*ψTEX28*–MW opsin–*ψTEX28*–MW opsin–*TEX28* gene array. They have inherited two copies of each of the *ψTEX28* and MW opsin genes as the result of further duplication events. Other males with colour blindness have lost a functional X-linked opsin gene. Their genome has undergone gene loss and reverted functionally to a NWM conformation [63].

4.4.4 Blood

In humans, the β-globin locus contains five genes that are responsible for making protein subunits of the oxygen carrier haemoglobin, and

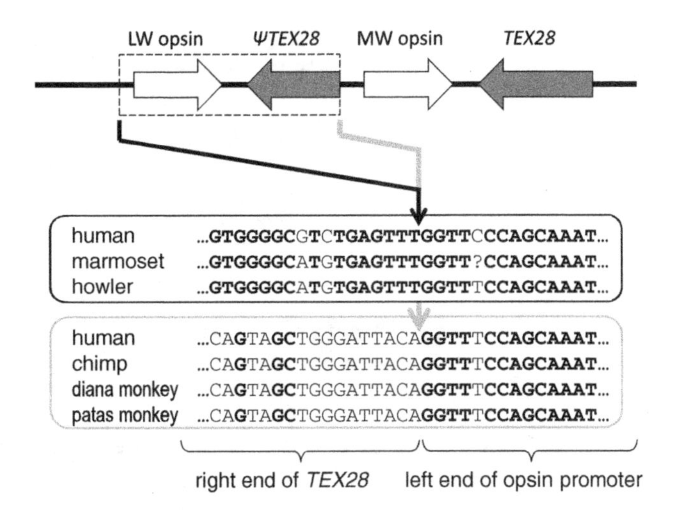

FIGURE 4.14. DNA SEQUENCES AT THE JUNCTION POINT BETWEEN THE *ΨTEX28* AND MW OPSIN GENES [62]
Upper box: sequences to the left of NWM (marmoset and howler monkey) opsin and human LW opsin genes. *Lower box*: sequences to the left of the MW opsin gene in humans, other apes and OWMs. Bases that are identical in all seven sequences are in *bold*.

the non-protein-coding η-globin pseudogene (designated ψη). These genes are scattered over some 60,000 bases of DNA in the order $\varepsilon - \gamma^1 - \gamma^2 - \psi\eta - \delta - \beta$, which is also the order in which they are expressed during development. Thus the ε gene is embryonic; the γ^1 and γ^2 genes are fetal; and the δ and β genes are expressed after birth. The gene family has arisen by duplication, and the mechanism of this has been detailed for the two γ genes [64].

Meiotic cell division is the process that generates gametes (egg and sperm cells). During this process, chromosome pairs line up side by side and exchange precisely equivalent lengths of chromosomal material. This process is called *homologous recombination*, and is effectively chromosome breakage with rejoining of ends from the partner chromosome. However, if highly similar sequences are present at two or more loci along the chromosome, then the chromosome pair may come together at non-equivalent sites. Exchange of

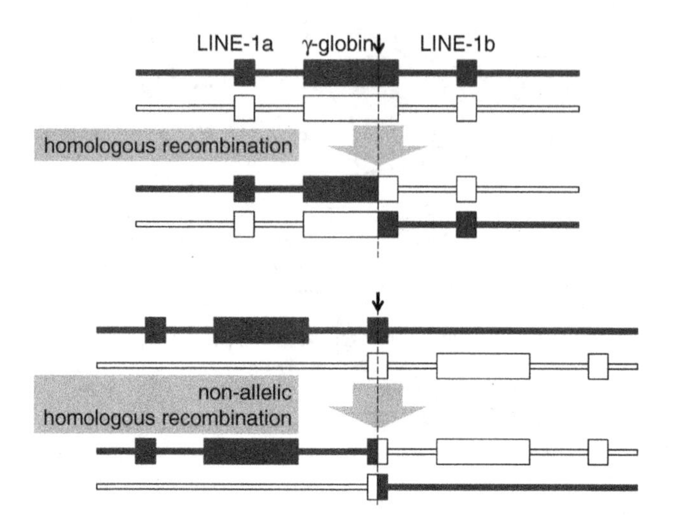

FIGURE 4.15. FORMATION OF A NEW γ-GLOBIN GENE
Homologous recombination (*upper diagram*): paired chromosomes align exactly. Equivalent segments on either side of the *dotted line* exchange places. Non-allelic homologous recombination (*lower diagram*): the LINE-1b element of the *grey chromosome* aligns with the LINE-1a element of the *white chromosome*. Non-equivalent segments on either side of the *dotted line* are exchanged, leading to gene duplication (upper chromosome) and deletion (lower chromosome.)

chromatin under these conditions will generate derivative chromosomes in which one gains, and the other loses, material. This is *non-allelic* homologous recombination or NAHR.

How does this relate to duplication of the γ-globin gene? Inspection of the gene locus indicates that the parental γ-globin gene was sandwiched between two LINE-1 elements (common to all apes and monkeys). The abnormal NAHR event occurred when the homologous chromosomes came together at non-equivalent LINE-1 elements: the left-hand LINE-1 element of one chromosome erroneously aligned with the right-hand LINE-1 element of the other. Recombination generated two products (Figure 4.15).

- One chromosome gained a chimaeric LINE-1 element (giving three in total) and an extra γ-globin gene. The new arrangement is: LINE-1→globin gene→chimaeric LINE-1→globin gene→LINE-1. This is the conformation retained in apes and monkeys.

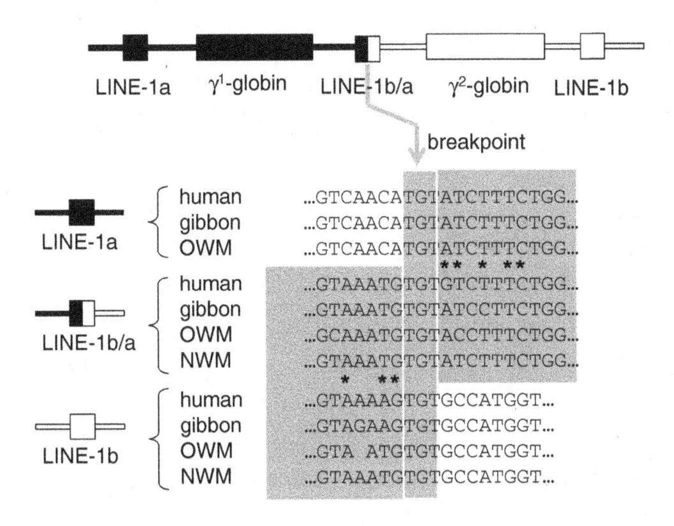

FIGURE 4.16. NON-ALLELIC HOMOLOGOUS RECOMBINATION: BIRTH OF THE γ-GLOBIN GENE PAIR [65]
Sequences are from the equivalent positions of three LINE-1 elements. *Asterisks* identify bases that are diagnostic for the LINE-1a element (*upper right*) and the LINE-1b element (*lower left*). These markers localise the junction point to the *shaded* TGT in the chimaeric LINE-1b/a.

- The reciprocal product of this unequal chromosome recombination event retained only one chimaeric LINE-1 element and no γ-globin gene. This truncated locus has been lost without trace.

Sequencing of the LINE elements has confirmed that NAHR generated the two γ-globin genes. The two original LINE-1 elements have similar but distinguishable sequences. The LINE-1 element situated between the two γ-globin genes is a chimaera containing diagnostic sequence variants from both parent LINE-1 elements [65]. The sequence of the left-hand portion is highly similar to the left-hand part of the original right-hand LINE-1b element; the sequence of the right-hand portion is highly similar to that of the right-hand part of the original left-hand LINE-1a element. And the breakpoint can be narrowed down to about three bases (Figure 4.16).

NAHR involving the two γ-globin genes continues in the human population. This has been demonstrated by analysis of the γ-globin gene locus in DNA from sperm cells and blood cells. NAHR generates variations in γ-globin copy numbers at a frequency of one copy number change per 100,000 DNA molecules. Some loci have one γ-globin gene (and have lost one gene); others have three or four genes (and have gained one or two relative to the normal condition). New copies of γ-globin genes are continuously generated. These, on occasion, in evolutionary time, persist and diverge in sequence. The end result is greater complexity and elaboration of function. Comparative analysis of the β-globin locus in eutherian mammals has shown that gene birth (and death) has occurred repeatedly during evolution [66].

4.4.5 *Immunity*

The immune system is a marvel of versatility. It operates to maintain a healthy relationship between the tissues of our own bodies and the legions of microbes, whether benign or sinister, that lurk in our environment. Microbes evolve rapidly, and the immune system must be able to respond effectively (on organismal and evolutionary timescales) to unanticipated new challenges. There is an arms race between the myriad pathogenic threats arrayed against our bodies, and our immune defences. Gene families involved in immunity appear to evolve rapidly relative to most other categories of genes.

One arm of the immune system is said to be *innate*: it is constituted of a stock inventory of cells and molecules that present a stereotypical set of defences to microbes. An example of the rapidly evolving components of innate immunity is provided by the populous *S100* gene family. These genes encode calcium-binding proteins that perform a diversity of protective roles. Our genome contains at least 18 *S100* genes. Of these, the *S100A7C* (*psoriasin*) and *S100A7A* (*koebnerisin*) gene pair are widely expressed in skin, and their respective proteins may normally act to restrain microbial invasion. (The skin is more than a passive barrier to microbes!) These proteins may also act to induce inflammation, a protective – although sometimes damaging – reaction to tissue injury and infection. Psoriasin itself was discovered through the inflammatory skin condition known as psoriasis [67].

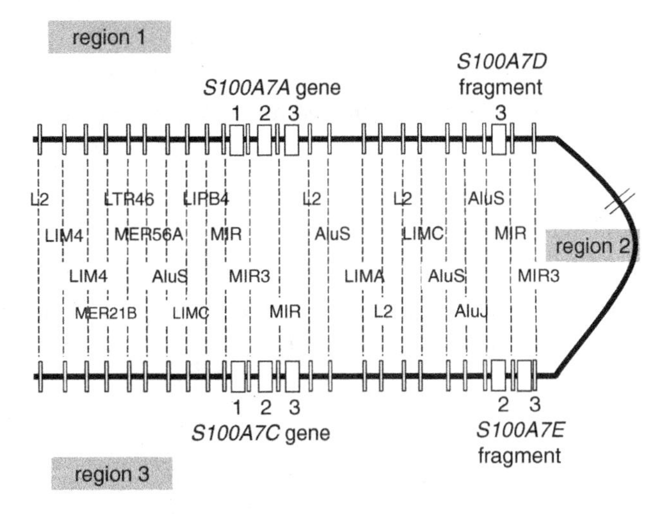

FIGURE 4.17. AN INVERTED REPEAT WITH *S100A7* GENE DUPLICATION [68] The two arms are in opposite orientation and are shown in a hairpin conformation to allow direct alignment. Twenty-three shared TEs demonstrate that regions 1 and 3 are copies. Younger TEs that were inserted after the duplication (and that are present in only one arm) are not shown. Not to scale.

The related psoriasin and koebnerisin genes arose as the result of the duplication of a large expanse of DNA. This duplication produced a 33,000-base segment (region 1) and a 31,000-base segment (region 3) separated by a linker of 11,000 bases. Besides the *S100A7* genes, 23 TEs are shared between regions 1 and 3 (Figure 4.17), demonstrating that the regions are duplicates. Regions 1 and 3 may be distinguished by the presence of a diagnostic rearrangement, and this genetic marker has been used to show that the two regions are present in chimpanzees as well as in humans. It follows that the duplication occurred before the chimpanzee lineage diverged from the human one.

This duplication generated two functional genes (*S100A7C* and *S100A7A*) where there had been only one before. Portions of two other genes (*S100A7D* and *S100A7E*) are also present within the duplicated fragments. It is likely that these copies did not survive the duplication process, and they persist as non-coding pseudogenes [68].

Defensins are another class of small proteins that are part of our antimicrobial arsenal. They neutralise viruses such as influenza virus and HIV. They kill bacteria and fungi, partly by damaging their cell membranes. They function as molecular signals that recruit or activate cellular components of the immune system, such as macrophages and mast cells. There are three sub-families of defensin genes, each of which is composed of numerous members. The defensin genes have proliferated by means of the familiar pattern of duplication and diversification [69].

The β-defensin genes are the oldest, and are found widely in vertebrates. Nearly 40 β-defensin genes in five clusters are distributed around the human genome. These clusters have counterparts in mouse and dog genomes, and so probably arose in the distant past through successive gene duplications. However, the number of copies of some of these genes is variable in the genomes of different people. For example, one cluster (containing at least three β-defensin genes) is present at a multiplicity of one to six copies per genome set [70].

As an aside, β-defensins may be used to repel more than microbes. Mammals are not in general venomous, but the platypus is a striking exception. During the mating season, the male platypus can use spurs on it back legs to deliver painful shots of venom to competitors or other intruders. The platypus has recruited a β-defensin gene to provide components of its venom. Gene duplication with divergence is the likely mechanism [71].

One of the β-defensin genes gave rise to the α-defensin gene sub-family, found only in mammals. The α-defensin genes have undergone birth-and-death changes during primate evolution. As with the *S100A7* genes, surrounding TEs have been used to identify orthologous genes in different species, and to track gene duplications [72].

The youngest sub-family, the θ-defensins, emerged out of the α-defensins. This third group is restricted to Old World primates, although θ-defensin proteins are not found in humans. The θ-defensins are small, cyclic proteins. They are formed when two small chains of amino acids are joined head to tail to form a closed circle. They are the only fully cyclic proteins made in mammals. There are six θ-defensin genes in the human genome, and they are of particular interest because they are all pseudogenes, transcribed into messenger RNAs but unable to direct the synthesis of proteins. They contain the same *stop* mutation, a C-to-T base substitution in codon 17, and this mutation is shared by the θ-defensin genes found in the genomes of chimps, bonobos and gorillas, and by one of several θ-defensin genes of orang-utans. We may conclude that the mutation arose in a great ape ancestor, and the unique pseudogene so generated was amplified to produce the multiple copies now present in the human genome [73].

What properties would the θ-defensin proteins possess if they could be regenerated? Chemical synthetic techniques have been used in the laboratory to make the functional peptides that would be specified by the human pseudogenes (had they not been disrupted by the *stop* mutation). The resulting proteins are called *retrocyclins*, and have been shown to suppress HIV infectivity. Subsequent work has generated fully novel retrocyclin derivatives, which neutralise the influenza virus in the test tube [74]. Scientists have tried another approach to activate the θ-defensin pseudogenes. A class of antibiotics (the aminoglycosides) can override the effect of *stop* mutations, and the use of such antibiotics has allowed θ-defensin proteins to be made by human cells possessing only the pseudogenes [75]. Perhaps θ-defensin genes may be resuscitated to once again take their place in our antiviral and antimicrobial armoury.

Expansions and contractions of gene families involved in the ever-evolving innate immune system are standard features of antimicrobial defences. These include gene families encoding receptors on natural killer cells, and TRIM and APOBEC3 antiviral proteins. Many more instalments may follow in this ongoing genetic drama [76].

The other branch of the immune system is designated as *adaptive*. This is a hugely sophisticated complex of cells and molecules that engages with the universe of biological molecules presented to us by our environment. Specific receptors provide highly discriminatory recognition of particular microbes and their products. The adaptive immune system progressively learns from and remembers such engagements through the lifespan of each individual, enabling us to survive in a world full of potentially destructive microbes. It has an amazing capacity to discriminate between what is *self* (normal components of the body) and what is *non-self* (biological molecules that originate from foreign organisms). And even when it comes to *non-self* structures, it is able to sense what is innocuous (and should be tolerated) and what is pathogenic (and should be resisted).

Genetic control of adaptive immunity is exerted by a vast tract of DNA known as the *major histocompatibility complex* (MHC), or in humans as the *human leukocyte antigen* (HLA) complex. The human MHC is located on the short arm of chromosome 6. The total sequence of the class I region (1.8 million bases in length) was reported in 1999, and shows a reiterated pattern of related genes and pseudogenes. This repeating pattern indicates that this locus was generated by multiple rounds of duplications. A detailed side-by-side comparison with the MHC region in the chimp genome indicates that both species share the same set of duplicated units. The generation of the MHC complex was essentially complete in a human–chimp ancestor. The main differences between the two species are 64 large (>100 bases) indels [77].

Conclusive evidence for this gene birth-by-duplication model arises from the fact that there is also a reiterated pattern of TEs across the MHC. Where the same TE (and its flanking sequences) recurs multiple times within the genome, we may be confident that the segment of DNA in which the TE is embedded has been duplicated from one original.

- The MHC class I region contains *alpha* and *beta* subregions, which comprise ten and four duplicated segments, respectively. Eleven of these segments contain a particular ERV16 sequence (a foundation member of the original duplicon).
- In the *beta* subregion, the *HLA-B* and *HLA-C* genes are known to have arisen as duplicate copies of one original, because their surrounding

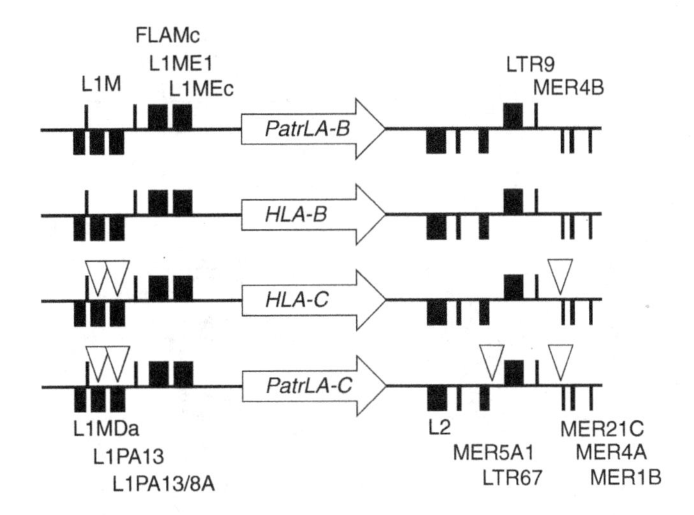

FIGURE 4.18. DUPLICATED *HLA-B* AND *HLA-C* GENES AND CHIMP ORTHOLOGUES [78]
An assemblage of TEs located near the human *HLA-B* and *-C* genes, and their chimp counterparts, *PatrLA-B* and *-C*. *Symbols above the line* represent TEs oriented from left to right; *symbols below the line* are oriented from right to left. *Triangles* indicate insertions. Not to scale.

sequences share a large assemblage of TEs, including 15 LINE-1 elements. The gene pair is also found in the chimp genome, and each of the units in the chimp possesses many of the same TEs as their human counterparts, including the same 15 LINE-1 elements. Some of these inserts are depicted in Figure 4.18. The duplication event occurred before the human and chimp lineages separated [78].

- In the *alpha* subregion, ten duplicons can be distinguished. The order in which TEs were added to duplicons indicates the order in which duplicons were formed. For two units known as the A and H duplicons (based on the presence of the *HLA-A* gene and the *HLA-H* pseudogene) shared TEs include those indicated in Figure 4.19. As before, the same two duplicons are present also in the chimp genome, and the same basic set of TEs is present in each chimp duplicon as in the human pair. These segments arose by a huge copy-and-paste event, and this occurred in an ancestor of humans and chimps [79].

The stages of development of the entire *alpha* subregion have been inferred (Figure 4.20). Duplicons were grouped into various

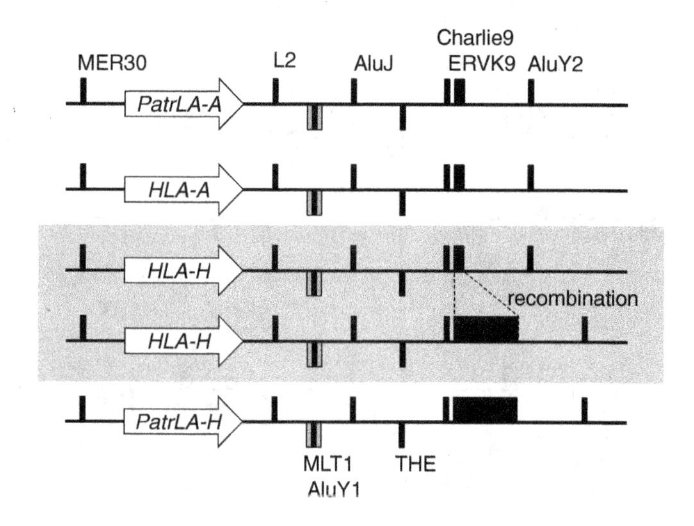

FIGURE 4.19. DUPLICATED *HLA-A* AND *HLA-H* GENES AND CHIMP ORTHOLOGUES [79]
An assemblage of TEs located near the human *HLA-A* and *-H* genes, and their chimp counterparts, *PatrLA-A* and *-H*. *Symbols above the line* represent TEs oriented from left to right; *those below* are oriented from right to left. Two schemes are shown for the *HLA-H* locus (*shaded*): the ERVK9 insert may be *either* full length (as in the chimp orthologue) *or* truncated by homologous recombination between the LTRs (as with the *HLA-A* gene). Not to scale.

classes on the basis of their diagnostic ERVs and TEs. The first four duplication steps were completed before the human, chimp and macaque lineages separated. Subsequent to the human–chimp and OWM divergence, the BAC three-duplicon unit was itself duplicated once in the lineage leading to apes, and extensively in the lineage leading to macaques. The *HLA-A* and *-H* genes (and their chimp counterparts; Figure 4.19) are found within the late-appearing C duplicons [80].

A second major group of MHC genes orchestrates adaptive immunity. Class II MHC genes include the *HLA-DRB* set of seven members in humans and chimps. As with the class I genes discussed above, a shared cohort of TEs has been used to demonstrate that these genes have multiplied from one precursor. An ancient L2 element, old

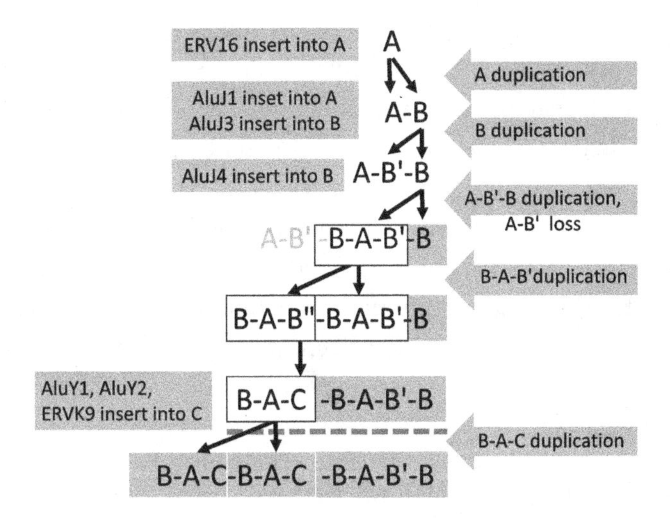

FIGURE 4.20. DEVELOPMENT OF THE ALPHA BLOCK OF THE MHC IN HUMANS AND CHIMPS [80]
The letters A, B and C represent duplicated segments of DNA. The same ten regions are present in both humans and chimps. All duplications *above the dotted line* were completed in an ancestor of humans, chimps and macaques.

Alu elements and several DNA transposons (of the MER116, MER53, MER20 and ZOMBI types), for example, are common to multiple *HLA-DRB* genes from multiple species (humans, chimps, OWMs and the marmoset, an NWM). The seven human genes are derivatives of one progenitor found in a simian ancestor. The family had grown to four by the time of an Old World primate ancestor [81].

The capacity of the adaptive immune system to monitor the enormous diversity of molecules that may enter our bodies (and potentially do mischief) is provided by protein receptors on lymphocytes. The repertoire of such *antigen receptors* has to be correspondingly diverse in order to provide adequate surveillance. The adaptive immune system has a range of strategies for generating an essentially unlimited variety of antigen receptors. One of these strategies is to have a large number of gene segments, each of which encodes a distinctive protein sequence contributing to a recognition-and-binding site for target molecules. In the *IGH* gene locus, there are more than 120 so-called *variable* gene segments. Approximately one-third of these are coding sequences. The remainder are pseudogenes. This entire genetic region is itself highly variable in the human population: each one of us may have a different set of variable gene segments. The variable gene array has been generated by segmental duplications, often following non-allelic homologous recombination [82]. Other receptor gene families that feature in adaptive immunity, including *IGLV* gene segments and immunoregulatory galectin genes (in which LINE-1 elements have driven gene duplication by NAHR), have expanded by the same mechanisms [83].

4.4.6 Master regulators of the genome

Genes that control the activities of other genes perform strategic roles in development. Many of these master genes exist in families. The mechanisms by which numerical expansion has occurred in particular families have to some extent been elucidated.

The *zinc finger* (*ZNF*) genes encode proteins that bind to DNA and that suppress the activity of other genes. The population of *ZNF* genes has undergone extensive change in recent evolutionary history. There are hundreds of members in the *Kruppel-ZNF* gene family, many of which are present in recently arising segmental duplications. Some appear to be human-specific; others are shared with chimps, others with all great apes and others with OWMs [84]. *Kruppel-ZNF* genes may have proliferated in bursts, and these periods of active gene birth may have coincided with the invasion of new types of ERVs into the genomes of our ancestors. This implies that new *Kruppel-ZNF* genes were recruited under the influence of natural selection because they acted to suppress ERV activity and so reduce the potentially disrupting effects that this activity may have caused [85].

The *FAM90A* family of genes encodes proteins with a single zinc finger domain. This feature suggests that such proteins have the capacity to bind to DNA or RNA, but their function is otherwise obscure. In the reference human genome, these genes comprise a family of 25 duplicated copies. Each duplicon is marked by an L1MB3 and an LTR5A element, demonstrating their origin from one progenitor gene bearing those signature TEs. Three of these genes have, in addition, gained upstream sequences from an unrelated gene as a result of a gene fusion event, and they are marked by the presence of AluSx and MIRb inserts. The expansion of the *FAM90A* family occurred in the hominoid (ape) lineage. In the human population, the number of gene copies possessed by different individuals is highly variable, extending over a tenfold range. Africans tend to have more copies than are found in the reference genome; East Asians tend to have fewer [86].

The *homeobox* genes encode proteins that function in the organisation of the body plan during embryonic development. These genes-that-control-genes multiplied to form a populous family in the distant past. At least a hundred members were already present in the genome of the last common ancestor of chordate animals (of which we vertebrates are the dominant issue). There are four collections of homeobox genes, together with conserved regulatory sequences, indicating that each of the four arrays has been derived from one original array [87]. The significance of this fourfold expansion will be explained later. Humans have 333 homeobox genes, of which 23% are pseudogenes. This gene family has been generally stable through primate history, with few duplications [88]. One of these is the *RHOXF2* gene on the X chromosome, which is active in male reproduction. It has been duplicated independently several times in Old World primate lineages. One of these duplication events occurred in a human–chimp ancestor [89].

However, subtle changes in homeobox function may have unsubtle consequences. We have noted that NANOG, a homeobox protein, may have an altered repertoire of target genes in primates because Alu elements have introduced new binding sites (Table 2.3). The *NANOGP8* processed pseudogene may upset cell regulation and promote cancer development (see Section 3.4). The *DUX4* gene is present in highly reiterated repeat arrays adjacent to the telomeres of chromosomes 4 and 10, and has the highest copy number of any protein-coding gene in our genome. Individual arrays vary from 11 to more than 150 gene copies. Extensive variation between generations is maintained by non-allelic homologous recombination. If the copy number of a chromosome 4 array decreases to less than 11, a muscle-wasting disease develops. This unstable arrangement has been inherited by all the great apes [90].

4.5 RETROGENEALOGY

LINE-1 elements use their encoded reverse transcriptase enzyme to copy and paste themselves throughout their resident genomes. However, these enzymes are promiscuous. They may associate with, and reverse-transcribe, any RNA molecules that present themselves too closely. They may, for example, attach themselves to messenger RNA molecules intended for protein-synthesising service. When LINE-1 enzymes commandeer messenger RNA molecules, they generate DNA copies of them and splice these into chromosomal DNA. The products are processed pseudogenes. These are inserted randomly with respect to DNA site, lack regulatory sequences and passively degenerate (Chapter 3). But this simplistic picture has been superseded by the finding that, against all expectations, many processed pseudogenes are preserved in a form that generates functional proteins [91].

But first a short dog story. Humans have produced dog breeds with short, stubby legs. Such *chondrodysplastic* dogs include the dachshund, corgi, basset hound, Scottish terrier and shih-tzu breeds. They share this distinctive morphology because their genomes possess a new copy of the *fgf4* gene that encodes a member of the fibroblast growth factor family, which is intimately involved in growth and development. The new gene copy arose by retrotransposition of the parent *fgf4* gene – through the unscheduled services of a LINE-1 element. It has all the features of a fully processed messenger RNA molecule that has been reverse-transcribed and inserted into the dog genome. It is bracketed by perfect duplications of the 13 base-long target site (AAGTCAGACAGAG) [92].

The retrotransposed *fgf4* gene exerts significant morphological effects that have been preserved by (artificial) selection. Dog breeders, with their particular ideas of what constitutes a desirable dog phenotype, have exploited a randomly arising trait, and have applied selective breeding to generate a range of mutant breeds. The presence of this reverse-transcribed novelty proves that dachshunds and Scottish terriers are derived ultimately from a shared ancestor.

4.5.1 Reverse-transcribed genes in primates

How many processed pseudogenes might be active in the human genome? One study identified nearly 4,000 reverse-transcribed genes in the human genome, and showed that 18% of these retained protein-coding capacity. Moreover, 30% were used as templates for RNA transcription. These genes also contained a lower-than-expected frequency of mutations that result in amino acid changes relative to mutations that do not result in amino acid changes. This tendency to conserve amino acid sequence is consistent with the idea that natural selection has been operating on a gene to maintain the integrity of its product. The retention of conserved open reading frames that are transcribed is strong evidence that these haphazardly acquired by-products of LINE-1-mediated reverse transcription are authentic genes. It has been estimated that, throughout primate evolution, the process of retrotransposition has generated one new gene every million years [93].

Reverse-transcribed genes that retain protein-coding function are called *retrogenes*. A succession of random events does, it may be concluded, generate new genes at an appreciable frequency. It must be emphasised that retrotransposed genes, whether they are crippled pseudogenes or have vital protein-coding (or other) roles are excellent phylogenetic markers. It is the elaborate, stepwise molecular *mechanism* of their generation that ensures their value as evolutionary markers. The presence or absence of *functionality* is irrelevant to the question of phylogenetic relatedness.

A subset of relatively young retrogenes has been investigated for their presence or absence in a panel of species representing

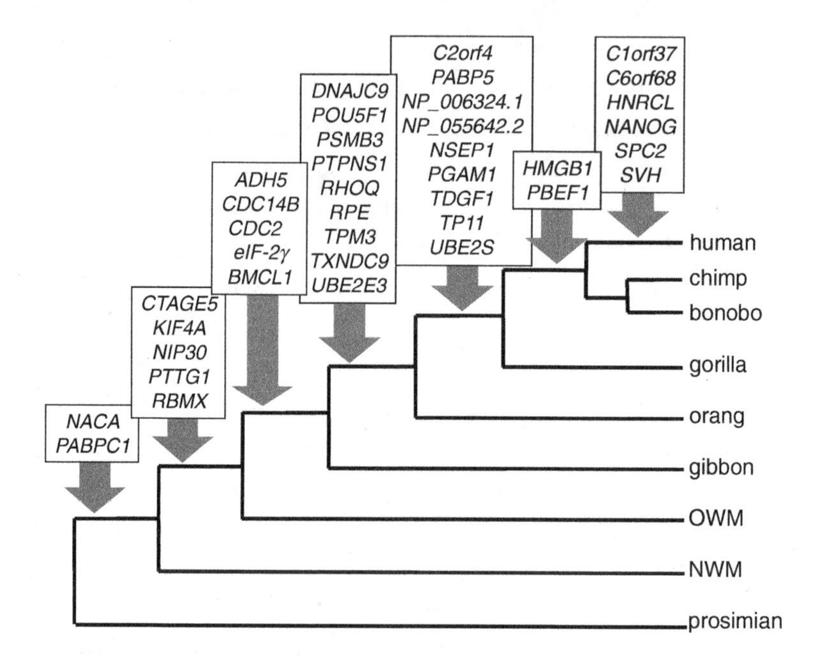

FIGURE 4.21. THE APPEARANCE OF REVERSE-TRANSCRIBED GENES DURING PRIMATE EVOLUTION [93]
Retrogenes are identified by the symbol of their parent gene.

primate groups (Figure 4.21). The order of appearance of these reverse-transcribed genes conforms to a familiar pattern. For example, two inserted genes are present in humans and chimps (but in no other species). These entered the primate germ-line in a human–chimp ancestor. Five are present in all apes and OWMs (but not in NWMs and prosimians). These arose in an ape–OWM progenitor. The distribution of inserts provides a perfect primate phylogeny. Shared retrogenes are magnificent markers of relatedness. We may add 'retrogenealogy' to the catalogue of markers that elucidate evolutionary history.

This work is not a flash in the pan. An independent study identified some 12,000 processed pseudogenes in the human genome, of which half showed some evidence of generating transcripts. Of these, 726 appeared to be serious contenders for being fully functional retrogenes. In a selection of 30 such genes, 10 were shown to be shared by

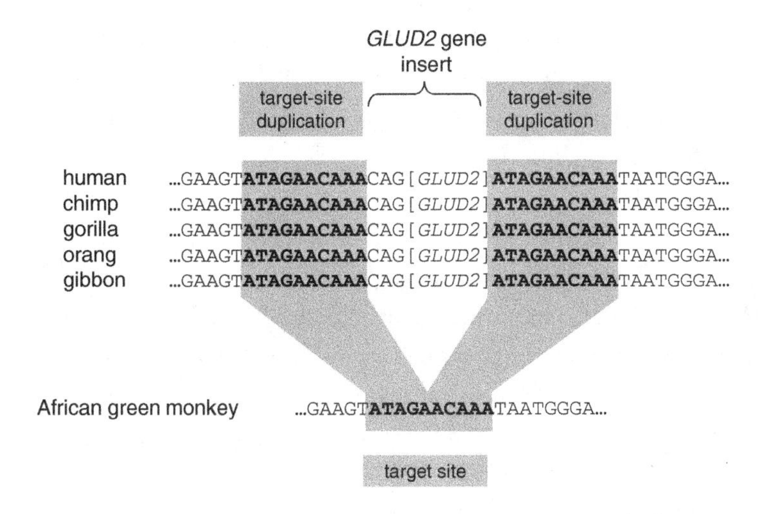

FIGURE 4.22. THE INSERTION SITE OF THE *GLUD2* RETROGENE [95]

humans and all other simians [94]. Retrotransposed genes typically function as single inserted units. In a significant minority of cases, however, they can join forces with nearby genes to form chimaeric transcripts. Or they can recruit new exons from flanking DNA to generate original transcriptional units.

The history of some primate-specific retrogenes has been elaborated. Humans and other apes possess two genes that encode the enzyme glutamate dehydrogenase. This enzyme metabolises glutamate, an important energy source. *GLUD1* encodes a widely expressed housekeeping enzyme. *GLUD2* is a processed, retrotransposed copy of *GLUD1*, and encodes a closely related enzyme that is expressed in the testis and brain. The insertion site of *GLUD2* is flanked by target-site duplications, as is typical for inserts that have been generated by LINE-1 endonucleases (Figure 4.22). The *GLUD2* insert is located in exactly the same site of the genome in all apes, and arose in an ancestor of the apes [95].

The sequences of the *GLUD2* retrogenes have been compared in all the species that possess them. From this comparison, the order of mutations arising since retrotransposition has been inferred. A burst of amino-acid-changing mutations occurred soon after the retrogene arose. Based on this work, researchers synthesised several long-extinct intermediate versions of the GLUD2 protein, and tested them for their behaviour in cells. One particular mutation (that converted glutamic acid into lysine) caused subtle changes in the intracellular location of the protein, enhancing its import into mitochondria [96].

The retrogene-encoded enzyme is expressed in Sertoli cells (in the testis) and in astrocytes (in the brain) and may be involved in supplying energy to germ cells, and in regulating the availability of glutamate (a neurotransmitter) to neurons, respectively. GLUD2 activity has been fine-tuned for an environment in which astrocytes are intimately associated with active neurons. It functions optimally under slightly more acidic conditions than does the parent enzyme [97]. Mutations that arose during the history of the GLUD2 enzyme have suppressed its basal (idling) catalytic activity and altered its sensitivity to inhibitors, activators and drugs [98].

It seems that one small step for a LINE-1 element has wrought one giant leap for hominoid-kind. But an unfortunate effect of possessing the *GLUD2* gene has also been uncovered. In men with Parkinson's disease, a rare variant (or polymorphism) in this gene (T1492G) is associated with disease onset at a younger age [99].

The *CDC14B* gene tells a similar story. This gene encodes an enzyme of the protein phosphatase class, and helps to control the late stages of mitosis. The *CDC14B* gene generates several transcripts, one of which was retrotransposed, in an ape ancestor, to form a new gene designated *CDC14Bretro*. The new gene accumulated multiple mutations, indicating that it acquired novel functions. The synthesis of hypothesised long-extinct intermediates shows that these mutations affected the localisation of the CDC14Bretro protein in cells. This hominoid-specific gene is expressed in testis and brain [100].

The formation of the *PIPSL* retrogene is a fascinating story of happenstance. It arose from the association of two precursor genes. The conventional *PIP5K1A* and *PSMD4* genes are adjacent to each other on chromosome 1. They encode a lipid kinase and a component

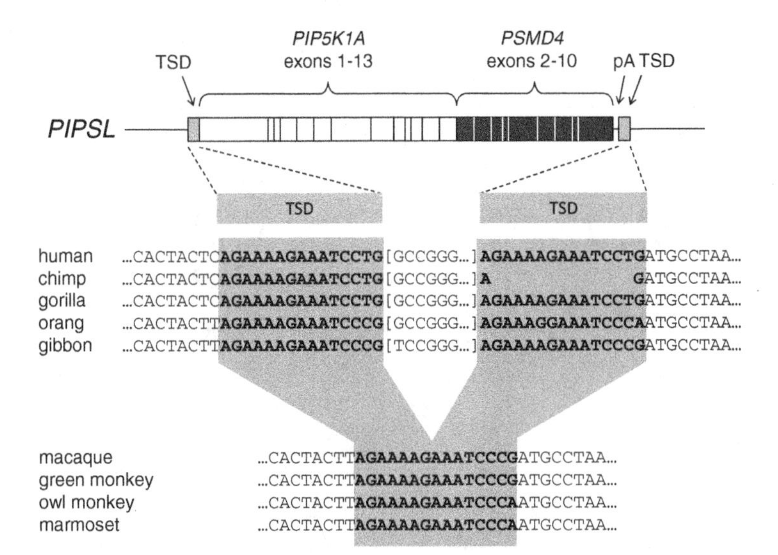

FIGURE 4.23. THE INSERTION SITE OF THE CHIMAERIC *PIPSL* RETROGENE [101]
pA, poly A sequence.

of the proteasome (a multi-protein complex that disposes of unwanted proteins). Transcription of the *PIP5K1A* gene occasionally continues into the *PSMD4* gene. The resulting messenger RNA contains the first 13 of the 15 *PIP5K1A* exons and the last nine of the ten *PSMD4* exons. One of these chimaeric RNA molecules has been retrotransposed, creating a ready-made chimaeric gene. As with the previous two examples, the hybrid *PIPSL* gene arose in an ape ancestor. All apes, but no monkeys, possess the insert (Figure 4.23).

A period of rapid diversification in the *PSMD4*-derived end of the new *PIPSL* gene was completed before the human and chimp lineages diverged. This indicates that the new gene was being honed for a new role by natural selection. This part of the new gene has also been conserved, whereas the *PIP5K1A*-derived part has degenerated in gorillas and gibbons. The gene is highly transcribed in the testis, and the protein is detectable at low levels [101].

There is an interesting twist to the retrogenic tale. The 5′ (left-hand) end of LINE-1 elements is often flipped 180° with respect to

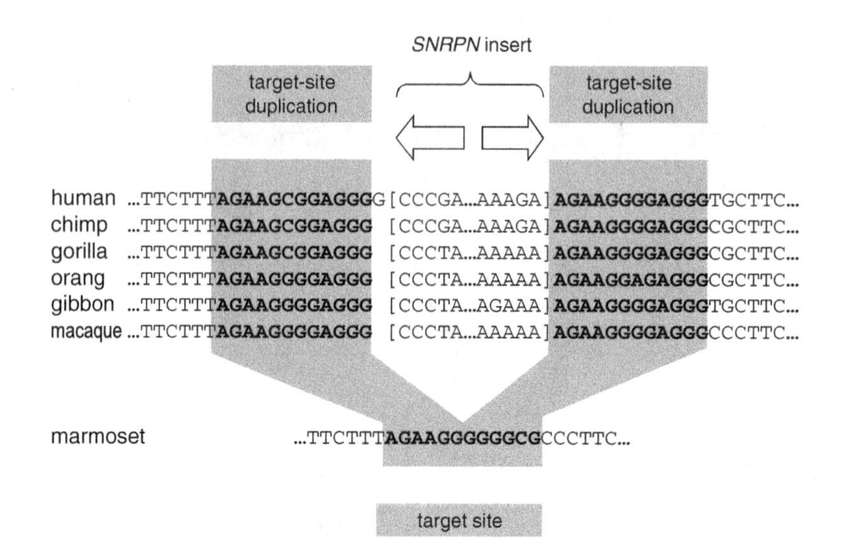

FIGURE 4.24. INSERTION SITE OF *SNRPN*, A CANDIDATE 5′-INVERTED RETROGENE [103]
Arrows indicate that the left-hand part of the gene is inverted with respect to the right-hand part.

the rest of the insert. This is known as *5′-inversion*, arises by a mechanism known as *twin priming* [102], and is another indicator of the sloppiness of the LINE-1-mediated retrotransposition process. A proportion of retrotransposed gene transcripts also show 5′-inversion – further evidence that processed pseudogenes are generated by the LINE-1 enzymatic machinery. Any protein-coding capacity in an inverted segment of DNA will be unrelated to the originally encoded protein sequence, and so 5′-inversion could be a mechanism for generating proteins containing wholly novel domains.

A computer search of the human genome identified six retrotransposed genes that are characterised both by the possession of 5′-inverted sequences and by their capacity to generate RNA transcripts. They are candidate retrogenes that include a copy of the *EIF3F* gene (shared by all apes) and of the *SNRPN* gene (shared by apes and OWMs; Figure 4.24). The *SNRPN* insert has retained the potential for making a protein in humans only [103]. The compounded randomness inherent in this process – the selection of a

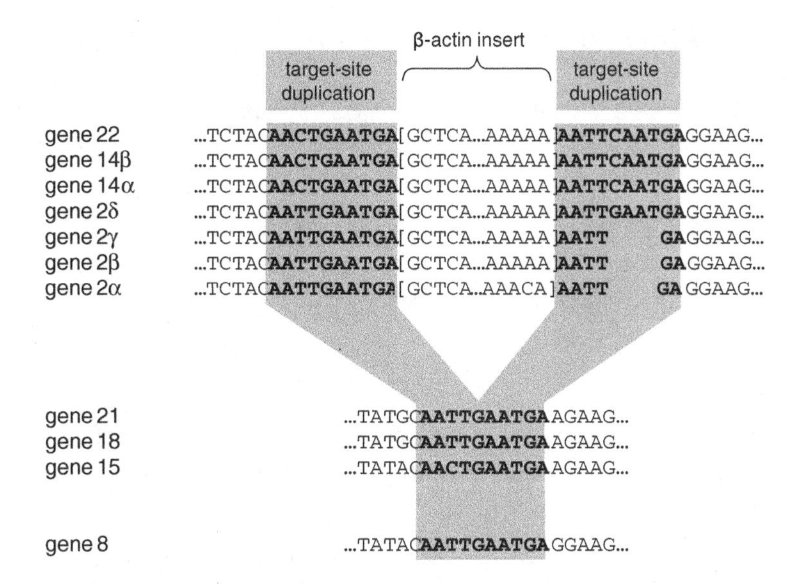

FIGURE 4.25. INSERTION SITE OF A β-ACTIN RETROGENE IN *POTE* GENES [104]

particular messenger RNA, its insertion into a particular site of the genome and its reorganisation by a particular 5′-inversion – entails that the products are irrefutable markers of descent from the founding ancestor.

A final 'retrogenealogical' example illustrates how gene families grow numerically. A family of genes, identified by the collective name of *POTE*, is present in the genomes of primates. There are at least 11 members in the human genome. Seven members of this family contain the insertion of β-actin gene sequences into the *POTE* coding region, creating a set of hybrid genes (Figure 4.25). The insertion site of the β-actin gene sequence, complete with target-site duplications, is the same in every case [104]. This multiply repeated insert establishes several things.

- One unique β-actin insertion event is now represented in seven genes. The one original gene into which the β-actin gene was inserted has undergone several duplication events to spawn a gene family.
- *POTE*-actin fusion genes are present in apes and OWMs. It follows that the original retrotransposition event occurred in a common ancestor of apes and OWMs.

- Several members of this gene family are transcribed, and POTE-actin chimaeric proteins are made in cells, indicating that new functionality has accrued from the situation. Indeed, POTE proteins are highly expressed in the testis, where they are associated with apoptotic cells. The actin-containing *POTE-2α* gene becomes more active when cells are exposed to apoptosis-inducing signals. The POTE-2α protein induces apoptosis, and may be a regulator of life-and-death decisions of cells [105]. *POTE-2α*, at least, is a card-carrying retrogene.

4.5.2 Reverse-transcribed genes in mammals

Primate-specific retrogenes are young in evolutionary terms. They have retained their target-site duplications, which provide unambiguous evidence for the TE-dependent (reverse transcriptase-endonuclease) mechanism of their generation. But our DNA also contains many retrogenes that appear to be much older. They are derived from processed transcripts. Introns have been removed. They are copies of identifiable parent genes. Each appears abruptly at a particular stage of genome history. Each has resisted degeneration, evidence that it has been maintained by natural selection. Computational surveys have enumerated at least 70 human retrogenes for which there is a mouse orthologue with an open reading frame [106]. These originated in a Euarchontoglires (primate–rodent) ancestor.

The *YY1* gene encodes a protein that belongs to the zinc finger family (see Section 4.4.6). The *YY1* gene is the proud parent of the *YY2* and *REX1* retrogenes, each of which encodes a protein that is itself a regulator of other genes. *YY2* was dropped, ready-made, into the fifth intron of another gene (*Mbtps2*), which is highly conserved in vertebrate species from humans to fish. Comparative studies show that both *YY2* and *REX1* were spliced into the genome of a germ-line cell from which primates, rodents and dogs (representing Boreoeutheria) have issued [107].

The YY1 and YY2 products each control the activity of a large number of genes, and although there is extensive overlap in their respective target gene sets, there is also a large degree of divergence. The *YY2* gene, despite its one-off copy-and-paste provenance, has probably exerted widespread effects on gene expression [108]. And the *REX1* product acts to maintain embryonic stem cells, at least in a murine experimental system. It also suppresses the activity of LTR retrotransposons – just as other zinc finger proteins have been hypothesised to do. There is irony in the fact that a gene generated by an unscheduled retrotransposition event now acts as a policeman to enforce constraints on other retrotransposition events [109].

Some genes are regulated by a process called *imprinting*. This means that the two copies (or alleles) are differentially tagged depending on whether they are inherited from the father or the mother. The activity of imprinted genes thus depends on their parent of origin. Some imprinted genes have arisen by the process of retrotransposition. The *MCTS2* retrogene is shared by humans and rodents, but not by dogs or cows, and dates from a Euarchontoglires ancestor. The *NAP1L5* and *INPP5F_V2* genes are shared by all these species (but not by opossums or chickens), and comparative data have shown that they were spliced into the germ-line of Boreoeutherian and eutherian ancestors, respectively [110].

These three retrogenes share a singular feature. Their parent genes are all located on the X chromosome, and retrotransposition has exported the copied versions to non-sex chromosomes (*autosomes*). Another striking example is provided by the *UTP14A* gene, which is also present on the X chromosome. Its product is required for the production of mature ribosomal RNA. It has spawned retrogenes on autosomal chromosomes on at least four separate occasions during mammalian history. In mice, the *Utp14b* retrogene is active in the testis and is essential for male fertility. This retrogene arose in a Euarchontoglires ancestor, but the human copy is a degenerated relic. However, humans and other primates possess an independently arising retrogene, *UTP14C*, which is also essential for male fertility. Cattle and elephants possess their own independent versions of this retrogene [111]. Random events generate analogous solutions when subjected to similar selective pressures. This situation is reminiscent of the repeated ERV exaptations in placental morphogenesis.

In some cases, copied genes evolve rapidly and provide wholly new functions. The *UBL4A* gene is also present on the X chromosome, and has been retrotransposed to create an autosomal copy, *Ubl4b*. This event occurred in an ancestor of eutherians and marsupials. In mice, the DNA sequence of the *Ubl4b* gene has diverged widely from its parent, but the respective small ubiquitin-like proteins specified by the parent and daughter genes are clearly related. The parental Ubl4a protein is expressed throughout sperm cell development, whereas the derivative Ubl4b copy is expressed at the later stages, suggesting that the two proteins perform different functions [112].

If this X-to-autosomal seeding of new genes is a general pattern, what selective pressure has been responsible for establishing such directionality? A clue comes from the observation that, during the development of male germ cells, the X chromosome is shut down. Perhaps the loss of expression of housekeeping genes on the X chromosome limits germ cell development. If this is so, then selective pressures would favour maintenance of the coding function of genes that have been copied from the X chromosome to autosomes; such gene products would be available during spermatogenesis.

Retrogenes arising on autosomes have occurred at a fairly constant rate throughout mammalian history (Table 4.1). However, retrogenes have been exported from [113] and imported into [114] the X chromosome (or its precursor) at one particular phase of history. This was the era of the eutherian ancestral lineage when the X chromosome was becoming differentiated as a sex chromosome. Retrogene export may have generated housekeeping genes on autosomes that remained active in developing male germ cells. Retrogene import may have served sex-specific roles. Either way, the retrogene traffic documents key stages in early mammalian history.

Some retrogenes are 'orphans': their parental genes have perished, leaving the daughter genes to carry on the family business. Such genes tend to be strongly conserved. Of 25 in the human

Table 4.1. *Retrogene traffic in mammals [113, 114]*

	Direction of retrogene traffic		
Ancestral lineage	Arising on autosome	Parent genes on X exporting retrogenes	Imported into X
Primate	39	4	6
Euarchontoglire	3	3	0
Eutheria	26	17	12
Eutherian–marsupial	15	1	1
Mammal	24	2	1

genome, three arose in a eutherian–marsupial progenitor, six in a mammal–bird ancestor, and others in ancestors we share with fish and with invertebrates (seven each) [115].

Families of genes produce small RNA molecules that do not encode proteins but that are involved in protein synthesis. These include the small nucleolar RNAs (snoRNAs) of the H/ACA and C/D box families. These gene families have been expanded by the proliferation of retrogenes. Most of the 200 H/ACA box retrogenes in the human genome have been inherited also by chimps, and many by mice [116]. The stepwise addition of new genes to the mammalian germ-line provides an ongoing mechanism for the development of new functions, and provides perfect signposts outlining mammalian evolution. These signposts are so persuasive because each of them arose by a complex and identifiable *mechanism*, a chain of molecular events that reflects the dynamic nature of the genome. Retrogenealogy is a spectacular science.

4.6 DNA TRANSPOSONS

Our genomes are littered with a shadowy class of jumping gene that we have ignored hitherto. These are the *DNA transposons*. They cut-and paste themselves around the genome, using an enzyme called a *transposase*. They propagate to high numbers because they coordinate their activity with cellular DNA replication [117]. Approximately

3% of our genome has been contributed by DNA transposons, and there are 380,000 individual inserts.

DNA transposons splice themselves into the genome in a manner that is random with respect to site. Indeed, scientists use DNA transposons (from fish and moths) as insertional mutagens (in mice) to discover genes that, when damaged, contribute to the development of cancer. This approach to finding potential cancer genes would work only if the mutagenic agent was indiscriminate in the way it targeted genes [118]. Such random behaviour indicates that DNA transposons should be powerful markers of common ancestry. If two species share the same insert at the same site, they must have inherited it from the ancestor in which that insert was added to the germ-line.

Whole-genome sequencing has revealed that essentially *all* the DNA transposons present in our genome are shared by OWMs. Each insert was already in place in an ancestor of apes and OWMs. We have 380,000 independent clonal markers of the monophylicity of Old World primates. Most of the individual inserts are present also in the genomes of non-primate mammals, and thus act as markers of remote eutherian ancestors [119]. Active DNA transposons long ago became extinct in primates, although they have been active much more recently in some species of bat [120].

But do DNA transposons contribute to gene formation? Scientists have created a catalogue of genes in eukaryotic organisms that might have originated in the dim past from genes encoding the transposase enzymes of DNA transposons [121]. Specific features showing the fingerprints of the original transposon have, of course, long ago disappeared.

What is highly likely, however, is that DNA transposons have generated genes that we possess. DNA transposons have been transmogrified into genes for microRNAs, those master regulators of gene expression. The *miR-1302* family of microRNAs was derived from a class of DNA transposon called MER53 elements, and appeared early in eutherian history. The genes that encode *miR-548* microRNAs are derived from a class of DNA transposons called *Made1* elements. Six

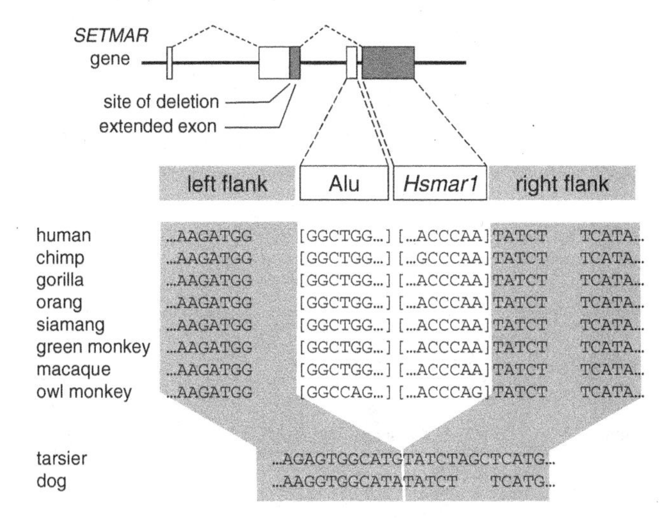

FIGURE 4.26. THE INSERTION SITES OF SIDE-BY-SIDE ALU AND *HSMAR1* ELEMENTS [124]
Top: The *SETMAR* gene, showing the two original exons, the extended exon (*shaded*), and the Alu and *Hsmar1* inserts. *Dotted inverted V-shapes* indicate splicing between the three exons. *Lower diagram*: *Shading* indicates sequences flanking the inserts (not TSDs as previously). Four bases were deleted to the left of the Alu element (CATG in the tarsier). The tarsier locus shows an AGC insertion to the right of the *Hsmar1* transposon, and a gap in the other sequences has been introduced to accommodate this.

of the seven *miR-548* genes in our genome are shared with relatives as distant as macaques. The seventh is shared with chimps. The *miR-548* microRNAs may contribute to the regulation of more than 10% of the genes in our genome [122]. In addition, many other classes of TE have contributed to our microRNA complement [123].

Genes of the *SET* family regulate the activities of genes. They encode enzymes that tag DNA-packaging proteins, called *histones*, with methyl groups. The methyl tags recruit other proteins that affect the conformation of chromatin, and hence the activities of genes. One particular *SET* gene has been modified by the nearby insertion of an Alu element and a DNA transposon of the *Hsmar1* type. Both of these arose in an ancestor of humans, other apes and monkeys (the anthropoid primates) (Figure 4.26).

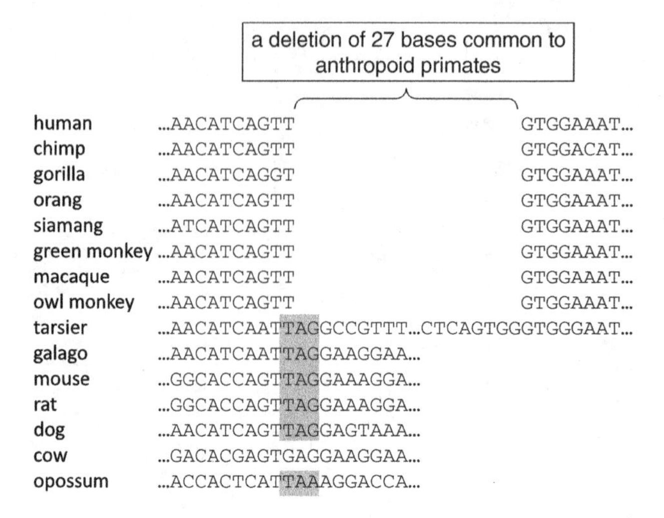

FIGURE 4.27. A 27-BASE DELETION, INHERITED BY ANTHROPOID PRIMATES, THAT REMOVED A TAG *STOP* CODON (*shaded*) [124]

The original *SET* gene and its new *Hsmar1* neighbour have formed a single chimaeric unit that is called the *SETMAR* gene. The left end is *SET*; the right end is *Hsmar1*. But an additional mutation was needed to enable this chimaeric gene to express a new protein. A 27-base deletion occurred in the right-hand end of the *SET* gene and removed a *stop* codon (TAG) and allowed the existing *SET* exon to extend rightwards, into what had been non-coding territory. This deletion is restricted to all anthropoid primates (Figure 4.27). The extended *SET* exon acquired the signals that allowed it to splice with the *Hsmar1* sequence. The transposon insertion and deletion demonstrate that (*enabling*) mutations do indeed contribute to new gene formation [124].

Cataloguing the genetic events that have assembled the *SETMAR* unit is the easy part. Deciphering the role of the novel protein is an open-ended project. The SETMAR protein performs a multitude of biochemical roles in cells. It appears to be involved in DNA repair, in the maintenance of genomic stability, and (potentially) in modulating the effectiveness of some anti-cancer drugs. In fact, the SETMAR protein enhances the efficiency of NHEJ, and so

Table 4.2. *Some DNA transposon-derived genes*

DNA transposon	Transposon gene	Cellular gene and function	Taxa possessing the gene(s)	Ref.
Hsmar1	transposase	*SETMAR*, DNA repair	Apes and monkeys	124
Charlie 10	transposase	*ZNF452*	Humans, cattle and dogs; Boreoeutherians (at least)	127
hAT group	transposase	*ZBED6*, regulator of development and growth	Eutherians; the transposon is present in all mammals	128
Pogo-like	transposase	*CENP-B*, a DNA-binding centromeric protein	Mammals	129
Crypton	tyrosine recombinase	Six genes in humans, from four initial colonisation events	Bony vertebrates	130
Harbinger	transposase, *MYB*-like	*HARBI1*, function unknown	Bony vertebrates	131
		NAIF1, induction of apoptosis	Bony vertebrates	
Transib	transposase	*RAG1* (recombination activating gene 1), generation of antibodies and antigen receptors	Jawed vertebrates	132

suppresses the formation of potentially cancer-causing chromosomal translocations [125]. But the new SETMAR protein does not only fit into multi-component DNA repair enzyme complexes. It has also become an integral part of signalling networks by which cells select between alternative responses to DNA damage [126]. Considering the fact that DNA transposons are disruptive mutagens, the stabilising effect of the SETMAR protein represents a dramatic volte-face.

A diverse anthology of such histories could be described. Many genes originated as parts of DNA transposons that were dropped willy-nilly into the genomes of distant ancestors. These genes perform a variety of functions (Table 4.2), but a common theme appears to be that their encoded proteins interact with DNA. This might be expected, as the originals were transposase enzymes that catalysed DNA rearrangements. The RAG1 recombinase underlies adaptive immunity, functioning to generate a practically infinite number of antigen receptor genes in lymphoid cells.

And finally, these genes appear to have arisen at vastly differing periods of history. The *SETMAR* gene is young in evolutionary terms, and its derivation from a transposase has been described at molecular resolution. The other genes are much older but, given the paradigm established by *SETMAR* evolution, it is natural to see them as other instances of exaptation – the domestication of genetic jetsam and flotsam. New genes arise from DNA transposons, just as they do from retrotransposons. Random events underlie the development of brand new genetic functionality in the context of lawful selection. And every such shared gene furnishes an incidental demonstration of monophylicity.

4.7 *DE NOVO* ORIGINS OF GENES

Many of our genes have been derived from TEs, or from gene copies embedded in segmental duplications, or from RNA molecules converted into DNA copies by promiscuous reverse transcriptases. But other genes appear to have arisen from scratch – against all expectations they have been cobbled together in a stepwise fashion from segments of the genomic wasteland. This mechanism of gene birth is described as origin *de novo* (or *from the beginning*). The products are *orphan* genes, found in one lineage only, lacking relatives in other lineages and contributing to lineage-specific characteristics [133].

Part of the versatility and complexity of genomes such as ours arises from the ability of one gene sequence to be copied into multiple alternative transcripts. A single transcriptional unit (simplistically called a *gene*) may generate multiple processed transcripts, each of which is composed of a different combination of exons. The variety of potential transcripts allows one gene sequence to specify the synthesis of multiple distinct proteins. Such a system lends itself to facile modification: during evolution, exons may be added to (or removed from) existing genes without disrupting existing function. We have noted how Alu or MIR elements have contributed to the phenomenon of exonisation (Chapter 2).

Exonisation *de novo* is exemplified by the gene encoding neuropsin. Two forms of this protein are produced. Type I is found in many species, but type II is expressed in the brains of humans only. The neuropsin gene acquired the ability to express the novel type II form only when a segment of intronic sequence was added to exon 3. An *enabling* mutation (the deletion of a C) created an open reading frame that (potentially) added 45 amino acids to the pre-existing type I form. This mutation is present in *all* ape species, and in *none* of eight species of monkey, indicating that it occurred in a hominoid ancestor (Figure 4.28). Frameshift mutations generally devastate coding sequences. This one *generated* a coding sequence [134]. A subsequent T-to-A mutation, in the lineage leading to humans only, generated the splice site needed to incorporate the extended reading frame into the neuropsin type II sequence [135].

Uric acid metabolism went haywire, independently, in ancestors of the great apes and of the gibbons, when the *UOX* gene was pseudogenised (Chapter 3). One of the disadvantages of increased concentrations of uric acid in the blood is the deposition of crystals in the kidneys, *uric acid nephrolithiasis*, a serious complication of gout. This condition is associated with a variant of the *ZNF365D* gene that encodes the talanin protein.

In the mouse genome, the DNA segment corresponding to *ZNF365D* is identifiable but generates no transcript. In the genomes

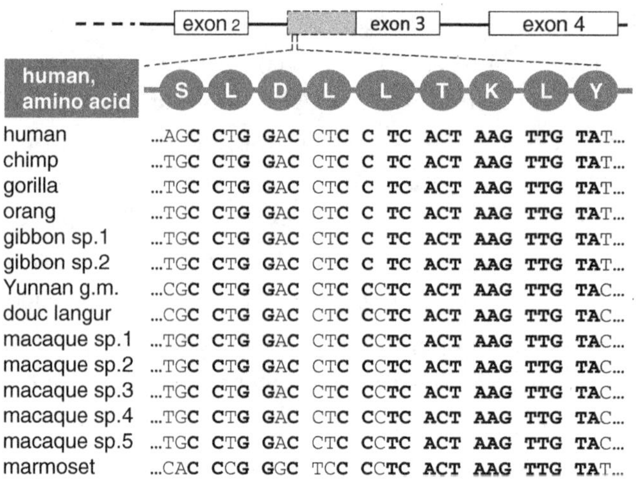

exon 2 exon 3 exon 4

human, amino acid	S L D L L T K L Y
human	...AGC CTG GAC CTC C TC ACT AAG TTG TAT...
chimp	...TGC CTG GAC CTC C TC ACT AAG TTG TAT...
gorilla	...TGC CTG GAC CTC C TC ACT AAG TTG TAT...
orang	...TGC CTG GAC CTC C TC ACT AAG TTG TAT...
gibbon sp.1	...TGC CTG GAC CTC C TC ACT AAG TTG TAT...
gibbon sp.2	...TGC CTG GAC CTC C TC ACT AAG TTG TAT...
Yunnan g.m.	...CGC CTG GAC CTC CCTC ACT AAG TTG TAC...
douc langur	...CGC CTG GAC CTC CCTC ACT AAG TTG TAC...
macaque sp.1	...TGC CTG GAC CTC CCTC ACT AAG TTG TAC...
macaque sp.2	...TGC CTG GAC CTC CCTC ACT AAG TTG TAC...
macaque sp.3	...TGC CTG GAC CTC CCTC ACT AAG TTG TAC...
macaque sp.4	...TGC CTG GAC CTC CCTC ACT AAG TTG TAC...
macaque sp.5	...TGC CTG GAC CTC CCTC ACT AAG TTG TAC...
marmoset	...CAC CCG GGC TCC CCTC ACT AAG TTG TAT...

FIGURE 4.28. AN ENABLING FRAMESHIFT MUTATION IN THE NEUROPSIN GENE [134]
The *shaded box* to the left of exon 3 indicates sequence that was recruited from intron 2. Amino acid sequence is represented by *white letters in grey circles*. The *oval identified as* L (leucine) locates the enabling frameshift mutation (loss of a C) that extended the reading frame.

of monkeys and non-human apes, the corresponding sequence is similar to that of humans, but contains *stop* codons and indels that preclude protein-coding capacity (Figure 4.29). All ape species, however, do generate a transcript, indicating that this sequence has acquired at least some of the features necessary to recruit and assemble transcriptional machinery. Only in humans is there a *ZNF365D* protein-coding gene sequence. Whatever function is assigned ultimately to this new gene, and the control of uric acid concentration has been hypothesised, the manner of its development establishes that accumulating mutations can assemble protein-coding genes from non-coding sequences [136].

Systematic searches through genome databases have identified three human genes for which the corresponding chimp and macaque base sequences are highly similar, but non-coding because of multiple disablements. Partial sequences of two of these are shown in

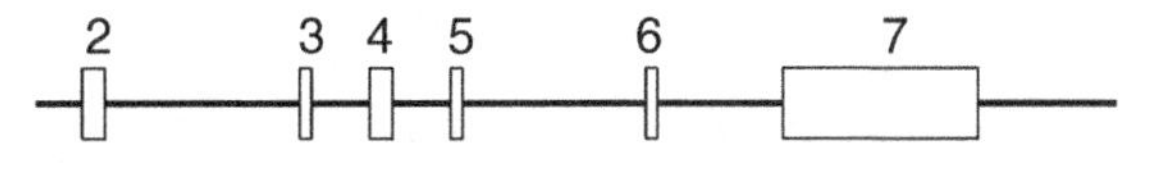

part of exon 4

human	...AGG AAA ACC CAA GTT TGG CGT TGG CAG TCA GGT AAT TCA TCA...
chimp	...AGG AAA ACC CAA GTT TGG CGT TGG CAG TCA GGT AAT TCA TCA...
gorilla	...AGG AAA ACC CAA GTT TGG CGT TGG CAG TCA GGT AAT TCA TCA...
orang	...AGG AAA ACC CAA GTT TGG TGT TGG CAG TCA GGT AAT TCA TCA...
gibbon	...AGG AAA ACC CAG GTT TGA CGT TGG TAG TCA GGT AAT TCA TCA...
baboon	...AGG AAA ACC TGA GTT TGG CGT TGG CAG TCA GGT AAT TCA TCA...
macaque	...AGG AAA ACC TGA GTT TGG CGT TGG CAG TCA GGT AAT TCA TCA...
marmoset	...AGG AAA ACC CAA GTT TGG CGT TGG CAG TCA GGT AGT TTC ATC...

part of exon 6

human	...AGA GAA AGT GTC		TCT ACA AGT...
chimp	...AGA GAA AGT GTC	TTC AGT CAG GAA CCC CAG TC	TCT ACA AGT...
gorilla	...AGA GAA AGT GTC	TTC AGT CAG GAA CCC CAG TC	TCT ACA AGT...
orang	...AGA GTA AGT GTC	TTC AGT CAG GAA CCC CAG TC	TCT ACA AGT...
gibbon	...AGA GAA AGT GTC	TTC AGT CAG GAA CCT AAG TC	TCT ACA AGT...
baboon	...AGA GAA AGT GTC	TTC AGT CAG GAA CCC CAG TC	TCT ACA AGT...
macaque	...AGA GAA AGT GTC	TTC AGT CAG GAA TCC CAG TC	TCT ACA AGT...

FIGURE 4.29. ORIGIN OF A PROTEIN-CODING GENE, *ZNF365D*, IN HUMANS [136]
The six protein-coding exons are depicted (*top*) together with partial sequences for exons 4 and 6. In exon 4, gibbon and monkey sequences contain *stop* codons and a T insertion. In exon 6, the human sequence contains a 20-base deletion generating an open reading frame.

Figure 4.30. The gene designated *C22orf45* in the human genome arose when a T-to-C mutation removed a *stop* codon that is retained in each of the other species. The *DNAH10OS* gene was formed in part when a ten-base insert generated an extensive protein-coding open reading frame [137]. Subsequent work has identified dozens of potential new genes generated in humans by enabling mutations and expressed as RNA transcripts and proteins [138], or present in primates for which the corresponding genetic regions in non-primate species are identifiable but lack the capacity to encode proteins [139].

Newly emergent genes tend to encode small proteins in which the coding sequences are present in only one exon. They tend to have a rapid rate of evolution, and to be expressed in a limited range of tissues (in particular reproductive tissues). In the vastness of the genome, small transcriptional units may be continuously mutating into

C22orf45 gene

human	...TGG AGA GGC CGA GTC CTC CC...
chimp	...TGG AGA GGC TGA GTC CTC CC...
gorilla	...TGG AGA GGC TGA GTC CTC CC...
orang	...TGG AGA GGC TGA GTC CTT CC...
gibbon	...TGC AGA GGC TGA GTC CTT CC...
macaque	...TGG AGA GGC TGA GTC CTC CC...

DNAH10OS gene

human	...CCCCAGGAATCCTCATTCCTGGGGCATCAA...
chimp	...CCCCAGGAAT GAGGCATCAA...
gorilla	...CCCCAGGAAT GAGGCATCAA...
orang	...CCCCAGGAAT GAGGCAGCAA...
gibbon	...CCCCAGGAAT GAGGCATCAA...
macaque	...CCCCAGGAAT GAGGCATCAA...

FIGURE 4.30. ORIGIN OF PROTEIN-CODING GENES IN HUMANS [137]
C22orf45 gene: an enabling T-to-C mutation that removed a *stop* codon (*shaded*) present in all other species (*upper diagram*). *DNAH10OS* gene: an enabling ten-base insert (*lower diagram*).

and out of existence. In those cases where reproductive success is enhanced, new constructions are preserved. Once subject to maintenance by selection, they can be elaborated and their functions extended with time.

Genes specifying non-coding RNAs may also arise *de novo*. MicroRNA genes have been appearing since before mammalian history. Especially rapid gene birth occurred in the ancestral lineages of both eutherian and hominoid mammals. Much microRNA gene birth has followed DNA duplications, but a family of such genes has appeared *de novo* in the human genome. The miR-941 molecules are expressed in stem cells and in the brain, and regulate cell signalling cascades. They are candidates for affecting lifespan [140].

Female eutherian mammals have two X chromosomes, whereas males have one. In order to maintain an equivalent dosage of gene transcripts arising from the X chromosome, one of the X chromosomes in females is randomly inactivated in each cell early in development. This inactivation is performed by coating the X chromosome with a long, non-coding RNA molecule called *XIST* (an acronym for

X (inactive)-specific transcript). This molecule acts like a glue to silence the X chromosome.

This mechanism of X chromosome inactivation operates only in eutherian mammals, and the *XIST* gene is found only in eutherians. The corresponding segment of the genome in marsupials, chickens, toads and fish does not have an *XIST* gene, and its place is occupied by another gene called *Lnx3*, which is missing in eutherians [141]. In the earliest eutherian mammals, the venerable old *Lnx3* gene was pseudogenised, and parts of the leftover sequence were salvaged to fabricate *XIST* (Figure 4.31, upper diagram). The old garage has been demolished, but some of its timbers have been scavenged to make the rustic garden gazebo.

Partial sequences of *XIST* exon 4 and *Lnx3* exon 4 are compared in Figure 4.31. The similarities are patent and extend well beyond the gene fragments shown. But in this short segment of DNA, three frameshift mutations are apparent in the eutherian sequences. These one- and two-base pair insertions may be seen either as *disabling* mutations (that scrambled the protein-coding capacity of *Lnx3* and contributed to its pseudogenisation) or as *enabling* mutations (that helped generate the brand new *XIST* gene). Either way, they are markers of eutherian monophylicity. Humans and moles really do have a common ancestor. The effect of the base changes on gene functionality, whether loss or gain (and in this case, both loss of *Lnx3* and gain of *XIST*), is irrelevant to the issue of phylogenetic relatedness.

But what is the source of the *XIST* exons that were not derived from recycled parts of the *Lnx3* gene? They include multiple bits and pieces identifiable as being TE-derived. The boundaries and target-site duplications of these have long disappeared, but fragments of ERVs, LINE elements, MIR elements and DNA transposons are readily discernable [142]. Recycling of gene sequences is not restricted to the *XIST* gene. Two other genes in this locus have also arisen by assimilating fragments of old genes, and by co-opting raw material imported in the form of TEs [143]. What might once have been dismissed as 'junk' DNA has been transmogrified into genetic riches.

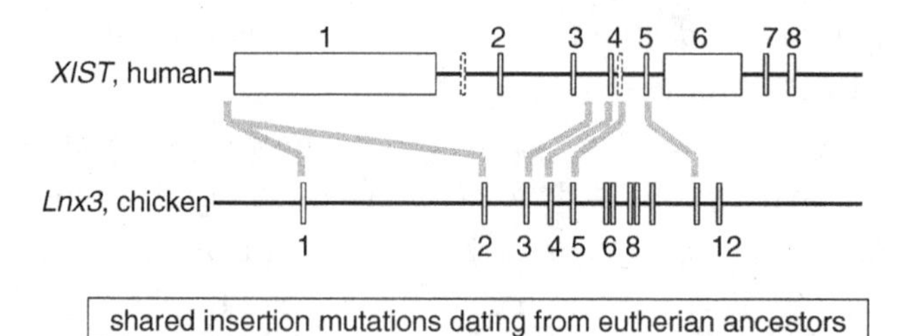

shared insertion mutations dating from eutherian ancestors

XIST, human ...GCATGAGGATCCTCCAGGGGAAAAGCTCACTACCACT...
XIST, mouse ...GCATGAGGATCCTCCAGGGGAATAGCTCACCACCACT...
XIST, rat ...GCATGAGGATCCTCCAGGGGAATAGCTCACCGCCACT...
XIST, vole ...GCATGAGGGTCCTCCAGGGGAATAGCTCACCAGCACT...
XIST, cow ...GCATGAGGATCCTCCAGGGGAAAGATTCACTACCACT...
XIST, dog ...GCATGAGGATCCTCCAGGGGAAAAGCTCACTACCACT...
XIST, mole ...GCTTGAGGACCCTCCAGGGGAAAAGCTCACTACCACT...
Lnx3, chicken ...G ATGAGAATAGTT GGGGGTAAGGACACG CCACT...

FIGURE 4.31. DERIVATION OF *XIST* GENE FROM *LNX3* GENE SEQUENCES [141, 142]
Upper diagram: Parts of the *XIST* gene derived from *Lnx3* are indicated by *grey lines*. *Dotted boxes* in *XIST* are exonised in other eutherians.
Lower diagram: a comparison of sequences from eutherian *XIST* exon 4 and chicken *Lnx3* exon 4.

Our elegant garden gazebo, which occupies pride of place on the front lawn, has been fabricated using timber salvaged from the old garage and scavenged from the municipal rubbish dump.

To perform its very particular role in cells, the *XIST* gene would have to be controlled by a suite of regulatory on/off mechanisms. Could a new gene and its regulatory circuits evolve concurrently? Evidence is growing that *XIST* has been incorporated into existing gene control networks. One key player is the *REX1* gene, the retrotransposed daughter of the *YY1* gene. *REX1* was spliced into the genome during the same epoch as that in which *XIST* originated. The new gene and its new regulator were assembled concurrently in pioneering eutherians [144].

The *XIST* gene encodes one of many long non-coding RNAs. The history of its formation will be one of many similar histories. The *HOTAIR* gene, for example, generates an RNA molecule that controls the activity of the body plan-specifying homeobox genes. *HOTAIR* regulates the regulators. It is absent from non-mammals and monotremes, and appeared *de novo* in eutherian–marsupial ancestors [145]. The recruitment of TEs, especially ERVs, as building blocks of long non-coding RNAs has been widespread [146]. And genes from which are transcribed regulated long non-coding RNAs (in OWMs) have been identified as the sources of 24 new protein-coding genes that have appeared in hominoids [147].

4.8 GENERATING GENES AND GENEALOGIES

A further mechanism of gene generation might be considered. This takes us back to genetic upheavals that occurred in deep time. Many gene families are organised around four members. It has been proposed that early in vertebrate development, many genes multiplied fourfold when the *entire genome* underwent two successive duplications: the first around the time when jawless fish arose and the second when cartilaginous fish arose. The whole-genome duplications were followed by widespread pruning of redundant genetic material.

This mechanism may be adduced in those cases where series of distantly related genes occur in the same order in different parts of the genome: so that 'gene A–gene B–gene C′ has its counterparts 'gene A′–gene B′–gene C″ elsewhere, with up to four related gene series. The fourfold reiteration of homeobox gene tandem arrays is a classical example [148]. The two-genome duplication scenario is reflected in various gene families.

- The α-actinin family is a select group with four members. Two of these are found in all cells, and contribute to the cytoskeleton – the protein network that holds cells together and enables them to migrate. The other two proteins, α-actinin-2 and -3, are present in muscle cells – although approximately a billion people lack α-actinin-3 because they share a particular mutation in the *ACTN3* gene. The people possessing the pseudogene do not have disease because α-actinin-2 substitutes for the scrambled α-actinin-3. But people with two copies of the *ACTN3* mutant gene are less likely to be elite athletes [149].

- Gene families that encode signalling proteins have been preferentially expanded. Of interest to cancer biologists is the four-gene family that encodes receptors for epidermal growth factors. These receptors play vital roles in embryonic development, wound healing and (when deranged by mutations) cancer pathogenesis [150]. The seven-member relaxin/insulin-like peptide gene family encodes hormones related to insulin, and probably expanded likewise. Subsequent losses or duplications occurred sporadically in particular lineages [151].
- The *ATP2A* genes, encoding calcium-transporting enzymes, comprise a candidate twice-duplicated family. The *ATP2A1* gene is of particular note because it has been the location of a segmental duplication that enabled it to gain an intron – a rare event. An uninterrupted exon, containing the sequence ...`AGGT`..., was duplicated to generate ...`AGgt` ... `agGT`... in which the (lower case) `gt` ... `ag` bases came to provide *splice donor* and *splice acceptor* sites, marking the termini of a new intron (Figure 4.32). These factors indicate that the singular events that gave rise to the intron occurred in an ancestor of humans and zebrafish [152].

The appearance of new introns in pre-existing genes may be exploited to provide phylogenetic markers. Such rare innovations have been documented in genes that are derived from TEs (and that do not possess introns to start with). Intron-forming events have occurred at a trickle over chordate history. Some arose in the last common ancestors of tetrapods, of amniotes and of eutherian–marsupial mammals. But some 20 domesticated TEs have generated 50 or more new introns in the ancestral line leading to eutherian mammals [153].

The study of gene families provides other phylogenetic markers from deep time. Gene sub-families appear sequentially, and the order of appearance can be used to deduce the order of relationship of animal taxa. Mammals possess at least a hundred *keratin-associated proteins* (KRTAPs) that contribute to the toughness of hair. The spawning of new KRTAP gene families has not been documented in primates; but the appearance of new families has been documented in both Boreoeutherian and eutherian–marsupial ancestors [154].

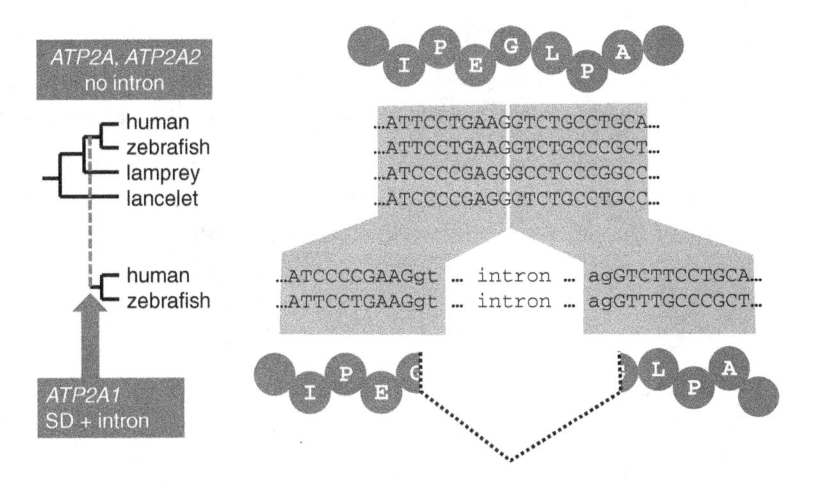

FIGURE 4.32. ACQUISITION OF AN INTRON IN THE *ATP2A1* GENE [152]
The *upper four sequences* represent a conserved region of the human, zebrafish (*ATP2A2*), lamprey and lancelet (*ATP2A*) genes. The *lower two sequences* are from the *ATP2A1* genes of humans and zebrafish, interrupted by an identically located intron. *Letters in circles* represent amino acids.

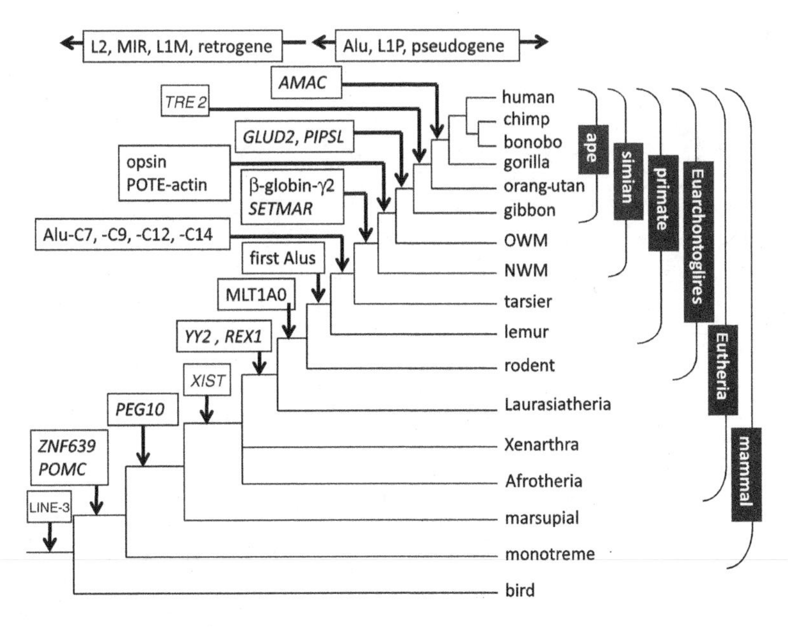

FIGURE 4.33. A SUMMARY OF OUR EVOLUTIONARY HISTORY

Do new genes arise? The selected empirical findings described herein have amply documented ways by which new genes have been fabricated, and multiple steps are often involved. Comparative analyses have identified ancient lineages in which these happenings occurred. By abstracting representative genetic markers from those described in the preceding chapters, we end up with a coherent and firmly established picture of our evolutionary history (Figure 4.33).

Epilogue: what really makes us human

We have reviewed aspects of the nascent and burgeoning field of comparative genomics. We have found that DNA sequences archive DNA history, and that our genome has arisen from those of (now extinct) progenitors by mechanisms that are intrinsic to genetic systems.

But it might be asserted that the interpretations placed upon the genetic data are dependent on scientists' evolutionistic presuppositions. When we discuss presuppositions, which we all possess (even if we are unaware of them), we have left science and entered the realm of metaphysics. I have sought to reflect exclusively on science in Chapters 1–4. My guiding presupposition is that the laws of nature are consistent (albeit incompletely known) – which is an underlying assumption fundamental to scientific practice. An Alu element that arises in the human germ-line today operates through basically the same biochemical mechanisms as did an Alu element that arose in an anthropoid ancestor.

What of the charge that evolution is a religion? The genetic considerations outlined are precisely those that prevail in the starkly technical environment of a genetics laboratory. We have considered only the molecules and mechanisms of heredity. We have not approached phylogenetics from the perspective of some arcane 'evolutionary' logic. On the contrary, my realisation that lineages of species may be defined by characteristic genetic fingerprints had compelling power *only* because, as a common-or-garden cell biologist, I was steeped in the principles by which lineages of cells are defined by the same familiar categories of heritable markers. The *casanova principle* is integral to heredity. If the conclusions drawn from Chapters 1–4 are invalid, then the genetic basis of such academic disciplines as developmental biology, virology, bacteriology, immunology, haematology and oncology must be questionable. A principled rejection of such conclusions necessitates a principled rejection of every branch of science that considers DNA in its purview.

Evolution, of course, may be transmogrified into a metaphysical system. This occurs when biological evolution is co-opted as the basis of a comprehensive world-view or of a system of ethics. But we have restricted ourselves to discussing the scientific evolutionary world-picture – upon which all of us, regardless of religion or world-view, should readily agree. Nevertheless, as people, we reflect also on what this genetic history might mean for our humanity. Does the comprehensive genetic story we have surveyed encompass all that needs to be said about our humanness?

Such reflection should attend to category boundaries and keep certain distinctions in mind. Biologically, we constitute a twig of the mammalian branch of the phylogenetic tree. It does not follow that ontologically, we are *merely* another one of the millions of species that comprise that tree. Physically we are apes, genetically continuous with lemurs and platypuses. It does not follow that metaphysically we are *nothing but* another kind of ape. Our distinctively human body plan is specified by our distinctively human gene complement. This is not the same thing as the dogmatic belief that we exist *only* to serve our genes, or that our life histories are genetically predetermined. People who would desist from monkey wars must distinguish between biological evolution and metaphysical extrapolations therefrom.

Genomic science has included a search for the genes 'that make us human'. That aim may engender anxiety, if not hostility, in people who see themselves as more than genes. More integrative assessments emphasise how humanness arises from the non-dissociable mix of genetics, environment and culture [1]. Our genes are necessary – but not sufficient – to account for our humanity. This may be illustrated by the inability of genetics alone to specify the function of two biological tissues that underlie our health and humanity. These tissues are responsible for immunity and conscious thought.

During the development of each individual, the immune and neural systems arise by evolutionary strategies involving Darwinian selection of cells into functional networks. Both immune and brain networks develop *only* in intimate interaction with the environment. The immune system requires cues from the microbial milieu in order to provide adequate surveillance for potential disturbances in the body. The neural system requires sensory input and social interaction with *persons* to develop into the substrate of a fully human personality. Neither system is genetically determined. The genes that specify immune and neural function require the tutelage of independent players in the outside world.

1 IMMUNE SYSTEMS

Our immune systems provide essential defence mechanisms against infectious diseases. Genes, of course, underlie all the functional potentialities of which our immune systems are capable (see Section 4.4.5). But genes encoding immune system components may malfunction in sterile germ-free environments. Those genes need to be schooled by products arising from the genes of other organisms if they are going to function appropriately. To be a healthy human is to integrate the action of one's own genes with those of innumerable others.

Humans throughout history have lived in intimate interaction with microbes. Some microbes are pathogenic, and these have always dominated our thinking about the microbial world. Words such as *plague, pestilence* and *pox* carry unhappy associations. But most microbes are fellow travellers, and we are unaware of their existence. They constitute the diverse commensal microflora that live innocuously in every nook and cranny of the surfaces of our bodies. The realisation is growing that many of them are beneficial to us. Their life-supporting effects are only now becoming apparent.

We frequently damage our skin. Cuts and abrasions are daily occurrences. The repair processes are in part regulated by bacteria that live in the skin: *Streptococcus epidermidis* is part of the repair process. It releases a product (lipoteichoic acid) that interacts with cellular TLR3 receptors and contributes to the control of inflammation and thence to the repair and ongoing integrity of the epidermal barrier that separates 'us' from the 'world out there'. For practical purposes these bugs are 'us' [2].

The same thing happens in the gut. Breaks in the lining of the intestine occur all the time. Repair is initiated in part by bacterial cell wall components called endotoxins (notorious for their damaging effects) interacting with cellular TLR4 receptors and inducing the synthesis of prostaglandins that orchestrate repair [3]. Indeed microbes in the intestine are implicated in the maintenance of structural, metabolic and immune normality by interacting with epithelial and immune cells [4]. Aggregates of lymphocytes and other cells mediating immunity (*lymphoid follicles*) are abundant in the wall of the intestine. They develop only in the presence of appropriate commensal bacteria and their released peptidoglycans [5].

It is likely that, during human evolution, our forebears and their microbial passengers worked out highly beneficial mutual relationships. Diverse communities of bacteria and worms lived with (pre-) humans in such a way as to be minimally provocative and disruptive. They became indispensable parts of our biology. We may rightly consider them to be 'old friends'. They secreted products, now dubbed *immunomodulins*, that acted as essential regulators of immune reactivity. These substances induce the production of a class of lymphocytes (regulatory T cells) that act as the 'fire extinguishers' of the immune system.

But the suspicion is growing among biomedical scientists that with urbanisation, affluence and increased hygiene-consciousness, we have distanced ourselves from many of our 'old friends'. In an ever more sanitised environment, their moderating influences have waned, and our immune systems have started to malfunction. This is an attractive explanation for the observation that, in wealthy over-developed countries, diseases of immunity are increasing in frequency. These include inflammatory bowel diseases, autoimmune diseases (type 1 diabetes, multiple sclerosis) and allergies (hay fever, asthma).

A spectrum of ideas that fall under the heading of the *hygiene* or *microbiota hypothesis* posit that these ailments of advanced civilisation are becoming more prevalent because our immune systems lack the tutelage offered by our former microbiological entourage [6].

Human genes encode the *potential* for an appropriately functioning immune system. Microbial genes allow that potential to be *realised.* An elaboration of this theme is that highly processed Western diets, deficient in fermentable plant fibre, affect the types of microbe in the gut and so indirectly promote asthma, autoimmune diseases and obesity [7].

The beneficial influences of 'old friends' have been shown experimentally in mice. The development of inflammatory bowel disease can be suppressed by the presence of a bacterium called *Bacteroides fragilis.* This microbe releases an immunomodulin (polysaccharide A) that promotes the function of regulatory T cells, those soothing moderators of immune reactivity [8]. Other bugs contribute similarly [9]. Pedigree *non-obese diabetic* mice spontaneously develop type 1 diabetes. Development of the disease can be suppressed if the mice are colonised with a group of human commensal bacteria, or if they are treated with material from one of several species of parasitic worms [10]. Experimental perturbations of the gut microflora affect the development of that inflammatory condition known as obesity [11].

Dramatic results have been obtained in humans as well. Patients suffering from inflammatory bowel diseases have been fed worm embryos as part of their medical treatment, and have subsequently shown clinical improvement [12]. Medical science is reaching new heights of sophistication with the transfer of faeces from the bowels of healthy people into the bowels of patients with chronic bowel infections, diabetes and ulcerative colitis. Bug therapy seems to help [13]. Patients with multiple sclerosis who were infected with worms during the course of their treatment had a much more favourable outcome than those who remained worm-free [14].

With respect to allergic disease, the safest environment is the microbially loaded context of the farm [15]. So could microbes be used to treat allergies? Some children at risk of developing eczema are protected if exposed to the commensal bacterium *Lactobacillus rhamnosus* during fetal development (by feeding the mother) and for the first two years of life [16]. Out-of-control immune systems really do seem to reform if reunited with 'old friends'.

People have widened the microbiota hypothesis. Synergising gut microbes may affect brain development and behaviour [17]. In overdeveloped countries, our estrangement from many species of commensal bacteria may promote the development of other inflammatory conditions including heart disease, cancer, neurodegeneration and depression [18]. Epidemiological evidence supports these speculations. As we age, our very personal assemblage of microbes reflects our diet and correlates with general vigour, muscle mass, mental outlook and our ability to live independently [19].

To be healthily human is to have an immune system trained by our microbial environment. Our own genes for immunity are necessary, but not sufficient, for optimal immune function. They must be *taught* how to function. They develop only by *experience*, in *relationship* with an invisible community of fellow travellers. Genes do not work in autonomous isolation.

2 NERVOUS SYSTEMS

Much work has been done to see whether the brains of humans and of other hominoids show different patterns of gene expression [20]. This fascinating question might tempt the unwary to presuppose a simple 'genes form brains' paradigm. But there is also the vital question of whether the brains of richly socialised humans and socially isolated humans might show differences in gene expression.

Just suppose we had identical twins: John enjoyed a rich network of family and social connections; George had lived alone in the jungle since infancy. Only John would have learned language and all that goes with it. Might these histories be reflected in their brain transcriptomes? Such gene expression experiments cannot be done, of course. Even if such genetically identical pairs existed, people's brains are not biopsied routinely.

Humans undergo delayed brain development (called *neoteny*) relative to other primates [21]. Some genes in the prefrontal cortex are expressed *later* in humans than in chimps. The difference in expression of these genes is maximal at about 10 years of age. This delay in brain development reflects an extended period of neuronal plasticity, allowing children a longer time to acquire knowledge and skills from caregivers. Suppose that George of the jungle's (human) genes for brain development operated in a chimp cultural *milieu*. His brain may have been configured very differently from John's. Genes make a human animal. Living in community may be required to make a human person.

The development of the brain is hugely flexible. It is not determined by genes alone. In a much-cited study, the brains of London taxi drivers were compared with those of non-taxi drivers, using an imaging method known as structural MRI [22]. The brains of London cabbies were shown to have distinctive anatomical features: the posterior hippocampus was increased in volume and the anterior hippocampus was reduced. These changes were correlated with the time spent driving taxis, suggesting that they developed in response to the experience of navigating around London. In other words, people did not become taxi drivers because they started off with distinctive hippocampi.

Similar studies have been performed on other groups of trained people. Musicians have distinctively developed structural features in their brains. These are located in the precentral gyrus, the corpus callosum and the cerebellum. The extent of these changes correlates with the age at which musical training commenced, indicating that the changes are a consequence, not a cause, of musicianship. Jugglers, university students cramming for exams and elite golfers have their own characteristic brain morphologies and functional characteristics. These changes reflect experience-dependent neural plasticity [23].

Autonomously acting genes are not the sole determinant of brain development. Genes function in environments. For many leading researchers, the old 'nature versus nurture' dichotomy is dead [24]. One cannot disentangle inherited and experiential influences on brain development. Simplistic either/or models have been replaced by new emphases on gene–environment interactions, genome regulation by experience, and brain plasticity. Genes function optimally in neural development only as they are instructed by multiple environmental cues: sensory stimulation, nurture, relationship, experience and training. Joan Stiles states that the 'emerging picture of brain development is of a dynamic and adaptive system that is constrained by *both* inherited factors and the experience of the organism' [25]. And if brain activity underlies mind, it follows that genes cannot fully determine how human minds develop.

2.1 *Critical periods*

Brain development in infancy is exquisitely sensitive to environmental influences. These influences may be physical (light, sound) or social (the non-quantifiable quality of attentive loving nurture). Appropriate stimuli are required during critical periods for the laying down of the relevant neural circuits. The outside world modifies the function of genes in nerves.

One might suppose that vision, at least, is hard-wired in the brain. But vision arises in infancy, when the appropriate brain circuits are formed in response to neural input from the eyes. If for any reason an eye is (reversibly) closed during a critical period in infancy, then the necessary neural circuits are not assembled, and vision is permanently impaired. This is the phenomenon of *amblyopia*, which affects up to 5% of the human population [26]. Conversely, people who are blind from birth, or who lose their vision during a critical period (the first three years of life), redeploy a part of the brain that usually responds to moving objects that are *seen*, so that it responds to moving objects that are *heard*. The middle temporal complex adapts to serve whatever modality of sensory reception feeds impulses to it [27].

Humans respond to faces – especially those that express fear. This may be part of a mechanism that alerts individuals to danger, and is probably functional by 7 months of age – just when infants become hazardously mobile. Particular neural circuits mediate this sensitivity to fearful faces. The basic circuitry is gene-specified, but requires real-life *experience* of people's faces to acquire its fully functional form. This experience-driven fine-tuning may be especially efficient during a sensitive period when individuals first encounter faces [28].

The development of the normal range of auditory sensitivity requires exposure to sounds of the corresponding range of wavelengths. Such exposures sculpt the requisite neuronal connections in the primary auditory cortex [29]. The acquisition of language from infancy requires neural circuitry that is laid down by experience. Exposure to language is required for the neural connections that allow babies to discriminate between phonemes (the fundamental units of speech), syllables, syntax and words. Such learning is maximally efficient during critical periods of development. It seems to be necessarily social [30]. There may be connections between language and thinking. We may think both in images (initially) and in words. One must suppose that to think in words requires the capacity for using language, which requires a cultural context. It follows that the capacity for human thought is at least partially dependent on social *milieu* [31]. Even the capacity to read (whether acquired as a child or an adult) affects the functional circuitry of the brain [32].

Multiple aspects of the social environment in which children are raised (parenting, neighbourhood influences, economic and socio-political context) affect early development. During critical periods in infancy, neurons become poised to respond to such inputs and form networks that support cognitive, sensory, muscular, emotional, behavioural and social competencies. Infancy is a critical period, when social inputs feature as lifelong determinants of health [33].

2.2 *Learning from neglect*

There is a new threat on the plains of Africa. Gang trouble. Family units have been disrupted, and groups of violent, antisocial adolescents are terrorising their neighbours. The gangs are composed of young male elephants and their victims are rhinos. Young elephants need their mothers and the input of extended family groups to be socialised. The culling of elephant herds has disrupted bonds between mothers and their offspring. It has led to the phenomenon of orphaned elephants that grow up to be delinquent – depressed, unpredictable and aggressive. Juvenile elephants that experience separation from their mothers, deprivation or trauma undergo perturbed brain development. The production of neurons, synapses and neurotransmitters is abnormal [34]. Elephants' genes produce optimally functioning elephants' brains only in the context of elephants' social networks.

Similar findings have accrued from studies of rodents, pigs and dogs. Early-life nurturing is essential for normal brain development. Maternal inattention, maternal separation or early weaning predisposes the offspring to later behavioural problems such as anxiety, aggressiveness and (in turn) poor mothering skills. Long-term learning and memory may be compromised. Such behavioural problems appear to reflect abnormalities in myelination, in the formation of dendritic spines, and in the production of neurons in the hippocampus. Males may be more vulnerable than females to behavioural damage from neglect [35]. In mice, there is a critical period (the fourth and fifth weeks) during which social interaction drives myelination and oligodendrocyte maturation in the medial prefrontal cortex, and establishes subsequent patterns of social exploration and memory [36].

Early-life anxiety generates long-term changes in the effects of glucocorticoid stress hormones. This is true whether one studies maternal inattention in rodents, or abuse and neglect of human infants [37]. Elevated stress hormones adversely modify brain development, and the consequences may extend (epigenetically) across generations. The neglect of young children deprives them of social and emotional inputs that are vital for their long-term development. Socioemotional deprivation has enduring effects on children's behaviour, as well as on the function and the very structure of their brains.

In one study, orphans who had spent most of their early lives in starkly impoverished institutions, and who were subsequently adopted and nurtured in families for several years, manifested normal executive functions (planning and decision-making) but performed poorly in at least some functions that are related to memory, attention and learning [38]. These observations suggest that there were persistent abnormalities in the circuitry of the prefrontal cortex and its connections with other areas.

Orphans who are institutionalised for the first few years of life, and who are given suboptimal levels of attention and care, often appear to be mentally retarded. When these children are adopted into supportive families, cognitive performance rapidly improves but subtle behavioural abnormalities may remain. The children tend to have reduced verbal skills, attention deficits and greater impulsivity. Sophisticated imaging techniques have shown that parts of these children's brains remain abnormal, both functionally (in terms of metabolic rate) [39] and structurally (in terms of the volume of cortical grey matter and the anatomy of connecting nerve tracts). In particular, a tract of nerve fibres known as the left uncinate fasciculus appears to be structurally altered [40].

Neglect of children is associated with altered anatomy, in adolescents, of the main nerve fibre tract in the brain, the corpus callosum [41]. Inadequate early-life nurturing predisposes people to an adulthood characterised by anxiety in social relationships. This is accompanied by alterations in neuronal structures, especially in brain areas implicated in emotional function [42]. Emotional neglect as children is associated with depression and reduced left hippocampal white matter volume as adults [43]. Genes alone do not determine the structure of white matter (nerve fibres) or grey matter (nerve cell bodies). Genes generate brains in the context of personal care.

Researchers try to distinguish the effects of socioemotional deprivation from other factors such as malnutrition and alcohol damage, although confounding factors cannot be rigorously excluded in human studies. Nevertheless, experimental studies on other primates, such as marmosets and macaques, designed to minimise confounding factors, have established that neglect is causally related to enduring abnormalities in behaviour, in brain structure and function, and (perhaps via the sympathetic nervous system) in long-term immune capability [44]. In captive macaques, smaller social networks (fewer animals per group) appear to lead to reductions in some grey matter areas and in certain neural networks (reduced coupling of activity in frontal and temporal cortex) [45].

Children from poor socioeconomic backgrounds experience more depression, anxiety, attention problems and conduct disorders than other children. The association between poverty and behavioural difficulties is multifactorial. Stressed parents may be less involved with their children, who may thus manifest neurological sequelae of neglect [46].

Adequate levels of care may be particularly important during a defined phase of brain development. Nurture may be needed during a *sensitive* or even *critical period*. Young children transferred from institutional neglect into caring foster families show a big catch-up in brain function and structure. This capacity to rebound declines as the age of transfer increases [47].

In summary, children's brains develop optimally in an environment characterised by stable emotional attachments with their caregivers. Human genes make healthy human persons in the social *milieu* of human love and attachment – a love that is learned and freely given, and that must be categorically different from genetically encoded instinctive altruism [48]. Genes, sculpted over millions of years of biological evolution, generate a neural primordium that attains its potential only in the context of personal qualities such as love and all the virtues that go with it. To be human is to have evolved to the stage at which genes provide the capacity for the expression of those qualities that constitute optimal levels of relationship, and that may be seen as not merely biological but as moral and spiritual.

3 FEATURES OF PERSONHOOD

Being fully human entails a capacity known as *mentalisation* [49]. Our own mental abilities allow us to perceive and interpret mental states in other people. As we observe how other people act, we can attribute to them feelings, thoughts, desires, goals and beliefs. It seems that we acquire such understandings as a 'developmental achievement'. Mentalisation arises from the earliest relationships that we form with our first caregivers. It is not a 'constitutional given'. We are not born with it. Our genes confer upon us the capacity to develop it, but do not specify it. Inadequate relationships with early-life caregivers may lead to abnormalities of mentalisation, and may be factors in the development of mental disease.

A somewhat more narrowly defined capacity is known as *theory of mind* (ToM). This is the ability to recognise that other people have a mind like one's own, and that their understanding of a situation might be different from one's own, as it depends on their own experience. Children who have experienced inadequate early-life nurturing show impaired ToM – a risk factor for enduring psychological problems. Improved caregiver input enhances the development of ToM in preschool children [50].

The acquisition of language may be important for the development of ToM [51]. This hypothesis arises from the study of a community of deaf people in Nicaragua. These people developed a sign language spontaneously and for the first time. The language has been developing progressively over several decades. Early versions of the language lacked signs for mental states (verbs for belief and knowledge, such as *think* and *know*), and people who possessed this basic language had only a limited understanding of how others might act in situations in which they lacked the observer's knowledge. The subsequent development of a more elaborate language with mental-state vocabulary conferred upon individuals the capacity to anticipate others' actions. Language, in other words, is important for the development of ToM. The acquisition of ToM is not innate, but depends on social interaction.

The role of culture and language in the development of ToM has been shown using functional MRI to identify those parts of the brain that are metabolically stimulated when people are involved in certain mental activities [52]. In both monolingual American and bilingual Japanese children, ToM is associated with the medial prefrontal cortex and precuneus. But other areas are language- or culture-dependent: the temporo-parietal junction (in Americans) and the inferior frontal gyrus (in Japanese). There was variation in brain areas used in ToM depending on whether Japanese people had learned English as children or as adults.

Functional MRI has shown that brain areas associated with ToM are involved also in religious thinking. Relational and religious thinking share at least some neural circuits [53]. This finding resonates with spiritual frameworks such as those that are central to Christian thought, which emphasises the psychosomatic unity of human being, eschews mind/spirit–body dualism and emphasises the centrality of relationship in religious experience [54].

Humans can orientate their actions according to long-term goals. *Intentionality* is the capacity to entertain goals (and the plans on how to attain them) that extend beyond current sensory experience. Philippe Rochat proposes that 'intentionality emerges as the by-product of reciprocal social exchanges' [55]. In this framework, intentionality arises from the *interpersonal* realm. It is not innate (genetically specified) but is *learned* from others.

Peter Steeves has proposed that the burgeoning consciousness of the human infant will not develop human intentionality on its own, but requires the presence of a 'significant other' who is human. To *develop* human intentionality, one must be treated as one who *has* human intentionality. To be human is to be treated by humans as human. Human caregivers attending to infants are responsible for enabling human persons [56]. Such an understanding of human development is seen in the Xhosa saying '*umuntu ngumuntu ngabantu*': a person becomes a person through persons [57]. Relationship is required for the realisation of human potential. Genes provide the potential for personhood, but cannot suffice to make us human. The marvel of the human genome is that it supports *social* learning [58]. Western individualism may be a distorted metaphysical lens through which to interpret the nature of humanness.

4 STORIES AND NARRATIVE IDENTITY

In a period spanning the 1940s to the 1960s, thousands of children were raised under dehumanising conditions in orphanages in Quebec, Canada. Known as the 'orphans of Duplessis' (after the State Premier of the time) many of them experienced emotional hardship during adulthood. For selected individuals, the interplay between childhood experience, natural temperament and adult influence has been explored by following their stories. Bob, for example, was shunted between several institutions, could recall few positive relationships with staff, and was often hit, verbally abused or locked up. But he refused to be cowed. He found solace in music and, as an adult, relief with supportive families. His background threw a long shadow over his employment, in which his desire for independence meant that he often struggled to meet the expectations of others [59].

Psychiatrists have used the power of stories to elucidate the effects of trauma, such as early-life neglect and abuse, on subsequent emotional and social development. Stories 'communicate experience, connect people to their past, and bring meaning to their present and future' [60]. People's stories may elucidate the basis of 'organic' diseases [61]. The ability to compose stories – narratives that integrate and interpret events – may be intrinsic to our humanity and to the normal workings of our minds. Narratives may be fundamental constituents of conscious thought and of the way in which people communicate their thoughts to others. One of the functions of conscious thought may be to construct experience as stories: internal 'movies' that simulate experience. Such narratives are vital ways of understanding the past and planning for the future [62].

The particularity of stories affects our personal development throughout life [63]. Research indicates that we are authors, constructing evolving stories of our lives. That is, we develop a *life story* or *narrative identity* that is integral to the concept of *self*. Parental discourse through infancy is essential for the development of the self-story in the child. We construct our narrative identities on the basis of stories learned from parents, family and the wider society. It has been suggested that

> stories are the substance of the self ... To witness the construction of situated stories is to understand the dynamic development of the self. Indeed, sharing stories is the mechanism through which people become selves. [64]

The importance of stories explains why, in the heady year during which the human genome study was published (2001), Alex Mauron could write:

> To be a human person means more than having a human genome, it means having a narrative identity of one's own. Likewise, membership in the human family involves a rich nexus of cultural links that cannot be reduced to taxonomy. On the question of human nature, we need a philosophical fresh start that cannot be provided by genomics alone. [65]

The flood of spectacular findings issuing from genomics research provides an ever-present temptation to disregard these cultural aspects of humanness.

Stories contribute to the formation of our world-views: the frameworks by which we perceive and interpret the world. We all possess a world-view, whether or not we are aware of it. Theologian Tom Wright suggests that a key feature of all world-views is the element of story. Our lives are 'grounded in and constituted by the implicit or explicit stories which humans tell themselves and one another'. Indeed world-views and their stories are normative: they claim to make sense of reality. Stories articulate, legitimate and support, modify, challenge, subvert or even destroy world-views [66].

Stories to which we are attentive effectively structure our experiences, guide and orient our lives, and shape our senses of virtue, value, vision and obligation [67]. It is stories that form and sustain the moral life: maxims and precepts are merely skeletal abstractions of the contents of stories. Stories give coherence to the intentionality that orients our lives, and so are important factors in moulding the consistency of our behaviour. That is, stories are character-forming. As Hauerwas states, '*theories* are meant to help you know the world without changing the world yourself' but '*stories* help you deal with the world by changing it through changing yourself' [68].

Genes cannot induce virtue in individuals, create loving and supportive families and communities, engender integrity in lawyers or even in scientists, generate just societies, or sustain democracies. It is the stories we assimilate that direct the courses of our lives. Prevailing stories may describe the inane world of celebrity culture. They may portray compunction-free strategies for amassing wealth. They may glamorise military conquest and the subjugation of others. Individuals and communities that regale themselves with such stories end up by re-enacting them.

But there are stories that take us by surprise, haunt, inspire and potentially ennoble us. We may stumble upon stories that are counterintuitive but that address us with compelling authority. One may be disturbed, excited or even enraged. This is where religiously motivated opposition to biological evolution seems so incongruous. No branch of biology can be expected to provide a formative, humanising narrative. Should not the constituents of the anti-evolution lobby instead direct their energies to the narrative summarised as: 'Love your enemies, do good to those who hate you' [69] – a perennial challenge issuing from a historical context of injustice, foreign militaristic oppression and seething revolutionary ferment?

This story is embodied in the injunction to 'love your neighbour as you love yourself' – where *neighbour* was defined as anyone to whom one might show compassion, regardless of the most-enduring or deep-seated traditions of tribal, cultural or political animosity

[70]. Holmes Rolston has said of an individual who fulfils this moral imperative [71]:

> Such a person acts, on the moral account, intending to benefit others at cost to himself, and, on the genetic account increasing the likelihood of the aided person's having offspring over one's having them.

This maximally inclusive understanding of neighbour-love represents an ethic that is foreign to the mindset of, and the institutions created by, *Homo economicus* or *H. industrialis*. It stands in radical contrast to the destructive stories and attitudes that legitimate self-interest and sustain long-lasting hostilities between communities.

This story inverted human values with the challenge, 'If one of you wants to be great, he must be the servant of the rest, and if one of you wants to be first, he must be the slave of all' [72]. This story transcends and contravenes the logic of genetics. It presents people with the moral choice to disregard their own genetic fitness (reproductive success) for the sake of that of non-reciprocating, even hostile, strangers. Science provides indicatives (the oceans are acidifying); stories provide imperatives (my consumptive lifestyle is selfish).

We should be alert for any tacit assumption that the possession of a human genome guarantees the perpetuation, let alone the continual progress, of human civilisation. The evolutionary geneticist Svante Paabo has argued as follows.

> We need to leave behind the view that the genetic history of our species is *the* history par excellence. We must realise that our genes are but one aspect of our history, and that there are many other histories that are even more important ... it is a delusion to think that genomics in isolation will ever tell us what it means to be human. To work toward that lofty goal, we need an approach that includes the cognitive sciences, primatology, the social sciences, and the humanities. [73]

Evolutionary geneticist Francisco Ayala has quipped that 'In matters of value, meaning and purpose, science has all the answers except the interesting ones.' The scientific story does not reveal the world of *persons*. Again, to quote Ayala: 'Successful as it is, and universally encompassing as its subject is, a scientific view of the world is hopelessly incomplete' [74]. If we are to defuse the evolution wars, scientists who pronounce on others' stories (such as the formative history of Christianity) must show the same informed openness to that history as they would require of a religious non-scientist who pronounces on biological history.

Our genomes embody a fascinating and highly enlightening story of our evolutionary development and biological history. This story has been explicated by rigorous empirical investigation. Perhaps there is a tendency among scientists, obsessively absorbed in their fields of study, to see humanity only in terms of those preoccupying academic disciplines. Professional tunnel vision may make us oblivious to our own formative narratives and uncritically dismissive of those of others. It is easy to forget that we are formed as persons by the particularities of human history, and these should be assessed on their own merits. A critical openness is needed when we investigate histories, for when it comes to being human, genetics is not the whole story.

References

PROLOGUE

1 Waller J, *Fabulous Science: Fact and Fiction in the History of Scientific Discovery* (Oxford: Oxford University Press, 2002), 181–4

2 Padian K (2008). Darwin's enduring legacy. *Nature* **451**, 632–4; Bowler BJ (2009). Darwin's originality. *Science* **323**, 223–6

3 Vollbrecht E and Sigmon B (2005). Amazing grass: developmental genetics of maize domestication. *Biochemical Society Transactions* **33**, 1502–6; Tian F, Stevens NM and Buckler ES IV (2009). Tracking footprints of maize domestication and evidence for a massive selective sweep on chromosome 10. *Proceedings of the National Academy of Sciences of the USA* **106** (Suppl. 1), 9979–86

4 National Academy of Sciences, *Science, Evolution, and Creationism* (Washington DC: National Academies Press, 2008), available at www.nap.edu/catalog/11876.html

5 Waller, *Fabulous Science*, 199–202

6 Polkinghorne J, *Science and Creation* (London: SPCK, 1988), 54

7 Jeeves M, quoted in Henry CF (ed.), *Horizons of Science: Christian Scholars Speak Out* (San Francisco: Harper & Row, 1978), 29

8 MacKay DM, *The Clockwork Image: A Christian Perspective on Science* (London: Intervarsity Press, 1974), 54

9 Spencer N, *God and Darwin* (London: SPCK, 2009), 48, 78–9, 82–4, 124

10 See http://ncse.com/files/pub/legal/kitzmiller/highlights/2005-12-20-Kitzmiller-decision.pdf; see also Editorial (2005). Dealing with design. *Nature* **434**, 1053; Brumfiel G (2005). Intelligent design: who has designs on your students' minds? *Nature* **434**, 1062; Leshner AI (2005). Redefining science. *Science* **309**, 221

11 Brooke JH (2008). Charles Darwin on religion: a statement prepared for the International Society for Science and Religion, available at www.issr.org.uk/issr-statements/charles-darwin-on-religion; Shapin S (2010). The Darwin Show. *London Review of Books* **32**, 3, available at www.lrb.co.uk/v32/n01/steven-shapin/the-darwin-show

12 Russel CA, *Cross-Currents: Interactions between Science and Faith* (Leicester: Intervarsity Press, 1985), 150

13 Livingstone DN, *Darwin's Forgotten Defenders* (Grand Rapids, MI: Eerdmans, 1987), 118

14 Including John Stek, Bruce Waltke and John Walton. See for example Van Till HJ, Snow RE, Stek JH and Young DA, *Portraits of Creation* (Grand Rapids, MI: Eerdmans, 1990)

15 Wenham GJ, *Word Biblical Commentary: Genesis 1–15* (Nashville: Thomas Nelson, 1987); Waltke BK and Fredricks CJ, *Genesis: A Commentary* (Grand Rapids, MI: Zondervan, 2001); Wenham GJ, *Exploring the Old Testament: Volume 1 The Pentateuch* (London: SPCK, 2003); Walton JH, *The Lost World of Genesis One* (Downers Grove, IL: Intervarsity Press, 2009)

16 Stott JRW, *Understanding the Bible* (London: Scripture Union, 1984), 48–9; Livingstone DN and Noll MA (2000). B.B. Warfield (1851–1921): a biblical inerrantist as evolutionist. *Isis* **91**, 283–304; Wright NT, www.ntwrightpage.com/sermons/Wisdom_Troubled_Time.htm; Bauckham R, *Bible and Ecology* (London: Darton, Longman & Todd, 2010), 158–60

17 Livingstone, *Defenders*, 102–5; Lindberg DC and Numbers RL, *God and Nature: Historical Essays on the Encounter between Christianity and Science* (Berkeley, CA: University of California Press, 1986), 376–7

18 Lindberg and Numbers, *God and Nature*, 402–3

19 Hunter GW, *A Civic Biology: Presented in Problems*. (New York: American Book Company, 1914), 196, 261–3

20 Spencer, *God and Darwin*, 87–98

21 For example, Wang HH, Isaacs FJ, Carr PA *et al.* (2009). Programming cells by multiplex genome engineering and accelerated evolution. *Nature* **460**, 894–8; Worsdorfer B, Woycechowsky KJ and Hilvert D (2011). Directed evolution of a protein container. *Science* **331**, 589–92

22 Haught J, in Conway Morris S (ed.), *The Deep Structure of Biology: Is Convergence Sufficiently Ubiquitous to Give a Directional Signal?* (West Conshohocken, PA: Templeton Foundation Press, 2008), 218

23 Polkinghorne, *Science and Creation*, 47–50; Burge T (2002). What else does physics tell us about God? *Science and Christian Belief* **14**, 79–80; Gingerich O, *God's Universe* (Cambridge, MA: Harvard University Press, 2006), 118–20; Conway Morris, *Deep Structure*, 228–30; Ewart P (2009). The necessity of chance: randomness, purpose and the sovereignty of God. *Science and Christian Belief* **21**, 111–31

24 Vermeij GJ (2006). Historical contingency and the purported uniqueness of evolutionary innovations. *Proceedings of the National Academy of Sciences of the USA* **103**, 1804–9

25 Conway Morris S, *Life's Solution: Inevitable Humans in a Lonely Universe* (Cambridge: Cambridge University Press, 2003); Conway Morris, *Deep Structure*, 46

26 Ohlrich C, *The Suffering God* (London: Triangle, 1982); Polkinghorne J, *The Way the World Is* (London: SPCK, 1983), 69–77; Rolston H III, *Genes, Genesis and God* (Cambridge: Cambridge University Press, 1999), 303–7

27 See Wright, ref. 16

28 For exceptions, see Dingli D and Nowak MA (2006). Infectious tumour cells. *Nature* **443**, 35–6; Belov K (2011). The role of the major histocompatibility complex in the spread of contagious cancers. *Mammalian Genome* **22**, 83–90

29 Katzir N, Rechavi G, Cohen GB *et al.* (1985). 'Retroposon' insertion into the cellular oncogene c-*myc* in canine transmissible venereal tumor. *Proceedings of the National Academy of Sciences of the USA* **82**, 1054–8; Murgia C, Pritchard JK, Kim SY *et al.* (2006). Clonal origin and evolution of a transmissible cancer. *Cell* **126**, 477–87; Rebbeck CA, Thomas R, Breen M *et al.* (2009). Origins and evolution of a transmissible cancer. *Evolution* **63**, 2340–9; Rebbeck CA, Leroi AM and Burt A (2011). Mitochondrial capture by a transmissible cancer. *Science* **331**, 303

30 Pearse A-M and Swift K (2006). Transmission of devil facial-tumour disease. *Nature* **439**, 549; Murchison EP, Tovar C, Hsu A *et al.* (2010). The Tasmanian devil transcriptome reveals Schwann cell origins of a clonally transmissible cancer. *Science* **327**, 84–7; Murchison EP, Schulz-Trieglaff OB, Ning Z *et al.* (2012). Genome sequencing and analysis of the Tasmanian devil and its transmissible cancer. *Cell* **148**, 780–91; for overviews, see Jones ME and McCallum H (2011). The devil's cancer. *Scientific American* **304**(6), 72–7; Belov K (2012). Contagious cancer: lessons from the devil and the dog. *BioEssays* **34**, 285–92

31 Gill P, Ivanov PL, Kimpton C *et al.* (1994). Identification of the remains of the Romanov family by DNA analysis. *Nature Genetics* **6**, 130–5; Ivanov PL, Wadhams MJ, Roby RK *et al.* (1996). Mitochondrial DNA sequence heteroplasmy in the Grand Duke of Russia Georgij Romanov establishes the authenticity of the remains of Tsar Nicholas II. *Nature Genetics* **12**, 417–20

32 Coble MD, Loreille OM, Wadhams MJ *et al.* (2009). Mystery solved: the identification of the two missing Romanov children using DNA analysis. *PLoS ONE* **4**, e4838; Rogaev EI, Grigorenko AP, Moliaka YK *et al.* (2009). Genomic identification in the historical case of the Nicholas II royal family. *Proceedings of the National Academy of Sciences of the USA* **106**, 5258–63

33 Rogaev EI, Grigorenko AP, Faskhutdinova G *et al.* (2009). Genotype analysis identifies the cause of the 'royal disease'. *Science* **326**, 817; Lannoy N and

Hermans C (2010). The 'royal disease' – haemophilia A or B? A haematological mystery is finally solved. *Haemophilia* **16**, 843–7

34 Coble MD (2011). The identification of the Romanovs: can we (finally) put the controversies to rest? *Investigative Genetics* **2**, 20

35 Avise JC, *Inside the Human Genome: A Case for Non-Intelligent Design* (Oxford: Oxford University Press, 2010); particularly recommended is Fairbanks DJ, *Relics of Eden: The Powerful Evidence of Evolution in Human DNA* (Amherst, MA, and New York: Prometheus Books, 2007); from a more theological perspective, see Alexander DR, *Creation and Evolution: Do We Have to Choose?* (Oxford: Monarch, 2008); Pattemore P, *Am I My Keepers Brother?: Human Origins from a Christian and a Scientific Perspective* (see www.amimykeepersbrother.com, 2011)

36 Finlay GJ (2003). *Homo divinus*: the ape that bears God's image. *Science and Christian Belief* **15**, 17–40; Finlay GJ (2008). Evolution as created history. *Science and Christian Belief* **20**, 67–89; Finlay GJ (2008). Human evolution: how random process fulfils divine purpose. *Perspectives on Science and Christian Faith* **60**(2), 103–14; Finlay GJ. The emergence of human distinctiveness: the genetic story, in Jeeves M (ed.), *Rethinking Human Nature: a Multidisciplinary Approach* (Grand Rapids, MI: Eerdmans, 2011)

1 RETROVIRAL GENEALOGY

1 Weinberg RA, *The Biology of Cancer* (New York: Garland Science, 2007)

2 Boccardo E and Villa LL (2007). Viral origins of human cancer. *Current Medicinal Chemistry* **14**, 2526–39; Dayaram T and Marriott SJ (2008). Effect of transforming viruses on molecular mechanisms associated with cancer. *Journal of Cellular Physiology* **216**, 309–14; Moore MS and Chang Y (2010). Why do viruses cause cancer? Highlights of the first century of human tumour virology. *Nature Reviews Cancer* **10**, 878–89

3 Rubin H (2011). The early history of tumor virology: Rous, RIF, and RAV. *Proceedings of the National Academy of Sciences of the USA* **108**, 14389–96

4 Mortreux F, Gabet A-S and Wattel E (2003). Molecular and cellular aspects of HTLV-1, associated leukemogenesis *in vivo*. *Leukemia* **17**, 26–38; Karpas A (2004). Human retroviruses in leukaemia and AIDS: reflections on their discovery, biology and epidemiology. *Biological Reviews of the Cambridge Philosophical Society* **79**, 911–33; Verdonck K, Gonzalez E, Van Dooren S *et al.* (2007). Human T-lymphotropic virus 1: recent knowledge about an ancient infection. *Lancet Infectious Diseases* **7**, 266–81; Boxus M and Willems L (2009). Mechanisms of HTLV-1 persistence and transformation. *British Journal*

of Cancer **101**, 1497–501; Goncalves DU, Proietti FA, Ribas JGR *et al.* (2010). Epidemiology, treatment, and prevention of human T-cell leukemia virus type 1-associated diseases. *Clinical Microbiological Reviews* **23**, 577–89

5 Leclercq I, Mortreux F, Cavrois M *et al.* (2000). Host sequences flanking the human T-cell leukemia virus type 1 provirus in vivo. *Journal of Virology* **74**, 2305–12

6 Chou KS, Okayama A, Su I-J *et al.* (1996). Preferred nucleotide sequence at the integration target site of human T-cell leukemia virus type I from patients with adult T-cell leukemia. *International Journal of Cancer* **65**, 20–4

7 Seiki M, Hattori S, Hirayama Y and Yoshida M (1983). Human adult T-cell leukemia virus: complete nucleotide sequence of the provirus genome integrated into leukemia cell DNA. *Proceedings of the National Academy of Sciences of the USA* **80**, 3618–22

8 Hacein-Bey-Abina S, Von Kalle C, Schmidt M *et al.* (2003). *LMO2*-associated clonal T cell proliferation in two patients after gene therapy for SCID-X1. *Science* **302**, 415–19; Deichmann A, Hasein-Bey-Abina S, Schmidt M *et al.* (2007). Vector integration is nonrandom and clustered and influences the fate of lymphopoiesis in SCID-X1 gene therapy. *Journal of Clinical Investigation* **117**, 2225–32; Kaiser J (2009). β-Thalassemia treatment succeeds, with a caveat. *Science* **326**, 1468–9

9 Hayward WS, Neel BG and Atrin SM (1981). Activation of a cellular *onc* gene by promoter insertion in ALV-induced lymphoid leukosis. *Nature* **290**, 475–80; Peters G, Lee AE and Dickson C (1984). Activation of cellular gene by mouse mammary tumour virus may occur early in mammary tumour development. *Nature* **309**, 273–5; Nusse R, Van Ooyen A, Rijsewijk F *et al.* (1985). Retroviral insertional mutagenesis in murine mammary cancer. *Proceedings of the Royal Society of London Series B* **226**, 3–13; Moules V, Pomier C, Sibon D *et al.* (2005). Fate of premalignant clones during the asymptomatic phase preceding lymphoid malignancy. *Cancer Research* **65**, 1234–43

10 Westaway D, Payne G and Varmus HE (1984). Proviral deletions and oncogene base-substitutions in insertionally mutagenized c-myc alleles may contribute to the progression of avian bursal tumors. *Proceedings of the National Academy of Sciences of the USA* **81**, 843–7

11 Vinokurova S, Wentzensen N, Einenkel J *et al.* (2005). Clonal history of papillomavirus-induced dysplasia in the female lower genital tract. *Journal of the National Cancer Institute* **97**, 1816–21; see also Luft F, Klaes R, Nees M *et al.* (2001). Detection of integrated papillomavirus sequences by ligation-mediated PCR (DIPS-PCR) and molecular characterization in cervical cancer cells. *International Journal of Cancer* **92**, 9–17

12 Ng IO-L, Guan X-Y, Poon R T-P *et al.* (2003). Determination of the molecular relationship between multiple tumour nodules in hepatocellular carcinoma differentiates multicentric origin from intrahepatic metastases. *Journal of Pathology* **199**, 345–53; Saigo K, Yoshida K, Ikeda R *et al.* (2008). Integration of hepatitis B virus DNA into the myeloid/lymphoid or mixed lineage leukaemia (MLL4) gene and rearrangements of MLL4 in human hepatocellular carcinoma. *Human Mutation* **29**, 703–8

13 Feng H, Shuda M, Chang Y and Moore PS (2008). Clonal integration of a polyomavirus in human Merkel cell carcinoma. *Science* **319**, 1096–100; Gandhi RK, Rosenberg AS and Somach SC (2009). Merkel cell polyomavirus: an update. *Journal of Cutaneous Pathology* **36**, 1327–9

14 Weiss RA (2006). The discovery of endogenous retroviruses. *Retrovirology* **3**, 67; Stoye JP (2012). Studies of endogenous retroviruses reveal a continuing evolutionary saga. *Nature Reviews Microbiology* **10**, 395–406; Feschotte C and Gilbert C (2012). Endogenous viruses: insights into viral evolution and impact on host biology. *Nature Reviews Genetics* **13**, 283–96

15 Lower R, Lower J and Kurth R (1996). The viruses in all of us: characteristics and biological significance of human endogenous retrovirus sequences. *Proceedings of the National Academy of Sciences of the USA* **93**, 5177–84; Bromham L (2002). The human zoo: endogenous retroviruses in the human genome. *Trends in Ecology and Evolution* **17**, 91–7; Bannert N and Kurth R (2004). Retroelements and the human genome: new perspectives on an old relation. *Proceedings of the National Academy of Sciences of the USA* **101** (Suppl. 2), 14572–9; Kurth R and Bannert N (2010). Beneficial and detrimental effects of human endogenous retroviruses. *International Journal of Cancer* **126**, 306–14

16 Smit AFA (1999). Interspersed repeats and other mementos of transposable elements in mammalian genomes. *Current Opinion in Genetics and Development* **9**, 657–63; Gifford R and Tristem M (2003). The evolution, distribution and diversity of endogenous retroviruses. *Virus Genes* **26**, 291; Mayer J and Meese E (2005). Human endogenous retroviruses in the primate lineage and their influence on host genomes. *Cytogenetic and Genome Research* **110**, 448–56; Blikstad V, Benachenhou F, Sperber GO and Blomberg J (2008). Evolution of human endogenous retroviral sequences: a conceptual account. *Cellular and Molecular Life Sciences* **65**, 3348–65; Blomberg J, Benachenhou F, Blikstad V *et al.* (2009). Classification and nomenclature of endogenous retroviral sequences (ERVs): problems and recommendations. *Gene* **448**, 115–23

17 Bonner TI, O'Connell C and Cohen M (1982). Cloned endogenous retroviral sequences from human DNA. *Proceedings of the National Academy of*

Sciences of the USA **79**, 4709–13; see also Mariani-Constantini R, Horn TM and Callahan R (1989). Ancestry of a human endogenous retrovirus family. *Journal of Virology* **63**, 4982–5

18 Johnson WE and Coffin JM (1999). Constructing primate phylogenies from ancient retrovirus sequences. *Proceedings of the National Academy of Sciences of the USA* **96**, 10254–60

19 Lindeskog M, Mager DL and Blomberg J (1999). Isolation of a human endogenous retroviral HERV-H element with an open *env* reading frame. *Virology* **258**, 441–50; de Parsival N, Casella J-F, Gressin L and Heidmann T (2001). Characterisation of the three HERV-H proviruses with an open envelope reading frame encompassing the immunosuppressive domain and evolutionary history in primates. *Virology* **279**, 558–69; Benit L, Calteau A and Heidmann T (2003). Characterization of the low-copy HERV-Fc family: evidence for recent integration in primates of elements with coding envelope genes. *Virology* **312**, 159–68

20 Barbulescu M, Turner G, Seaman MI *et al.* (1999). Many human endogenous retrovirus K (HERV-K) proviruses are unique to humans. *Current Biology* **9**, 861–8

21 Turner G, Barbulescu M, Su M *et al.* (2001). Insertional polymorphisms of full-length endogenous retroviruses in humans. *Current Biology* **11**, 1531–5; Jha AR, Pillai SK, York VA *et al.* (2009). Cross-sectional dating of novel haplotypes of HERV-K 113 and HERV-K 115 indicate these proviruses originated in Africa before *Homo sapiens*. *Molecular Biology and Evolution* **26**, 2617–26

22 Jha AR, Nixon DF, Rosenberg MG *et al.* (2011). Human endogenous retrovirus K106 (HERV-K106) was infectious after the emergence of anatomically modern humans. *PLoS ONE* **6**, e20234

23 Agoni L, Golden A, Guha C and Lenz J (2012). Neandertal and Denisovan retroviruses. *Current Biology* **22**, R437–8

24 Subramanian RP, Wildschutte JH, Russo C and Coffin JM (2011). Identification, characterization, and comparative genomic distribution of the HERV-K (HML-2) group of human endogenous retroviruses. *Retrovirology* **8**, 90

25 Lebedev YB, Belonovitch OS, Zybrova NV *et al.* (2000). Differences in HERV-K LTR insertions in orthologous loci of humans and the great apes. *Gene* **247**, 265–77; Sverdlov ED (2000). Retroviruses and primate evolution. *BioEssays* **22**, 161–71; Kurdyukov SG, Lebedev YB, Artamonova II *et al.* (2001). Full-sized HERV-K (HML-2) human endogenous retroviral LTR sequences on human chromosome 21: map locations and evolutionary history. *Gene* **273**, 51–61; Nadezhdin EV, Lebedev YB, Glazkova DV *et al.* (2001). Identification of paralogous HERV-K LTRs on human chromosomes 3, 4, 7 and 11 containing clusters of olfactory receptor genes. *Molecular Genetics and Genomics* **265**, 820–5

26 Sin H-S, Koh E, Taya M *et al.* (2011). A novel Y chromosome microdeletion with the loss of an endogenous retrovirus-related, testis-specific transcript in the AZFb region. *Journal of Urology* **186**, 1545–52

27 Subramanian *et al.* (2011), ref. 24; see also Hughes JF and Coffin JM (2004). Human endogenous retrovirus K solo-LTR formation and insertional polymorphisms: implications for human and viral evolution. *Proceedings of the National Academy of Sciences of the USA* **101**, 1668–72; Belshaw R, Watson J, Katzourakis A *et al.* (2007). Rate of recombinational deletion among human endogenous retroviruses. *Journal of Virology* **81**, 9437–42

28 Huh J-W, Kim D-S, Ha H-S *et al.* (2007). Formation of a new solo-LTR of the human endogenous retrovirus H family in human chromosome 21. *Molecules and Cells* **22**, 360–3

29 Choi J, Koh E, Matsui F *et al.* (2008). Study of azoospermia factor-a deletion caused by homologous recombination between the human endogenous retroviral elements and population-specific alleles in Japanese infertile males. *Fertility and Sterility* **89**, 1177–82

30 Hughes JF and Coffin JM (2001). Evidence for genomic rearrangements mediated by human endogenous retroviruses during primate evolution. *Nature Genetics* **29**, 487–9; Hughes JF and Coffin JM (2005). Human endogenous retroviral elements as indicators of ectopic recombination events in the primate genome. *Genetics* **171**, 1183–94

31 Lindeskog M, Medstrand P, Cunningham AA and Blomberg J (1999). Coamplification and dispersion of adjacent human endogenous retroviral HERV-H and HERV-E elements: presence of spliced hybrid transcripts in normal leukocytes. *Virology* **244**, 219–29

32 Jamain S, Girondot M, Leroy P *et al.* (2001). Transduction of the human gene *FAM8A1* by endogenous retrovirus during primate evolution. *Genomics* **78**, 38–49

33 International Human Genome Sequencing Consortium (2001). Initial sequencing and analysis of the human genome. *Nature* **409**, 860–921; Li W-H, Gu Z, Wang H and Nekrutenko A (2001). Evolutionary analyses of the human genome. *Nature* **409**, 847–9

34 Chimpanzee Sequencing and Analysis Consortium (2005). Initial sequence of the chimpanzee genome and comparison with the human genome. *Nature* **437**, 69–87

35 Prufer K, Munch K, Hellmann I *et al.* (2012). The bonobo genome compared with the human and chimpanzee genomes. *Nature* **486**, 527–31 (for ERVs and TEs, see Supplementary Information)

36 Ventura M, Catacchio CR, Alkan C *et al.* (2011). Gorilla genome structural variation reveals evolutionary parallelisms with chimpanzee. *Genome*

Research **21**, 1640–9; for a genome analysis (without reference to ERVs), see Scally A, Dutheil JY, Hillier LW *et al.* (2012). Insights into hominid evolution from the gorilla genome sequence. *Nature* **483**, 169–75

37 Locke DP, Hillier LW, Warren WC *et al.* (2011). Comparative and demographic analysis of orang-utan genomes. *Nature* **469**, 529–33 (for ERVs, see Supplementary Information)

38 Rhesus Macaque Genome Sequencing and Analysis Consortium (2007). Evolutionary and biomedical insights from the rhesus macaque genome. *Science* **316**, 222–34; Han K, Konkel MK, Xing J *et al.* (2007). Mobile DNA in Old World monkeys: a glimpse through the rhesus macaque genome. *Science* **316**, 238–40

39 Meyer M, Kircher M, Gansauge M-T *et al.* (2012). A high-coverage genome sequence from an archaic Denisovan individual. *Science* **338**, 222–6

40 Green RE, Krause J, Briggs AW *et al.* (2012). A draft sequence of the Neandertal genome. *Science* **328**, 710–22

41 Kijima TE and Innan H (2010). On the estimation of the insertion time of retrotransposable elements. *Molecular Biology and Evolution* **27**, 896–904

42 de Parsival N and Heidmann T (2005). Human endogenous retroviruses: from infectious elements to human genes. *Cytogenetic and Genome Research* **110**, 318–32

43 Herve CA, Forrest G, Lower R *et al.* (2004). Conservation and loss of the *ERV3* open reading frame in primates. *Genomics* **83**, 940–3

44 Aagaard L, Villesen P, Kjeldbjerg AL and Pedersen FS (2005). The approximately 30-million-year-old ERVPb1 envelope gene is evolutionarily conserved among hominoids and Old World monkeys. *Genomics* **86**, 685–91; Kjeldbjerg AL, Villesen P, Aagaard L and Pedersen FS (2008). Gene conversion and purifying selection of a placenta-specific ERV-V envelope gene during simian evolution. *BMC Evolutionary Biology* **8**, 266; for ERV-Pb1 envelope fusogenic capacity, see Aagaard L, Bjerregaard B, Kjeldbjerg AL *et al.* (2012). Silencing of endogenous envelope genes in human choriocarcinoma cells shows that envPb1 is involved in heterotypic cell fusions. *Journal of General Virology* **93**, 1696–9

45 Bonnaud B, Beliaeff J, Bouton O *et al.* (2005). Natural history of the ERVWE1 endogenous retroviral locus. *Retrovirology* **2**, 57

46 Mi S, Lee X, Li X *et al.* (2000). Syncytin is a captive retroviral envelope protein involved in placental morphogenesis. *Nature* **403**, 785–9

47 Tolosa JM, Schjenken JE, Clifton VL *et al.* (2012). The endogenous retroviral envelope protein syncytin-1 inhibits LPS/PHA-stimulated cytokine responses in human blood and is sorted into placental exosomes. *Placenta* **33**, 933–41; Fahlbusch FB, Ruebner M, Volkert G *et al.* (2012). Corticotropin-releasing

hormone stimulates expression of leptin, 11beta-HSD2 and syncytin-1 in primary human trophoblasts. *Reproductive Biology and Endocrinology* **10**, 80

48 Prudhomme S, Oriol G and Mallet F (2004). A retroviral promoter and a cellular enhancer define a bipartite element which controls *env* ERVWE1 placental expression. *Journal of Virology* **78**, 12157–68; Matsuura K, Jigami T, Taniue K *et al.* (2011). Identification of a link between Wnt/β-catenin signalling and the cell fusion pathway. *Nature Communications* **2**, 548; Ruebner M, Langbein M, Strissel PL *et al.* (2012). Regulation of the human endogenous retroviral syncytin-1 and cell–cell fusion by the nuclear hormone receptors PPARγ/RXRα in placentogenesis. *Journal of Cellular Biochemistry* **113**, 2383–96

49 Muir A, Lever AML and Moffett A (2006). Human endogenous retrovirus-W envelope (syncytin) is expressed in both villous and extravillous trophoblast populations. *Journal of General Virology* **87**, 2067–71; Hayward MD, Potgens AJG, Drewlo S *et al.* (2007). Distribution of human endogenous retrovirus type W receptor in normal human villous placenta. *Pathology* **39**, 406–12

50 Mi *et al.* (2000), ref. 46; Mallet F, Bouton O, Prudhomme S *et al.* (2004). The endogenous locus ERVWE1 is a bona fide gene involved in hominoid placental physiology. *Proceedings of the National Academy of Sciences of the USA* **101**, 1731–6

51 Caceres M, NISC Comparative Sequencing Program and Thomas JW (2006). The gene of retroviral origin syncytin 1 is specific to hominoids and is inactive in Old World monkeys. *Journal of Heredity* **97**, 100–6; see also Bonnaud B, Bouton O, Oriol G *et al.* (2004). Evidence of selection on the domesticated ERVWE1 *env* retroviral element involved in placentation. *Molecular Biology and Evolution* **21**, 1895–901

52 Søe K, Andersen TL, Hobolt-Pederson A-S *et al.* (2011). Involvement of human endogenous retroviral syncytin-1 in human osteoclast fusion. *Bone* **48**, 837–46

53 Malassine A, Blaise S, Handschuh K *et al.* (2007). Expression of the fusogenic HERV-FRD Env glycoprotein (syncytin 2) in human placenta is restricted to villous cytotrophoblastic cells. *Placenta* **28**, 185–91; Esnault C, Priet S, Ribet D *et al.* (2008). A placenta-specific receptor for the fusogenic, endogenous retrovirus-derived, human syncytin-2. *Proceedings of the National Academy of Sciences of the USA* **105**, 17532–7

54 Blaise S, de Parseval N, Benit L and Heidmann T (2003). Genome-wide screening for fusogenic human endogenous retrovirus envelopes identifies syncytin 2, a gene conserved on primate evolution. *Proceedings of the National Academy of Sciences of the USA* **100**, 13013–18; Mangeney M, Renard M, Schlecht-Louf G *et al.* (2007). Placental syncytins: genetic disjunction between the fusogenic

and immunosuppressive activity of retroviral envelope proteins. *Proceedings of the National Academy of Sciences of the USA* **104**, 20534–9

55 Tolosa *et al.* (2012), ref. 47; see also Kammerer U, Germeyer A, Stengel S *et al.* (2011). Human endogenous retrovirus K (HERV-K) is expressed in villous and extravillous cytotrophoblast cells of the human placenta. *Journal of Reproductive Immunology* **91**, 1–8

56 Heidmann O, Vernochet C, Dupressoir A and Heidmann T (2009). Identification of an endogenous retroviral envelope gene with fusogenic activity and placenta-specific expression in the rabbit: a new 'syncytin' in a third order of mammals. *Retrovirology* **6**, 107; Vernochet C, Heidmann O, Dupressoir A *et al.* (2011). A syncytin-like endogenous retroviral envelope gene of the guinea pig specifically expressed in the placenta junctional zone and conserved in Caviomorpha. *Placenta* **32**, 885–92; Cornelis G, Heidmann O, Bernard-Stoecklin S *et al.* (2012). Ancestral capture of *syncytin-Car1*, a fusogenic endogenous retroviral *envelope* gene involved in placentation and conserved in Carnivora. *Proceedings of the National Academy of Sciences of the USA* **109**, E432–41

57 Dunlap KA, Palmarini M, Varela M *et al.* (2006). Endogenous retroviruses regulate periimplantation placental growth and differentiation. *Proceedings of the National Academy of Sciences of the USA* **103**, 14390–5; Baba K, Nakaya Y, Shojima T *et al.* (2011). Identification of novel endogenous betaretroviruses which are transcribed in bovine placenta. *Journal of Virology* **85**, 1237–45

58 Dupressoir A, Vernochet C, Bawa O *et al.* (2009). Syncytin-A knockout mice demonstrate the critical role in placentation of a fusogenic, endogenous retrovirus-derived, envelope gene. *Proceedings of the National Academy of Sciences of the USA* **106**, 12127–32; Dupressoir A, Vernochet C, Harper F *et al.* (2011). A pair of co-opted retroviral envelope syncytin genes is required for formation of the two-layered murine placental syncytiotrophoblast. *Proceedings of the National Academy of Sciences of the USA* **108**, E1164–73

59 Dunlap *et al.* (2006), ref. 57; Black SG, Arnaud F, Palmarini M and Spencer TE (2010). Endogenous retroviruses in endogenous differentiation and placental development. *American Journal of Reproductive Immunology* **64**, 255–64

60 Kudaka W, Oda T, Jinno Y *et al.* (2008). Cellular localization of placenta-specific human endogenous retrovirus (HERV) transcripts and their possible implication in pregnancy-induced hypertension. *Placenta* **29**, 282–9; Langbein M, Strick R, Strissel PL *et al.* (2008). Impaired cytotrophoblast cell–cell fusion is associated with reduced syncytin and increased apoptosis in patients with placental dysfunction. *Molecular Reproduction and Development* **75**, 175–83; Ruebner M, Strissel PL, Langbein M *et al.* (2010). Impaired cell fusion and differentiation in placentae from patients with intrauterine growth restriction

correlate with reduced levels of HERV envelope genes. *Journal of Molecular Medicine* **88**, 1143–56; Vargas A, Toufaily C, LeBellego F *et al.* (2011). Reduced expression of both syncytin 1 and syncytin 2 correlates with severity of pre-eclampsia. *Reproductive Sciences* **18**, 1085–91

61 Mallasine A, Frendo JL, Blaise S *et al.* (2008). Human endogenous retrovirus-FRD envelope protein (syncytin 2) expression in normal and trisomy 21-affected placenta. *Retrovirology* **5**, 6

62 Lynch C and Tristem M (2003). A co-opted *gypsy*-type LTR-retrotransposon is conserved in the genomes of humans, sheep, mice, and rats. *Current Biology* **13**, 1518–23; Brandt J, Veith AM and Volff J-N (2005). A family of neofunctionalized Ty3/gypsy retrotransposon genes in mammalian genomes. *Cytogenetic and Genome Research* **110**, 307–17; Volff J-N (2009). Cellular genes derived from Gypsy/Ty3 retrotransposons in mammalian genomes. *Annals of the New York Academy of Sciences* **1178**, 233–43

63 Clark MB, Janicke M, Gottesbuhren U *et al.* (2007). Mammalian gene *PEG10* expresses two reading frames by high efficiency –1 frameshifting in embryonic-associated tissues. *Journal of Biological Chemistry* **282**, 37359–69

64 Ono R, Nakamura K, Inoue K *et al.* (2006). Deletion of *Peg10*, an imprinted gene acquired from a retrotransposon, causes early embryonic lethality. *Nature Genetics* **38**, 101–6; Lim AL, Ng S, Leow SCP *et al.* (2012). Epigenetic state and expression of imprinted genes in umbilical cord correlates with growth parameters in human pregnancy. *Journal of Medical Genetics* **49**, 689–97

65 Suzuki S, Ono R, Narita T *et al.* (2007). Retrotransposon silencing by DNA methylation can drive mammalian genomic imprinting. *PLoS Genetics* **3**, e55

66 Sekita Y, Wagatsuma H, Nakamura K *et al.* (2008). Role of transposon-derived imprinted gene, *RTL1*, in the feto-maternal interface of mouse placenta. *Nature Genetics* **40**, 243–8

67 Edwards CA, Mungall AJ, Matthews L *et al.* (2008). The evolution of the DLK1-DIO3 imprinted domain in mammals. *PLoS Biology* **6**, e135

68 Knox K and Baker JC (2008). Genomic evolution of the placenta using co-option and duplication and divergence. *Genome Research* **8**, 695–705; Rawn SM and Cross JC (2008). The evolution, regulation, and function of placental-specific genes. *Annual Review of Cell and Developmental Biology* **24**, 159–81

69 Sugimoto J and Schust DJ (2009). Human endogenous retroviruses and the placenta. *Reproductive Sciences* **16**, 1023–33; Kaneko-Ishino T and Ishino F (2010). Retrotransposon silencing by DNA methylation contributed to the evolution of placentation and genomic imprinting in mammals. *Development, Growth and Differentiation* **52**, 533–43; Dupressoir A, Lavialle C and Heidmann T

(2012). From ancestral infectious retroviruses to bona fide cellular genes: role of the captured syncytins in placentation. *Placenta* **33**, 663–71

70 Piriyapongsa J, Polavarapu N, Borodovsky M and McDonald J (2007). Exonization of the LTR transposable elements in human genome. *BMC Genomics* **8**, 291

71 Huh JW, Kim TH, Yi JM *et al.* (2006). Molecular evolution of the periphilin gene in relation to human endogenous retrovirus M element. *Journal of Molecular Evolution* **62**, 730–7

72 Ling J, Pi W, Bollag R *et al.* (2002). The solitary long terminal repeats of ERV-9 endogenous retroviruses are conserved during primate evolution and possess enhancer activities in embryonic and hematopoietic cells. *Journal of Virology* **76**, 2410–23; Pi W, Zhu X, Wu M *et al.* (2010). Long-range function of an intergenic retrotransposon. *Proceedings of the National Academy of Sciences of the USA* **107**, 12992–7

73 Dunn CA, van de Lagemaat LN, Baillie GJ and Mager DL (2005). Endogenous retrovirus long terminal repeats as ready-to-use mobile promoters: the case of primate *β3GAL- T5*. *Gene* **364**, 2–12

74 Schulte AM and Wellstein A (1998). Structure and phylogenetic analysis of an endogenous retrovirus inserted into the human growth factor gene pleiotrophin. *Journal of Virology* **72**, 6065–72; Ball M, Carmody M, Wynne F *et al.* (2009). Expression of pleiotrophin and its receptors in human placenta suggests roles in trophoblast life cycle and angiogenesis. *Placenta* **30**, 649–53

75 Bieche I, Laurent A, Laurendeau I *et al.* (2003). Placenta-specific INSL4 expression is mediated by a human endogenous retroviral element. *Biology of Reproduction* **68**, 1422–9

76 Landry J-R and Mager DL (2003). Functional analysis of the endogenous retroviral promoter of the human endothelin B receptor gene. *Journal of Virology* **77**, 7459–66

77 Cohen CJ, Rebollo R, Babovic S *et al.* (2011). Placenta-specific expression of the IL-2 receptor B subunit from an endogenous retroviral promoter. *Journal of Biological Chemistry* **286**, 35543–52

78 Emera D, Casola C, Lynch VJ *et al.* (2012). Convergent evolution of endometrial prolactin expression in primates, mice, and elephants through the independent recruitment of transposable elements. *Molecular Biology and Evolution* **29**, 39–47; Emera D and Wagner DP (2012). Transformation of a transposon into a derived prolactin promoter with function during human pregnancy. *Proceedings of the National Academy of Sciences of the USA* **109**, 11246–51

79 Huh J-W, Ha H-S, Kim D-S and Kim H-S (2008). Placenta-restricted expression of LTR-derived *NOS3*. *Placenta* **29**, 602–8

80 Wang T, Zeng J, Lowe CB *et al.* (2007). Species-specific endogenous retroviruses shape the transcriptional network of the human tumor suppressor protein p53. *Proceedings of the National Academy of Sciences of the USA* **104**, 18613–18

81 Beyer U, Moll-Rocek J, Moll UM and Dobbelstein M (2011). Endogenous retrovirus drives hitherto unknown proapoptotic p63 isoforms in the male germ line of humans and great apes. *Proceedings of the National Academy of Sciences of the USA* **108**, 3624–9

82 Conley AB, Piriyapongsa J and Jordan IK (2008). Retroviral promoters in the human genome. *Bioinformatics* **24**, 1563–7

83 van de Lagemaat LN, Landry J-R, Mager DL and Medstrand P (2003). Transposable elements in mammals promote regulatory variation and diversification of genes with specialized functions. *Trends in Genetics* **19**, 530–6; Cohen CJ, Lock WM and Mager DL (2009). Endogenous retroviral LTRs as promoters for human genes: a critical assessment. *Gene* **448**, 105–14

84 Barbulescu M, Turner G, Su M *et al.* (2001). A HERV-K provirus in chimpanzees, bonobos and gorillas, but not humans. *Current Biology* **11**, 779–83

85 Scally *et al.* (2012), ref. 36

86 Maksakova IA, Romanish MT, Gagnier L *et al.* (2006). Retroviral elements and their hosts: insertional mutagenesis in the mouse germ-line. *PLoS Genetics* **2**, e2

87 Theodorou V, Kimm MA, Boer M *et al.* (2007). MMTV insertional mutagenesis identifies genes, gene families, and pathways involved in mammary cancer. *Nature Genetics* **39**, 759–69; Ishiguro T, Avila H, Lin SY *et al.* (2010). Gene trapping identifies chloride channel 4 as a novel inducer of colon cancer cell migration, invasion and metastasis. *British Journal of Cancer* **102**, 774–82; Dail M, Li Q, McDaniel A *et al.* (2010). Mutant *Ikzf1*, $Kras^{G12D}$, and *Notch1* cooperate in T lineage leukemogenesis and modulate responses to targeted agents. *Proceedings of the National Academy of Sciences of the USA* **107**, 5106–11

88 Deichmann *et al.* (2007), ref. 8

89 Barbulescu *et al.* (1999), ref. 20

90 Hughes and Coffin (2001), ref. 30

91 Lee YN and Bieniasz PD (2007). Reconstitution of an infectious endogenous retrovirus. *PLoS Pathogens* **3**, e10

92 Barbulescu *et al.* (2001), ref. 84

93 Dewannieux M, Harper F, Richaud A *et al.* (2006). Identification of an infectious progenitor for the multiple copy HERV-K human endogenous retroelements. *Genome Research* **16**, 1548–56
94 Turner *et al.* (2001), ref. 21
95 Agoni *et al.* (2012), ref. 23
96 Lapuk AV, Khil PP, Lavrentieva IV *et al.* (1999). A human endogenous retrovirus-like (HERV) LTR formed more than 10 million years ago due to an insertion of HERV-H LTR into the 5′ LTR of HERV-K is situated on human chromosomes 10, 19, and Y. *Journal of General Virology* **80**, 835–9
97 Young GR, Eksmond U, Salcedo R *et al.* (2012). Resurrection of endogenous retrovirus in antibody-deficient mice. *Nature* **491**, 774–8
98 Stoye (2012), ref. 14
99 Dewannieux *et al.* (2006), ref. 93; Lee and Bieniasz (2007), ref. 91; Brady T, Lee YN, Ronen K *et al.* (2009). Integration target site selection by a resurrected human endogenous retrovirus. *Genes and Development* **23**, 633–42
100 Hanke K, Kramer P, Seeher S *et al.* (2009). Reconstitution of the ancestral glycoprotein of a human endogenous retrovirus-K and modulation of its functional activity by truncation of the cytoplasmic domain. *Journal of Virology* **83**, 12790–800
101 Kraus B, Boller K, Reuter A and Schnierle BS (2011). Characterization of the human endogenous retrovirus K gag protein: identification of protease cleavage sites. *Retrovirology* **8**, 21
102 For ERV-Kcon sequence, see Lee and Bieniasz (2007), ref. 91 (Supporting Information); for K106 and K110 sequences, see Barbulescu *et al.* (1999), ref. 20, available at www.ncbi.nlm.nih.gov/nuccore/AF164621.1 and www.ncbi.nlm.nih.gov/nuccore/AF164618.1; for regulatory elements, see Jha *et al.* (2011), ref 22; Fuchs NV, Kraft M, Tondera C *et al.* (2011). Expression of the human endogenous retrovirus (HERV) group HML-2/HERV-K does not depend on canonical promoter elements but is regulated by transcription factors Sp1 and Sp3. *Journal of Virology* **85**, 3436–48
103 Horie M, Honda T, Suzuki Y *et al.* (2010). Endogenous non-retroviral RNA elements in mammalian genomes. *Nature* **463**, 84; Katzourakis A and Gifford RJ (2010). Endogenous viral elements in animal genomes. *PLoS Genetics* **6**, e1001191; Feschotte and Gilbert (2012), ref. 14
104 Liu H, Fu Y, Xie J *et al.* (2011). Widespread endogenization of densoviruses and parvoviruses in animal and human genomes. *Journal of Virology* **85**, 9863–76
105 Arbuckle JH, Medveczky MM, Luka J *et al.* (2010). The latent human herpesvirus-6A genome specifically integrates in telomeres of human chromosomes

in vivo and in vitro. *Proceedings of the National Academy of Sciences of the USA* **107**, 5563–8

2 JUMPING GENEALOGY

1 Kapitonov VV, Pavlicek A and Jurka J. Anthology of repetitive DNA, in Meyers RA (ed.), *Encyclopedia of Molecular Cell Biology and Molecular Medicine Volume 1* (Weinheim: Wiley-VCH Verlag, 2004), 251–305; Kazazian HH (2004). Mobile elements: drivers of genome evolution. *Science* **303**, 1626–32; Jurka J, Kapitonov VV, Kohany O and Jurka MV (2007). Repetitive sequences in complex genomes: structure and evolution. *Annual Review of Genomics and Human Genetics* **8**, 241–59; Wicker T, Sabot F, Hua-Van A *et al.* (2007). A unified classification system for eukaryotic transposable elements. *Nature Reviews Genetics* **8**, 973–82
2 Kramerov DA and Vassetzky NS (2011). Origin and evolution of SINEs in eukaryotic genomes. *Heredity* **107**, 487–95
3 Giordano J, Ge Y, Gelfand Y *et al.* (2007). Evolutionary history of mammalian transposons determined by genome-wide defragmentation. *PLoS Computational Biology* **3**, e137
4 Deininger PL and Batzer MA (2002). Mammalian retroelements. *Genome Research* **12**, 1455–65; Kazazian HH (2007). Progress in understanding the biology of the human mutagen LINE-1. *Human Mutation* **28**, 527–39; Goodier JL and Kazazian HH (2008). Retrotransposons revisited: the restraint and rehabilitation of parasites. *Cell* **135**, 23–35; Cordaux R and Batzer MA (2009). The impact of retrotransposons on human genome evolution. *Nature Reviews Genetics* **10**, 691–703
5 Khan H, Smit A and Boissinot S (2006). Molecular evolution and tempo of amplification of human LINE-1 retrotransposons since the origin of primates. *Genome Research* **16**, 78–87; Kordis D, Lovšin N and Gubenšek F (2006). Phylogenomic analysis of the L1 retrotransposons in Deuterostomia. *Systematic Biology* **55**, 886–901; Waters PD, Dobigny G, Waddell PJ and Robinson TJ (2007). Evolutionary history of LINE-1 in the major clades of placental mammals. *PLoS ONE* **1**, e258
6 Giordano *et al.* (2007), ref. 3
7 Eickbush TH and Jumburuthugoda VK (2008). The diversity of retrotransposons and the properties of their reverse transcriptases. *Virus Research* **134**, 221–34
8 Cordaux R, Hedges DJ and Batzer MA (2004). Retrotransposition of *Alu* elements: how many sources? *Trends in Genetics* **20**, 464–7; Bennett EA, Keller H, Mills RE *et al.* (2008). Active *Alu* retrotransposons in the human genome. *Genome Research* **18**, 1875–83

9 Price AL, Eskin E and Pevzner PA (2004). Whole-genome analysis of Alu repeat elements reveals complex evolutionary history. *Genome Research* **14**, 2245–52; Churakov G, Grundmann N, Kuritzin A *et al.* (2010). A novel web-based TinT application and the chronology of the primate Alu retroposon activity. *BMC Evolutionary Biology* **10**, 376

10 Wang HH, Xing J, Grover D *et al.* (2005). SVA elements: a hominid-specific retroposon family. *Journal of Molecular Biology* **354**, 994–1007; Hancks DC and Kazazian HH (2010). SVA retrotransposons: evolution and genetic instability. *Seminars in Cancer Biology* **20**, 234–45

11 Baillie JK, Barnett MW, Upton KR *et al.* (2011). Somatic retrotransposition alters the genetic landscape of the human brain. *Nature* **479**, 534–7; Tyekucheva S, Yolken RH, McCombie WR *et al.* (2011). Establishing the baseline level of repetitive element expression in the human cortex. *BMC Genomics* **12**, 495

12 Kaneko H, Dridi S, Tarallo V *et al.* (2011). DICER1 deficit induces *Alu* toxicity in age-related macular degeneration. *Nature* **471**, 325–30; Tarrallo V, Hirano Y, Gelfand BD *et al.* (2012). DICER1 loss and Alu RNA induce age-related macular degeneration via the NLRP3 inflammasome and MyD88. *Cell* **149**, 847–59

13 Wallace NA, Belancio VP and Deininger PL (2008). L1 mobile element expression causes multiple types of toxicity. *Gene* **419**, 75–81; Belancio VP, Roy-Engel AM, Pochampally RR and Deininger P (2010). Somatic expression of LINE-1 elements in human tissues. *Nucleic Acids Research* **38**, 3909–22

14 Han JS and Boeke JD (2004). A highly active synthetic mammalian retrotransposon. *Nature* **429**, 314–18; van den Hurk JAJM, Meij IC, del Carmen Seleme M *et al.* (2007). L1 retrotransposition can occur early in human embryonic development. *Human Molecular Genetics* **16**, 1587–92; see also Bogerd HP, Wiegand HL, Hulme AE *et al.* (2006). Cellular inhibitors of long interspersed element 1 and Alu retrotransposition. *Proceedings of the National Academy of Sciences of the USA* **103**, 8780–5; Chiu Y-L, Witkowska HE, Hall SC *et al.* (2006). High-molecular-mass APOBEC3G complexes restrict Alu retrotransposition. *Proceedings of the National Academy of Sciences of the USA* **103**, 15588–93; Bennett *et al.* (2008), ref. 8; Hancks DC, Goodier JL, Mandal PK *et al.* (2011). Retrotransposition of marked SVA elements by human L1s in cultured cells. *Human Molecular Genetics* **20**, 3386–400

15 Muotri AR, Chu VT, Marchetto MCN *et al.* (2005). Somatic mosaicism in neuronal precursor cells mediated by L1 retrotransposition. *Nature* **435**, 903–10; An W, Han JS, Wheelan SJ *et al.* (2006). Active retrotransposition by a synthetic L1 element in mice. *Proceedings of the National Academy of Sciences of the USA* **103**, 18662–7; Okudaira N, Goto M, Yanobu-Takanashi R *et al.* (2011). Involvement of retrotransposition of long interspersed nucleotide element-1

in skin carcinogenesis induced by 7, 12-dimethylbenz[*a*]anthracene and 12-*O*-tetradecanylphorbol-13-acetate. *Cancer Science* **102**, 2000–6

16 Miki Y, Nishisho I, Horii A *et al.* (1991). Disruption of the APC gene by a retrotransposal insertion of L1 sequence in a colon cancer. *Cancer Research* **52**, 643–5

17 Lee E, Iskow R, Yang L *et al.* (2012). Landscape of somatic retrotransposition in human cancers. *Science* **337**, 967–71

18 Cordaux R, Hedges DJ, Herke SW and Batzer M (2006). Estimating the retrotransposition rate of human Alu elements. *Gene* **373**, 134–7; Xing J, Zhang Y, Han K *et al.* (2009). Mobile elements create structural variation: analysis of a complete human genome. *Genome Research* **19**, 1516–26; Huang CRL, Schneider AM, Lu Y *et al.* (2010). Mobile interspersed repeats are major structural variants in the human genome. *Cell* **141**, 1171–82; Rouchka E, Montoya-Durango DE, Stribinskis V *et al.* (2010). Assessment of genetic variation for the LINE-1 retrotransposon from next generation sequence data. *BMC Bioinformatics* **11** (Suppl 9), S12; Ewing AD and Kazazian HH (2011). Whole-genome resequencing allows detection of many rare LINE-1 insertion alleles in humans. *Genome Research* **21**, 985–90

19 Iskow RC, McCabe MT, Mills RE *et al.* (2010). Natural mutagenesis of human genomes by endogenous retrotransposons. *Cell* **141**, 1253–61; Hormozdiari F, Alkan C, Ventura M *et al.* (2011). *Alu* repeat discovery and characterization within human genomes. *Genome Research* **21**, 840–9

20 Chen J-M, Stenson PD, Cooper DN and Ferec C (2005). A systematic study of LINE-1 endonuclease-dependent retrotranspositional events causing human genetic disease. *Human Genetics* **117**, 411–27; Callinan PA and Batzer MA (2006). Retrotransposable elements and human disease. *Genome Dynamics* **1**, 104–15; Chen J-M, Ferec C and Cooper DN (2006). LINE-1 endonuclease-dependent retrotranspositional events causing human genetic disease: mutation detection bias and multiple mechanisms of target gene disruption. *Journal of Biomedicine and Biotechnology* **2006**(1), Article ID 56182; Belancio VP, Hedges DJ and Deininger P (2008). Mammalian non-LTR retrotransposons: for better or worse, in sickness and in health. *Genome Research* **18**, 343–58; Belancio VP, Deininger PL and Roy-Engel AM (2009). LINE dancing in the human genome: transposable elements and disease. *Genome Medicine* **1**, 97; Burns KH and Boeke JD (2012). Human transposon tectonics. *Cell* **149**, 740–52

21 Wallace MR, Andersen LB, Saulino AM *et al.* (1991). A de novo Alu insertion results in neurofibromatosis type 1. *Nature* **353**, 864–6

22 Gallus GN, Cardaioli E, Rufa A *et al.* (2010). Alu-element insertion in an *OPA1* intron sequence associated with autosomal dominant optic atrophy. *Molecular Vision* **16**, 178–83

23 Watanabe M, Kobayashi K, Jin F *et al.* (2005). Founder SVA retrotransposal insertion in Fukuyama-type congenital muscular dystrophy and its origin in Japanese and Northeast Asian populations. *American Journal of Medical Genetics Part A* **138**, 344–8; Taniguchi-Ikeda M, Kobayashi K, Kanagawa M *et al.* (2011). Pathogenic exon-trapping by SVA retrotransposon and rescue in Fukuyama muscular dystrophy. *Nature* **478**, 127–31

24 Takasu M, Hayashi R, Maruya E *et al.* (2007). Deletion of entire *HLA-A* gene accompanied by an insertion of a retrotransposon. *Tissue Antigens* **70**, 144–50

25 Machado PM, Brandao RD, Cavaco BM *et al.* (2007). Screening for a *BRCA2* rearrangement in high-risk breast/ovarian cancer families: evidence for a founder effect and analysis of the associated phenotypes. *Journal of Clinical Oncology* **25**, 2027–34

26 Bouchet C, Vuillaumier-Barrot S, Gonzales M *et al.* (2007). Detection of an *Alu* insertion in the *POMT1* gene from three French Walker–Warburg syndrome families. *Molecular Genetics and Metabolism* **90**, 93–6

27 Konkel MK and Batzer MA (2010). A mobile threat to genome stability: the impact of non-LTR retrotransposons upon the human genome. *Seminars in Cancer Biology* **20**, 211–21

28 Katzir *et al.* (1985), see Prologue, ref. 29

29 Ovchinnikov I, Rubin A and Swergold GD (2002). Tracing the LINEs of human evolution. *Proceedings of the National Academy of Sciences of the USA* **99**, 10522–7; Mathews L, Chi SY, Greenberg N *et al.* (2003). Large differences between LINE-1 amplification rates in the human and chimpanzee lineages. *American Journal of Human Genetics* **72**, 739–48

30 The data are from the whole-genome studies, see Chapter 1, refs. 33–8. The detailed data for the LINE-1, LINE-2 and Alu elements were from the dataset of Prufer *et al.* (2012), see Chapter 1, ref. 35; see also Mills RE, Bennett EA, Iskow RC *et al.* (2006). Recently mobilized transposons in the human and chimpanzee genomes. *American Journal of Human Genetics* **78**, 671–9; Mills RE, Bennett EA, Iskow RC and Devine SE (2007). Which transposable elements are active in the human genome? *Trends in Genetics* **23**, 183–91; Lee J, Cordaux R, Han K *et al.* (2007). Different evolutionary fates of recently integrated human and chimpanzee LINE-1 retrotransposons. *Gene* **390**, 18–27

31 Buzdin A, Ustyugova S, Gogvadze E *et al.* (2002). A new family of chimeric retrotranscripts formed by a full copy of U6 small nuclear RNA fused to the 3′ terminus of L1. *Genomics* **80**, 402–6; Buzdin A, Gogvadze E, Kovalskaya E *et al.* (2003). The human genome contains many types of chimeric retrogenes generated through in vivo RNA recombination. *Nucleic Acids Research* **31**, 4385–90; Gogvadze EV, Buzdin AA and Sverdlov ED (2005). Multiple template

switches on LINE-directed reverse transcription: the most probable formation mechanism for the double and triple chimeric retroelements in mammals. *Russian Journal of Bioorganic Chemistry* **31**, 74; Buzdin A, Gogvadze E and Lebrun M-H (2007). Chimeric retrogenes suggest a role for the nucleolus in LINE amplification. *FEBS Letters* **581**, 2877–82

32 Hasnaoui M, Doucet AJ, Meziane O and Gilbert N (2009). Ancient repeat sequence derived from U6 snRNA in primate genomes. *Gene* **448**, 139–44

33 Ling J, Zhang L, Jin H *et al.* (2004). Dynamic retrotransposition of ERV-9 LTR and L1 in the β-globin gene locus during primate evolution. *Molecular Phylogenetics and Evolution* **30**, 867–71

34 Hamdi H, Nishio H, Zielinski R and Dugaiczyk A (1999). Origin and phylogenetic distribution of *Alu* DNA repeats: irreversible events in the evolution of primates. *Journal of Molecular Biology* **289**, 861–71

35 Ajala AR, Almeida SS, Rangel M *et al.* (2012). Association of *ACE* gene insertion/deletion polymorphism with birth weight, blood pressure levels, and ACE activity in healthy children. *American Journal of Hypertension* **7**, 827–32

36 Martinez J, Dugaiczyk LJ, Zielinski R and Dugaiczyk A (2001). Human genetic disorders: a phylogenetic perspective. *Journal of Molecular Biology* **308**, 587–96

37 Gibbons R and Dugaiczyk A (2005). Phylogenetic roots of Alu-mediated rearrangements leading to cancer. *Genome* **48**, 160–7

38 Hamdi HK, Nishio H, Travis J *et al.* (2000). *Alu*-mediated phylogenetic novelties in gene regulation and development. *Journal of Molecular Biology* **299**, 931–9

39 Salem A-H, Ray DA, Xing J *et al.* (2003). Alu elements and hominid phylogenetics. *Proceedings of the National Academy of Sciences of the USA* **100**, 12787–91; Xing J, Salem A-H, Hedges DJ *et al.* (2003). Comprehensive analysis of two Alu Yd subfamilies. *Journal of Molecular Evolution* **57** (Suppl 1), S76–89; Salem A-H, Ray DA, Hedges DJ *et al.* (2005). Analysis of the human Alu Ye lineage. *BMC Evolutionary Biology* **5**, 18

40 Hedges DJ, Callinan PA, Cordaux R *et al.* (2004). Differential *Alu* mobilization and polymorphism among the human and chimpanzee lineages. *Genome Research* **14**, 1068–75

41 Sheikh TH and Deininger PL (1996). The role and amplification of the HS Alu subfamily founder gene. *Journal of Molecular Evolution* **42**, 15–21; Han K, Xing J, Wang H *et al.* (2005). Under the genomic radar: the stealth model of *Alu* amplification. *Genome Research* **15**, 655–64; Walker JA, Konkel MK, Ullmer B *et al.* (2012). Orangutan *Alu* quiescence reveals possible source element: support for ancient backseat drivers. *Mobile DNA* **3**, 8

42 See Chapter 1, refs. 33–8

43 Lee J, Ha J, Son S-Y and Han K (2012). Human genome deletions generated by SVA-associated events. *Comparative and Functional Genomics* **2012**, Article ID 807270; see also Chapter 1, ref. 38

44 See Chapter 1, refs. 33–8

45 Kouprina N, Pavlicek A, Mochida GH *et al.* (2004). Accelerated evolution of the *ASPM* gene controlling brain size begins prior to human brain expansion. *PLoS Biology* **2**, e126; Pavlicek A, Noskov VN, Kouprina N *et al.* (2004). Evolution of the tumor suppressor *BRCA1* locus in primates: implications for cancer predisposition. *Human Molecular Genetics* **13**, 2737–51

46 Crouau-Roy B and Clisson I (2000). Evolution of an Alu DNA element of type Sx in the lineage of primates and the origin of an associated tetranucleotide microsatellite. *Genome* **43**, 642–8

47 Schmitz J, Ohme M and Zischler H (2001). SINE insertions in cladistic analyses and the phylogenetic affiliations of *Tarsius bancanus* to other primates. *Genetics* **157**, 777–84; Schmitz J, Roos C and Zischler H (2005). Primate phylogeny: molecular evidence from retroposons. *Cytogenetic and Genome Research* **108**, 26–37; for striking confirmation of simian-tarsier monophylicity, see Hartig G, Churakov G, Warren WC *et al.* (2013). Retrophylogenomics place tarsiers on the evolutionary branch of anthropoids. *Scientific Report* **3**, 1756

48 Kuryshev VY, Skryabin BV, Kremerskothen J *et al.* (2001). Birth of a gene: locus of neuronal BC200 snmRNA in three prosimians and human BC200 pseudogenes as archives of change in the *Anthropoidea* lineage. *Journal of Molecular Biology* **309**, 1049–66

49 Schmitz J and Zischler H (2003). A novel family of tRNA-related SINEs in the colugo and two new retrotransposable markers separating dermopterans from primates. *Molecular Phylogenetics and Evolution* **28**, 341–9

50 Liu GE, Alkan C, Jiang L *et al.* (2009). Comparative analysis of *Alu* repeats in primate genomes. *Genome Research* **19**, 876–85

51 Kojima KK (2010). Alu monomer revisited: recent generation of Alu monomers. *Molecular Biology and Evolution* **28**, 13–15

52 Park S-J, Huh J-W, Kim Y-H *et al.* (2012). Intron retention and TE exonization events in *ZRANB2*. *Comparative and Functional Genomics* **2012**, 170208

53 Schmitz J, Ohme M, Suryobroto B and Zischler H (2002). The colugo (*Cynocephalus variegatus*, Dermoptera): the primates' gliding sister? *Molecular Biology and Evolution* **19**, 2308–12

54 Levin HL and Moran JV (2011). Dynamic interactions between transposable elements and their hosts. *Nature Reviews Genetics* **12**, 615–27

55 Levy A, Schwartz S and Ast G (2010). Large-scale discovery of insertion hotspots and preferential integration sites of human transposed elements. *Nucleic Acids Research* **38**, 1515–30

56 Wimmer K, Callens T, Wernstedt A and Messiaen L (2011). The *NF1* gene contains hotspots for L1 endonuclease-dependent *de novo* insertion. *PLoS Genetics* **11**, e1002371

57 Cantrell MA, Filanoski BJ, Ingermann AR *et al.* (2001). An ancient retrovirus-like element contains hot spot for SINE insertion. *Genetics* **158**, 769–77; for preferred integration sites in human disease, see Levin and Moran (2011), ref. 54

58 Vincent BJ, Myers JS, Ho HJ *et al.* (2003). Following the LINEs: an analysis of primate genomic variation at human-specific LINE-1 insertion sites. *Molecular Biology and Evolution* **20**, 1338–48; Ho HJ, Ray DA, Salem A-H *et al.* (2005). Straightening out the LINEs: LINE-1 orthologous loci. *Genomics* **85**, 201–7

59 Roy-Engel AM, Carroll ML, El-Sawy M *et al.* (2002). Non-traditional *Alu* evolution and primate genomic diversity. *Journal of Molecular Biology* **316**, 1033–40; Salem A-H, Kilroy GE, Watkins WS *et al.* (2003). Recently integrated Alu elements and human genomic diversity. *Molecular Biology and Evolution* **20**, 1349–61

60 Salem *et al.* (2003), ref. 39

61 van de Lagemaat L, Gagnier L, Medstrand P and Mager DL (2005). Genomic deletions and precise removal of transposable elements mediated by short identical DNA segments in primates. *Genome Research* **15**, 1243–9

62 Ray DA, Xing J, Salem A-H and Batzer MA (2006). SINEs of a *nearly* perfect character. *Systematic Biology* **55**, 928–35; Ray DA (2007). SINEs of progress: mobile elements applications to molecular ecology. *Molecular Ecology* **16**, 19–33

63 Meyer TJ, McLain AT, Oldenburg JM *et al.* (2012). An Alu-based phylogeny of gibbons (Hylobatidae). *Molecular Biology and Evolution* **29**, 3441–50

64 Xing J, Wang H, Han K *et al.* (2005). A mobile element-based phylogeny of Old World monkeys. *Molecular Phylogenetics and Evolution* **37**, 872–80; Xing J, Wang H, Zhang Y *et al.* (2007). A mobile element-based evolutionary history of guenone (tribe Cercopithecini). *BMC Biology* **5**, 5; Li J, Han K, Xing J *et al.* (2009). Phylogeny of the macaques (Cercopithecidae: *Macaca*) based on *Alu* elements. *Gene* **448**, 242–9

65 Singer SS, Schmitz J, Schwiegk C and Zischler H (2002). Molecular cladistic markers in New World monkey phylogeny (Platyrrhini, Primates). *Molecular Phylogenetics and Evolution* **26**, 490–501; Ray DA, Xing J, Hedges DJ *et al.* (2005). *Alu* insertion loci and platyrrhine primate phylogeny. *Molecular Phylogenetics and Evolution* **35**, 117–26

66 Roos C, Schmitz J and Zischler H (2004). Primate jumping genes elucidate strepsirrhine phylogeny. *Proceedings of the National Academy of Sciences of the USA* **101**, 10650–4; McLain AT, Meyer TJ, Faulk C *et al.* (2012). An *Alu*-based phylogeny of lemurs (Infraorder: Lemuriformes). *PLoS ONE* **7**, e44035

67 Xing J, Witherspoon DJ, Ray DA *et al.* (2007). Mobile DNA elements in primate and human evolution. *American Journal of Physical Anthropology* **134** (Suppl 45), 2–19

68 Herke SW, Xing J, Ray DA *et al.* (2006). A SINE-based dichotomous key for primate identification. *Gene* **390**, 39–51

69 Murphy WJ, Eizirik E, O'Brien SJ *et al.* (2001). Resolution of the early placental mammal radiation using Bayesian phylogenetics. *Science* **294**, 2348–51; Binina-Emonds ORP, Cardillo M, Jones KE *et al.* (2007). The delayed rise of present-day mammals. *Nature* **446**, 507–12; Springer MS and Murphy WJ (2007). Mammalian evolution and biomedicine: new views from phylogeny. *Biological Reviews of the Cambridge Philosophical Society* **82**, 375–92; Wildman DE, Uddin M, Opazo JC *et al.* (2007). Genomics, biogeography, and the diversification of placental mammals. *Proceedings of the National Academy of Sciences of the USA* **104**, 14395–400; Prasad AB, Allard MW, NISC Comparative Sequencing Program and Green ED (2008). Confirming the phylogeny of mammals by use of large comparative sequence data sets. *Molecular Biology and Evolution* **25**, 1795–808; Meredith RW, Janecka JE, Gatesy J *et al.* (2011). Impacts of the Cretaceous terrestrial revolution and the KPg extinction on mammal diversification. *Science* **334**, 521–4; Stringer MS, Meredith RW, Janecka JE and Murphy WJ (2011). The historical biogeography of Mammalia. *Philosophical Transactions of the Royal Society of London Series B, Biological Sciences* **366**, 2478–502

70 Thomas JW, Touchman JW, Blakesley RW *et al.* (2003). Comparative analyses of multi-species sequences from targeted genomic regions. *Nature* **424**, 788–93 (Phylogenetic Analyses, Supplementary Information 5)

71 Hughes DC (2000). MIRs as agents of mammalian gene evolution. *Trends in Genetics* **16**, 60–2

72 Lin L, Jiang P, Shen S *et al.* (2009). Large-scale analysis of exonised mammalian-wide interspersed repeats in primate genomes. *Human Molecular Genetics* **18**, 2204–14

73 Silva JC, Shabalina SA, Harris DG *et al.* (2003). Conserved fragments of transposable elements in intergenic regions: evidence for widespread recruitment of MIR- and L2-derived sequences within the mouse and human genomes. *Genetical Research* **82**, 1–18; Zhu L, Swergold GD and Seldin MF (2003). Examination of sequence homology between human chromosome 20 and the mouse genome: intense conservation of many genomic elements. *Human Genetics* **113**, 60–70

74 Kriegs JO, Churakov G, Jurka J *et al.* (2007). Evolutionary history of 7SL RNA-derived SINEs in Supraprimates. *Trends in Genetics* **23**, 158–61

75 Armiger V, Erlandsson R, Pielberg G *et al.* (2003). Comparative sequence analysis of the PRKAG3 region between human and pig: evolution of repetitive sequences and potential new exons. *Cytogenetic and Genome Research* **102**, 163–72

76 Lowe CB, Bejerano G and Haussler D (2007). Thousands of human mobile element fragments undergo strong purifying selection near developmental genes. *Proceedings of the National Academy of Sciences of the USA* **104**, 8005–10

77 Lindblad-Toh K, Garber M, Zuk O *et al.* (2011). A high-resolution map of human evolutionary constraint using 29 mammals. *Nature* **478**, 476–82

78 Kriegs JO, Churakov G, Kiefmann M *et al.* (2006). Retroposed elements as archives for the evolutionary history of placental mammals. *PLoS Biology* **4**, e91

79 Nishihara H, Hasegawa M and Okada N (2006). Pegasoferae, an unexpected mammalian clade revealed by tracking ancient retroposon insertions. *Proceedings of the National Academy of Sciences of the USA* **103**, 9929–34 (see Supporting Information)

80 Yu L and Zhang Y-P (2005). Evolutionary implications of multiple SINE insertions in an intronic region from diverse mammals. *Mammalian Genome* **16**, 651–60

81 Smith AM, Sanchez M-J, Follows GA *et al.* (2008). A novel mode of enhancer evolution: the *TAL1* stem cell enhancer recruited a MIR element to specifically boost its activity. *Genome Research* **18**, 1422–32

82 Kriegs *et al.* (2006), ref. 78

83 Blanchette M, Green ED, Miller W and Haussler D (2004). Reconstructing large regions of an ancestral mammalian genome in silico. *Genome Research* **14**, 2412–23

84 Murphy WJ, Pringle TH, Crider TA *et al.* (2007). Using genomic data to unravel the root of the placental mammal tree. *Genome Research* **17**, 413–21

85 Churakov G, Kriegs JO, Baertsch R *et al.* (2009). Mosaic retroposon insertion patterns in placental mammals. *Genome Research* **19**, 868–75

86 Nishihara H, Maruyama S and Okada N (2009). Retroposon analysis and recent geological data suggest near-simultaneous divergence of the three superorders of mammals. *Proceedings of the National Academy of Sciences of the USA* **106**, 5235–40

87 Data from Kriegs *et al.* (2006), ref. 78; Nishihara *et al.* (2006), ref. 79; Kriegs *et al.* (2007), ref. 74; Churakov *et al.* (2009), ref. 85; Nishihara *et al.* (2009), ref. 86; data in dark grey boxes are from refs. 88–96

88 Krull M, Petrusma M, Makalowski W *et al.* (2007). Functional persistence of exonised mammalian-wide interspersed repeat elements (MIRs). *Genome Research* **17**, 1139–45

89 Santangelo AM, de Souza FSJ, Franchini LF *et al.* (2007). Ancient exaptation of a CORE-SINE retroposon into a highly conserved mammalian neuronal enhancer of the proopiomelanocortin gene. *PLoS Genetics* **3**, e166; Alfoldi J, De Palma F, Grabherr M *et al.* (2011). The genome of the green anole lizard and a comparative analysis with birds and mammals. *Nature* **477**, 587–91 (Supplementary Table 18); Franchini LF, Lopez-Leal R, Nasif S *et al.* (2011). Convergent evolution of two mammalian neuronal enhancers by sequential exaptation of unrelated retroposons. *Proceedings of the National Academy of Sciences of the USA* **108**, 15270–5

90 Kamal M, Xie X and Lander ES (2006). A large family of ancient repeat elements in the human genome is under strong selection. *Proceedings of the National Academy of Sciences of the USA* **103**, 2740–5; see also Lowe *et al.* (2007), ref. 76

91 Bejerano G, Lowe CB, Ahituv N *et al.* (2006). A distal enhancer and an ultraconserved exon are derived from a novel retroposon. *Nature* **441**, 87–90

92 Lowe CB, Bejerano G, Salama SR and Haussler D (2010). Endangered species hold clues to human evolution. *Journal of Heredity* **101**, 437–47

93 Nishihara H, Smit AFA and Okada N (2006). Functional sequences derived from SINEs in the mammalian genome. *Genome Research* **16**, 864–74; Xie X, Kamal M and Lander ES (2006). A family of conserved noncoding elements derived from an ancient transposable element. *Proceedings of the National Academy of Sciences of the USA* **103**, 11659–64; Hirakawa M, Nishihara H, Kanehisa M and Okada N (2009). Characterization and evolutionary landscape of AmnSINE1 in Amniota genomes. *Gene* **441**, 100–10

94 Sasaki T, Nishihara H, Hirakawa M *et al.* (2008). Possible involvement of SINEs in mammalian-specific brain formation. *Proceedings of the National Academy of Sciences of the USA* **105**, 4220–5; Okada N, Sasaki T, Shimogori T and Nishihara H (2010). Emergence of mammals by emergency: exaptation. *Genes to Cells* **15**, 801–12

95 Gentles AJ, Wakefield MJ, Kohany O *et al.* (2007). Evolutionary dynamics of transposable elements in the short-tailed opossum *Monodelphis domestica*. *Genome Research* **17**, 992–1004

96 Elisaphenko EA, Kolesnikov NN, Shevchenko AI *et al.* (2008). A dual origin of the *XIST* gene from a protein-coding gene and a set of transposable elements. *PLoS ONE* **3**, e2521

97 Jurka J, Bao W, Kojima KK *et al.* (2012). Different groups of repetitive elements preserved in mammals correspond to different periods of regulatory innovation in vertebrates. *Biology Direct* **7**, 36

98 Alfoldi *et al.* (2011), ref. 89

99 Churakov G, Sadasivuni MK, Rosenbloom KR *et al.* (2010). Rodent evolution: back to the root. *Molecular Biology and Evolution* **27**, 1315–26; Kriegs JO, Zemann A, Churakov G *et al.* (2010). Retroposon insertions provide insights into deep lagomorph evolution. *Molecular Biology and Evolution* **27**, 2678–81

100 Hallstrom BM, Schneider A, Zoller S and Janke A (2011). A genomic approach to examine the complex evolution of laurasiatherian mammals. *PLoS ONE* **6**, e28199

101 Shimamura M, Yasue H, Ohshima K *et al.* (1997). Molecular evidence from retroposons that whales form a clade within even-toed ungulates. *Nature* **388**, 666–70; Shedlock AM and Okada N (2000). SINE insertions: powerful tools for molecular systematics. *BioEssays* **22**, 148–60; Wong K (2002). The mammals that conquered the seas. *Scientific American* **286**(5), 70–9

102 Nikaido M, Matsuno F, Hamilton H *et al.* (2001). Retroposon analysis of major cetacean lineages: the monophyly of toothed whales and the paraphyly of river dolphins. *Proceedings of the National Academy of Sciences of the USA* **98**, 7384–9; Sasaki T, Nikaido M, Wada S *et al.* (2006). *Balaenoptera omurai* is a newly discovered baleen whale that represents an ancient evolutionary lineage. *Molecular Phylogenetics and Evolution* **41**, 40–52; Chen Z, Xu S, Zhou K and Yang G (2011). Whale phylogeny and rapid radiation events revealed using novel retrotransposed elements and their flanking sequences. *BMC Evolutionary Biology* **11**, 314; Meredith RW, Gatesy J, Cheng J and Springer MS (2011). Pseudogenisation of the tooth gene enamelysin (*MMP20*) in the common ancestor of extant baleen whales. *Proceedings, Biological Sciences/ The Royal Society* **278**, 993–1002

103 Nijman IJ, van Tessel P and Lenstra JA (2002). SINE retrotransposon during the evolution of the pecoran ruminants. *Journal of Molecular Evolution* **54**, 9–16; Nilsson M, Klassert D, Bertelsen M *et al.* (2012). Activity of ancient RTE retroposons during the evolution of cows, spiral-horned antelopes and nilgais (Bovinae). *Molecular Biology and Evolution* **29**, 2885–8

104 Nishihara H, Satta Y, Nikaido M *et al.* (2005). A retroposon analysis of Afrotherian phylogeny. *Molecular Biology and Evolution* **22**, 1823–33; Moller-Krull M, Delsuc F, Churakov G *et al.* (2007). Retroposed elements and their flanking regions resolve the evolutionary history of xenarthran mammals (armadillos, anteaters, and sloths). *Molecular Biology and Evolution* **24**, 2573–82

105 Nilsson MA, Churakov G, Sommer M *et al.* (2010). Tracking marsupial evolution using archaic genomic retroposon insertions. *PLoS Biology* **8**, e1000436

106 Watanabe M, Nikaido M, Tsuda TT *et al.* (2006). The rise and fall of the CR1 subfamily in the lineage leading to penguins. *Gene* **365**, 57–66; Kaiser VB, van Tuinen M and Ellegren H (2007). Insertion events of CR1 retrotransposable elements elucidate the phylogenetic branching order in galliform birds. *Molecular Biology and Evolution* **24**, 338–47

107 Suh A, Paus M, Kiefmann M *et al.* (2011). Mesozoic retroposons reveal parrots as the closest living relatives of passerine birds. *Nature Communications* **2**, 443

108 Han K-L, Braun EL, Kimball RT *et al.* (2011). Are transposable element insertions homoplasy free? An examination using the avian tree of life. *Systematic Biology* **60**, 375–86

109 Shedlock AM, Takahashi K and Okada N (2004), SINEs of speciation: tracking lineages with retroposons. *Trends in Ecology and Evolution* **19**, 545–53

110 Kuryshev *et al.* (2001), ref. 48; see also Martignetti JA and Brosius J (1993). BC200 RNA: a neural RNA polymerase III product encoded by a monomer *Alu* element. *Proceedings of the National Academy of Sciences of the USA* **90**, 11563–7; Duning K, Buck F, Barnekow A and Kremerskothen J (2008). SYNCRIP, a component of dendritically localised mRNPs, binds to the translation regulator BC200 RNA. *Journal of Neurochemistry* **105**, 351–9

111 Parrott AM, Tsai M, Batchu P *et al.* (2011). The evolution and expression of the snaR family of small non-coding RNAs. *Nucleic Acids Research* **39**, 1485–500; Parrott AM and Mathews MB (2011). The evolution and consequences of snaR family transposition in primates. *Mobile Genetic Elements* **1**, 291–5

112 Courseaux A and Nahon J-L (2001). Birth of two chimeric genes in the *Hominidae* lineage. *Science* **291**, 1293–7; Schmieder S, Darre-Toulemonde F, Arguel M-J *et al.* (2008). Primate-specific spliced *PMCHL* RNAs are non-protein coding in human and macaque tissues. *BMC Evolutionary Biology* **8**, 330

113 Li C-Y, Zhang Y, Wang Z *et al.* (2010). A human-specific *de novo* protein-coding gene associated with human brain functions. *PLoS Computational Biology* **6**, e1000734

114 Wu M, Li L and Sun Z (2007). Transposable element fragments in protein-coding regions and their contributions to human functional proteins. *Gene* **401**, 165–71; Schmitz J and Brosius J (2011). Exonization of transposed elements: a challenge and opportunity for evolution. *Biochimie* **93**, 1928–34

115 Damert A, Lower J and Lower R (2004). Leptin receptor isoform 219.1: an example of protein evolution by LINE-1-mediated human-specific retrotransposition of a coding SVA element. *Molecular Biology and Evolution* **21**, 647–51

116 Mola G, Vela V, Fernandez-Figueras MI *et al.* (2007). Exonization of *Alu*-generated splice variants in the survivin gene of human and non-human primates. *Journal of Molecular Biology* **366**, 1055–63

117 Krull M, Brosius J and Schmitz J (2005). Alu-SINE exonization: en route to protein-coding function. *Molecular Biology and Evolution* **22**, 1702–11

118 Huh J-W, Kim Y-H, Lee S-R *et al.* (2010). Four different ways of alternative transcripts generation mechanism in *ADRA1A* gene. *Genes and Genetic Systems* **85**, 65–73

119 Moller-Krull M, Zemann A, Roos C *et al.* (2008). Beyond DNA: RNA editing and steps toward Alu exonization in primates. *Journal of Molecular Biology* **382**, 601–9

120 Singer SS, Mannel DN, Hehlgans T *et al.* (2004). From 'junk' to gene: curriculum vitae of a primate receptor isoform gene. *Journal of Molecular Biology* **341**, 883–6

121 Lee J-R, Huh J-W, Kim D-S *et al.* (2009). Lineage-specific evolutionary events on SFTPB gene: Alu recombination-mediated deletion (ARMD), exonization, and alternative splicing events. *Gene* **435**, 29–35

122 Park *et al.* (2012), ref. 52

123 Lin L, Shen S, Tye A *et al.* (2008). Diverse splicing patterns of exonised Alu elements in human tissues. *PLoS Genetics* **4**, e1000225

124 Hasler J and Strub K (2006). Alu elements as regulators of gene expression. *Nucleic Acids Research* **34**, 5491–7; Feschotte C (2008). Transposable elements and the evolution of regulatory networks. *Nature Reviews Genetics* **9**, 397–405

125 Pheasant M and Mattick JS (2007). Raising the estimate of functional human sequences. *Genome Research* **17**, 245–53; Petherick A (2008). The production line. *Nature* **454**, 1042–5; Mattick JS (2011). The central role of RNA in human development and cognition. *FEBS Letters* **585**, 1600–16

126 Mikkelsen TS, Wakefield MJ, Aken B *et al.* (2007). Genome of the marsupial *Monodelphis domestica* reveals innovation in non-coding sequences. *Nature* **447**, 167–77

127 Lowe CB and Haussler D (2012). 29 mammalian genomes reveal novel exaptations of mobile elements for likely regulatory functions in the human genome. *PLoS ONE* **7**, e43128

128 Thurman RE, Rynes E, Humbert R *et al.* (2012). The accessible chromatin landscape of the human genome. *Nature* **489**, 75–82 (Supplementary Material, Figures 4 and 5, and Table 3)

129 Conley AB, Miller WJ and Jordan IK (2008). Human *cis* natural antisense transcripts initiated by transposable elements. *Trends in Genetics* **24**, 53–6;

Faulkner GL, Kimura Y, Daub CO *et al.* (2009). The regulated retrotransposon transcriptome of mammalian cells. *Nature Genetics* **41**, 563

130 Bourque G, Leong B, Vega VB *et al.* (2008). Evolution of the mammalian transcription factor binding repertoire via transposable elements. *Genome Research* **18**, 1752–62; Polavarapu N, Marino-Ramirez L, Landsman D *et al.* (2008). Evolutionary rates and patterns for human transcription factor binding sites derived from repetitive DNA. *BMC Genomics* **9**, 226

131 Polak P and Domany E (2006). Alu elements contain many binding sites for transcription factors and may play a role in regulation of developmental processes. *BMC Genomics* **7**, 133

132 Gombart AF, Saito T and Koeffler HP (2009). Exaptation of an ancient Alu short interspersed element provides a highly conserved vitamin D-mediated innate immune response in humans and primates. *BMC Genomics* **10**, 321

133 Laperriere D, Wang T-T, White JH and Mader S (2007). Widespread Alu repeat-driven expansion of consensus DR2 retinoic acid response elements during primate evolution. *BMC Genomics* **8**, 23; Mason CE, Shu F-J, Wang C *et al.* (2010). Location analysis for the estrogen receptor-α reveals binding to diverse ERE sequences and widespread binding within repetitive DNA elements. *Nucleic Acids Research* **38**, 2355–68; Bolotin E, Chellappa K, Huang-Verslues W *et al.* (2011). Nuclear receptor HNF4alpha binding sequences are widespread in Alu repeats. *BMC Genomics* **12**, 560

134 Zemojtel T, Kielbasa SM, Arndt PF *et al.* (2009). Methylation and deamination of CpGs generate p53-binding sites on a genomic scale. *Trends in Genetics* **25**, 63–6

135 Zemojtel T, Kielbasa SM, Arndt PF *et al.* (2011). CpG deamination creates transcription factor binding sites with high efficiency. *Genome Biology and Evolution* **3**, 1304–11

136 Antonaki A, Demetriades C, Polyzos A *et al.* (2011). Genomic analysis reveals a novel NK-κB binding site in Alu repetitive elements. *Journal of Biological Chemistry* **286**, 38768–82

137 Pandey R, Mandal AK, Jha V and Mukerji M (2011). HSF binding in Alu repeats expands its involvement in stress through an antisense mechanism. *Genome Biology* **12**, R117

138 Lynch VJ, Leclerc RD, May G and Wagner GP (2011). Transposon-mediated re-wiring of gene regulatory networks contributed to the evolution of pregnancy in mammals. *Nature Genetics* **43**, 1154–9

139 Johnson R, Gamblin RJ, OOi L *et al.* (2006). Identification of the REST regulon reveals extensive transposable element-mediated binding site duplication. *Nucleic Acids Research* **34**, 3862–77

140 Schmidt D, Schwalie PC, Wilson MD *et al.* (2012). Waves of retrotransposon expansion remodel genome organization and CTCF binding in multiple mammalian lineages. *Cell* **148**, 335–48

141 Reviewed in ref. 20

142 Han K, Lee J, Meyer TJ *et al.* (2008). L1 recombination-associated deletions generate human genomic variation. *Proceedings of the National Academy of Sciences of the USA* **105**, 19366–71

143 Sen SK, Han K, Wang J *et al.* (2006). Human genomic deletions mediated by recombination between Alu elements. *American Journal of Human Genetics* **79**, 41–53

144 Han K, Lee J, Meyer TJ *et al.* (2007). Alu recombination-mediated structural deletions in the chimpanzee genome. *PLoS Genetics* **3**, 1939–49

145 Wang X, Mitra N, Secundino I *et al.* (2012). Specific inactivation of two immunomodulatory *SIGLEC* genes during human evolution. *Proceedings of the National Academy of Sciences of the USA* **109**, 9935–40

146 Zhang R, Wang Y-Q and Su B (2008). Molecular evolution of a primate-specific microRNA family. *Molecular Biology and Evolution* **25**, 1493–502; Lehnert S, Van Loo P, Thilakarathne PJ *et al.* (2009). Evidence for co-evolution between human microRNAs and Alu-repeats. *PLoS ONE* **4**, e4456

147 Lee J, Han K, Meyer TJ *et al.* (2008). Chromosomal inversions between human and chimpanzee lineages caused by retrotransposons. *PLoS ONE* **3**, e4047

148 Kelkar YD, Eckert KA, Chiaromonte F and Makova KD (2011). A matter of life or death: how microsatellites emerge in and vanish from the human genome. *Genome Research* **21**, 2038–48; Ahmed M and Liang P (2012). Transposable elements are a significant contributor to tandem repeats in the human genome. *Comparative and Functional Genomics* **2012**, Article ID 947089

149 Kurosaki T, Ueda S, Ishida T *et al.* (2012). The unstable CCTG repeat responsible for myotonic dystrophy type 2 originates from an *AluSx* element insertion into an early primate genome. *PLoS ONE* **7**, e38379 (Supporting Information, Figure S4)

150 Thompson R, Zoppis S and McCord B (2012). An overview of DNA typing methods for human identification: past, present, and future. *Methods in Molecular Biology* **830**, 3–16

151 Huda A and Jordan IK (2009). Epigenetic regulation of mammalian genomes by transposable elements. *Annals of the New York Academy of Sciences* **1178**, 276–84

152 Oliver KR and Greene WK (2009). Transposable elements: powerful facilitators of evolution. *BioEssays* **31**, 703–14; Oliver KR and Green WK (2011).

Mobile DNA and the TE-thrust hypothesis: supporting evidence from the primates. *Mobile DNA* **2**, 8

153 Hagan CR, Sheffield RF and Rudin CM (2003). Human Alu element retrotransposition induced by genotoxic stress. *Nature Genetics* **3**, 219–20; Stribinskis V and Ramos KS (2006). Activation of human long interspersed nuclear element 1 retrotransposition by benzo(*a*)pyrene, an ubiquitous environmental carcinogen. *Cancer Research* **66**, 2616–20; Teneng I, Stribinskis V and Ramos KS (2007). Context-specific regulation of *LINE-1*. *Genes to Cells* **12**, 1101–10; Okudaira N, Iijima K, Koyama T *et al.* (2010). Induction of long interspersed nucleotide element-1 (L1) retrotransposition by 6-formylindolo[3,2-*b*]carbazole (FICZ), a tryptophan photoproduct. *Proceedings of the National Academy of Sciences of the USA* **107**, 18487–92; Giorgi G, Marcantonio P and Del Re B (2011). LINE-1 retrotransposition in human neuroblastoma cells is affected by oxidative stress. *Cell and Tissue Research* **346**, 383–91; Okudaira *et al.* (2011), ref. 15; Teneng I, Montoya-Durango DE, Quertermous JL *et al.* (2011). Reactivation of L1 retrotransposon by benzo(*a*)pyrene involves complex genetic and epigenetic regulation. *Epigenetics* **6**, 355–67

154 Ishizaka Y, Okudaira N, Tamura M *et al.* (2012). Modes of retrotransposition by long interspersed element-1 by environmental factors. *Frontiers in Microbiology* **3**, Article 191

155 Zeh DW, Zeh JA and Ishida Y (2009). Transposable elements and an epigenetic basis for punctuated equilibria. *BioEssays* **31**, 715–26; Rebollo R, Horard B, Hubert B and Vieira C (2010). Jumping genes and epigenetics: towards new species. *Gene* **454**, 1–7

156 Britten RJ (2010). Transposable element insertions have strongly affected human evolution. *Proceedings of the National Academy of Sciences of the USA* **107**, 19945–8

157 Mattick (2011), ref. 125

158 de Koning AP, Gu W, Castoe TA *et al.* (2011). Repetitive elements may comprise over two-thirds of the human genome. *PLoS Genetics* **7**, e1002384

159 Werren JH (2011). Selfish genetic elements, genetic conflict, and evolutionary innovation. *Proceedings of the National Academy of Sciences of the USA* **108** (Suppl. 2), 10863–70

3 PSEUDOGENEALOGY

1 Interview (1998). Dr. Charles Dinarello elected to the U.S. National Academy of Sciences. *International Cytokine Society Newsletter* **6**, 1

2 Drayna D (2005). Founder mutations. *Scientific American* **294**(4), 78–85

3 Eiberg H, Troelsen J, Nielsen M *et al.* (2008). Blue eye colour in humans may be caused by a perfectly associated founder mutation in a regulatory element located within the *HERC2* gene inhibiting *OCA2* expression. *Human Genetics* **123**, 177–87; Sturm RA, Duffy DL, Zhao ZZ *et al.* (2008). A single SNP in an evolutionary conserved region within intron 86 of the *HERC2* gene determines human blue–brown eye colour. *American Journal of Human Genetics* **82**, 424–31

4 Dalbagni DG, Zhi-Ping R, Herr H *et al.* (2001). Genetic alterations in TP53 in recurrent urothelial cancer: a longitudinal study. *Clinical Cancer Research* **7**, 2797–801; Denzinger S, Mohren K, Knuechel R *et al.* (2006). Improved clonality analysis of multifocal bladder tumors by combination of histopathologic organ mapping, loss of heterozygosity, fluorescence in situ hybridization, and p53 analysis. *Human Pathology* **37**, 143–51

5 Hafner C, Knuechel R, Stoehr R and Hartman A (2002). Clonality of multifocal urothelial carcinomas: 10 years of molecular genetic studies. *International Journal of Cancer* **101**, 1–6

6 Gerlinger M, Rowan AJ, Horswell S *et al.* (2012). Tumour heterogeneity and branched evolution revealed by multiregion sequencing. *New England Journal of Medicine* **366**, 883–92

7 Greaves LC, Preston SL, Tadrous PJ *et al.* (2006). Mitochondrial DNA mutations are established in human colonic stem cells, and mutated clones expand by crypt fission. *Proceedings of the National Academy of Sciences of the USA* **103**, 714–19

8 Leedham SJ, Graham TA, Oukrif D *et al.* (2009). Clonality, founder mutations, and field cancerization in human ulcerative colitis-associated neoplasia. *Gastroenterology* **136**, 542–50; Salk JJ, Salipante SJ, Risques RA *et al.* (2009). Clonal expansions in ulcerative colitis identify patients with neoplasia. *Proceedings of the National Academy of Sciences of the USA* **106**, 20871–6; Hayashi H, Miyagi Y, Sekiyama A *et al.* (2012). Colorectal small cell carcinoma in ulcerative colitis with identical rare p53 gene mutation to associated adenocarcinoma and dysplasia. *Journal of Crohn's and Colitis* **6**, 112–15

9 Jones S, Chen W-D, Parmigiani G *et al.* (2008). Comparative lesion sequencing provides insights into tumor evolution. *Proceedings of the National Academy of Sciences of the USA* **105**, 4283–8

10 Campbell PJ, Pleasance ED, Stephens PJ *et al.* (2008). Subclonal phylogenetic structures in cancer revealed by ultra-deep sequencing. *Proceedings of the National Academy of Sciences of the USA* **105**, 13081–6; Shah SP, Morin RD, Khattra J *et al.* (2009). Mutational evolution in a lobular breast tumour, profiled at single nucleotide resolution. *Nature* **461**, 809–13; Campbell PJ, Yachida

S, Mudie LJ *et al.* (2010). The patterns and dynamics of genetic instability in metastatic pancreatic cancer. *Nature* **467**, 1109–13; Yachida S, Jones S, Bozic I *et al.* (2010). Distant metastasis occurs late during the genetic evolution of pancreatic cancer. *Nature* **467**, 1114–17; Tao Y, Ruan J, Yeh A-H *et al.* (2011). Rapid growth of a hepatocellular carcinoma and the driving mutations revealed by cell-population genetic analysis of whole-genome data. *Proceedings of the National Academy of Sciences of the USA* **108**, 12042–7

11 Navin N, Kendal J, Troge J *et al.* (2011). Tumour evolution inferred by single-cell sequencing. *Nature* **472**, 90–4

12 Greaves M (2010). Cancer stem cells? Back to Darwin. *Seminars in Cancer Biology* **20**, 65–70; Stratton MR (2011). Exploring the genomes of cancer cells: progress and promise. *Science* **331**, 1553–8; Greaves M and Maley CC (2012). Clonal evolution in cancer. *Nature* **481**, 306–13

13 Lieber MR (2008). The mechanism of human nonhomologous end joining. *Journal of Biological Chemistry* **283**, 1–5

14 Pace II JK, Sen SK, Batzer MA and Feschotte C (2009). Repair-mediated duplication by capture of proximal chromosome DNA has shaped vertebrate genome evolution. *PLoS Genetics* **5**, e1000469 (Supplementary Information, Figure S1)

15 Thomas EE, Srebro N, Sebat J *et al.* (2004). Distribution of short paired duplications in primate genomes. *Proceedings of the National Academy of Sciences of the USA* **101**, 10349–54

16 Morrish TA, Gilbert N, Myers JS *et al.* (2002). DNA repair mediated by endonuclease-independent LINE-1 retrotransposition. *Nature Genetics* **31**, 159–65

17 Sen SK, Huang CT, Han K and Batzer M (2007). Endonuclease-independent insertion provides an alternative pathway for L1 retrotransposition in the human genome. *Nucleic Acids Research* **35**, 3741–51

18 Srikanta D, Sen SK, Huang CT *et al.* (2009). An alternative pathway for *Alu* retrotransposition suggests a role in DNA double-strand break repair. *Genomics* **93**, 205–12

19 Willett-Brozick JE, Savul SA, Richey LE and Baysal BE (2001). Germ line insertion of mtDNA at the breakpoint junction of a reciprocal constitutional translocation. *Human Genetics* **109**, 216–23

20 Millar DS, Tysoe C, Lazarou LP *et al.* (2010). An isolated case of lissencephaly caused by the insertion of a mitochondrial genome-derived DNA sequence into the 5′ untranslated region of the *PAFAH1B1* (L1S1) gene. *Human Genomics* **4**, 384–93

21 Bensasson D, Zhang D-X, Hartl DL and Hewitt GM (2001). Mitochondrial pseudogenes: evolution's misplaced witness. *Trends in Ecology and Evolution*

16, 314–21; Hazkani-Covo E, Zeller RM and Martin W (2010). Molecular poltergeists: mitochondrial DNA copies (*numts*) in sequenced nuclear genomes. *PLoS Genetics* **6**, e1000834

22 Tourmen Y, Baris O, Dessen P *et al.* (2002). Structure and chromosomal distribution of human mitochondrial pseudogenes. *Genomics* **80**, 71–7; Woischnik M and Moraes CT (2002). Pattern of organization of human mitochondrial pseudogenes in the nuclear genome. *Genome Research* **12**, 885–93; Ricchetti M, Tekaia F and Dujon B (2004). Continued colonisation of the human genome by mitochondrial DNA. *PLoS Biology* **2**, e273; Gherman A, Chen PE, Teslovich TM *et al.* (2007). Population bottlenecks as a potential major shaping force of human genome architecture. *PLoS Genetics* **3**, e119; Simone D, Calabrese FM, Lang M *et al.* (2011). The reference human nuclear mitochondrial sequences compilation validated and implemented on the UCSC genome browser. *BMC Genomics* **12**, 517

23 Zischler H, Geisert H and Castresana J (1998). A hominoid-specific nuclear insertion of the mitochondrial D-loop: implications for reconstructing ancestral mitochondrial sequences. *Molecular Biology and Evolution* **15**, 463–9; Zischler H (2000). Nuclear integrations of mitochondrial DNA in primates; inference of associated mutational events. *Electrophoresis* **21**, 531–6; Schmitz J, Ohme M and Zischler H (2002). The complete mitochondrial sequence of *Tarsius bancanus*: evidence for an extensive nucleotide compositional plasticity of primate mitochondrial DNA. *Molecular Biology and Evolution* **19**, 544–53; Ovchinnikov IV and Kholina OI (2010). Genome digging: insight into the mitochondrial genome of Homo. *PLoS ONE* **5**, e14278

24 Ricchetti *et al.* (2004), ref. 22; Hazkani-Covo E and Graur D (2007). A comparative analysis of *numt* evolution in human and chimpanzee. *Molecular Biology and Evolution* **24**, 13–8; Hazkani-Covo E and Covo S (2008). *Numt*-mediated double-strand break repair mitigates deletions during primate genome evolution. *PLoS Genetics* **4**, e1000237; Lang M, Sazzini M, Calabrese FM *et al.* (2012). Polymorphic numts trace human population relationships. *Human Genetics* **131**, 757–71; Calabrese FM, Simone D and Attimonelli M (2012). Primates and mouse numts in the UCSC genome browser. *BMC Bioinformatics* **13** (Suppl. 4), S15 (figures kindly supplied by Dr F Calabrese, personal communication).

25 Jensen-Seaman MI, Wildschutte JH, Soto-Calderon ID and Anthony NM (2009). A comparative approach shows differences in patterns of numt insertion during hominoid evolution. *Journal of Molecular Evolution* **68**, 688–99; Hazkani-Covo E (2009). Mitochondrial insertions into primate nuclear genomes suggest the use of *numts* as a tool for phylogeny. *Molecular Biology and Evolution* **26**, 2175–9; Soto-Calderon ID, Lee EJ, Jensen-Seaman MI and Anthony NM (2012).

Factors affecting the relative abundance of nuclear copies of mitochondrial DNA (numts) in hominoids. *Journal of Molecular Evolution* **75**, 102–11; Tsuji J, Frith MC, Tomii K and Horton P (2012). Mammalian numt insertion is non-random. *Nucleic Acids Research* **40**, 9073–88 (see Supplementary Material)

26 Yunis JJ and Prakash O (1986). The origin of man: a chromosomal pictorial legacy. *Science* **215**, 1525–30

27 Kehrer-Sawatzki H and Cooper DN (2007). Structural differences between the human and chimpanzee genomes. *Human Genetics* **120**, 759–78

28 IJdo JW, Baldini A, Ward DC *et al.* (1991). Origin of human chromosome 2: an ancestral telomere–telomere fusion. *Proceedings of the National Academy of Sciences of the USA* **88**, 9051–5

29 Fan Y, Linardopoulou E, Friedman C *et al.* (2002). Genomic structure and evolution of the ancestral chromosome fusion site in 2q13–2q14.1 and paralogous regions on other human chromosomes. *Genome Research* **12**, 1651–62; Hillier LW, Graves TA, Fulton RS *et al.* (2005). Generation and annotation of the DNA sequences of human chromosomes 2 and 4. *Nature* **434**, 724–31

30 Azzalin CM, Nergadze SG and Giulotto E (2001). Human intrachromosomal telomeric repeats: sequence organization and mechanisms of origin. *Chromosoma* **110**, 75–82; Nergadze SG, Rocchi M, Azzalin CM *et al.* (2004). Insertion of telomeric repeats at intrachromosomal break sites during primate evolution. *Genome Research* **14**, 1704–10; Nergadze SG, Santagostino MA, Salzano A *et al.* (2007). Contribution of telomerase RNA retrotranscription to DNA double-strand break repair during mammalian genome evolution. *Genome Biology* **8**, R260

31 Nagy R, Sweet K and Eng C (2004). Highly penetrant cancer syndromes. *Oncogene* **23**, 6445–70

32 Harbour JW (1998). Overview of RB gene mutations in patients with retinoblastoma: implications for clinical genetic screening. *Ophthalmology* **105**, 1442–7; Balogh K, Patócs A, Majnik J *et al.* (2004). Genetic screening methods for the detection of mutations responsible for multiple endocrine neoplasia type 1. *Molecular Genetics and Metabolism* **83**, 74–81; Ferla R, Calo V, Cascio S *et al.* (2007). Founder mutations in *BRCA1* and *BRCA2* genes. *Annals of Oncology* **18** (Suppl. 6), vi93–8; Walsh T, Casadei S, Coats KH *et al.* (2006). Spectrum of mutations in *BRCA1, BRCA2, CHEK2,* and *TP53* in families at high risk of breast cancer. *Journal of the American Medical Association* **295**, 1379–88; Lindstrom E, Shimokawa T, Toftgard R and Zaphiropoulos PG (2006). PTCH mutations: distribution and analysis. *Human Mutation* **27**, 215–9; Nordstrom-O'Brien M, van der Luijt RB and van Rooijen E (2010). Genetic analysis of von Hippel–Lindau disease. *Human Mutation* **31**, 521–37

33 Garritano S, Gemignani F, Palmeri EI *et al.* (2009). Detailed haplotype analysis at the *TP53* locus in p.R337H mutation carriers in the population of Southern Brazil: evidence for a founder effect. *Human Mutation* **31**, 143–50

34 Zheng D and Gerstein MB (2007). The ambiguous boundary between genes and pseudogenes: the dead rise up, or do they? *Trends in Genetics* **23**, 219–24; Harrison PM, Hegyi H, Balasubramian S *et al.* (2002). Molecular fossils in the human genome: identification and analysis of the pseudogenes in chromosomes 21 and 22. *Genome Research* **12**, 272–80; Zheng D, Frankish A, Baertsch R *et al.* (2007). Pseudogenes in the ENCODE regions: consensus annotation, analysis of transcription, and evolution. *Genome Research* **17**, 839–51

35 Baguley BC and Finlay GJ (1988). Derivatives of amsacrine: determinants required for high activity against the Lewis lung carcinoma. *Journal of the National Cancer Institute* **80**, 195–9; Baguley BC and Finlay GJ (1988). Relationship between the structure of analogues of amsacrine and their degree of cross-resistance to adriamycin-resistant P388 leukaemia cells. *European Journal of Cancer and Clinical Oncology* **24**, 205–10

36 Neff MW, Robertson KR, Wong AK *et al.* (2004). Breed distribution and history of canine *mdr1–1Δ*, a pharmacogenetic mutation that marks the emergence of breeds from the collie lineage. *Proceedings of the National Academy of Sciences of the USA* **101**, 11725–30

37 Balasubramanian S, Habegger L, Frankish A *et al.* (2011). Gene inactivation and its implications for annotation in the era of personal genomics. *Genes and Development* **25**, 1–10; MacArthur DG, Balasubramanian S, Frankish A *et al.* (2012). A systematic survey of loss-of-function variants in human protein-coding genes. *Science* **335**, 823–8

38 Rompler H, Schulz A, Pitra C *et al.* (2005). The rise and fall of the chemoattractant receptor GPR33. *Journal of Biological Chemistry* **280**, 31068–75; Bohnekamp J, Boselt I, Saalbach A *et al.* (2010). Involvement of the chemokine-like receptor GPR33 in innate immunity. *Biochemical and Biophysical Research Communications* **396**, 272–7

39 Galvani AP and Novembre J (2005). The evolutionary history of the *CCR5-Δ32* HIV-resistance mutation. *Microbes and Infection* **7**, 302–9

40 Mercereau-Puijalon O and Menard D (2010). *Plasmodium vivax* and the Duffy antigen: a paradigm revisited. *Transfusion clinique et biologique* **17**, 176–83

41 Xue Y, Daly A, Yngvadottir B *et al.* (2006). Spread of an inactive form of caspase-12 in humans is due to recent positive selection. *American Journal of Human Genetics* **78**, 659–70

42 Hermel E and Klapstein KD (2011). A possible mechanism for maintenance of the deleterious allele of human *CASPASE-12*. *Medical Hypotheses* **77**, 803–6;

Hervella M, Plantinga TS, Alonso S *et al.* (2012). The loss of functional caspase-12 in Europe is a pre-Neolithic event. *PLoS ONE* 7, e37022

43 Wang *et al.* (2012), Chapter 2, ref. 145

44 Chou H-H, Takematsu H, Diaz S *et al.* (1998). A mutation in human CMP–sialic acid hydroxylase occurred after the *Homo–Pan* divergence. *Proceedings of the National Academy of Sciences of the USA* **95**, 11751–6; Hayakawa T, Satta Y, Gagneux P *et al.* (2001). *Alu*-mediated inactivation of the human CMP-N-acetylneuraminic acid hydroxylase gene. *Proceedings of the National Academy of Sciences of the USA* **98**, 11399–404

45 Hedlund M, Padler-Karavani V, Varki NM and Varki A (2008). Evidence for a human-specific mechanism for diet and antibody-mediated inflammation in carcinoma progression. *Proceedings of the National Academy of Sciences of the USA* **105**, 18936–41; Taylor RE, Gregg CJ, Padler-Karavani V *et al.* (2010). Novel mechanism for the generation of human xeno-autoantibodies against the nonhuman sialic acid *N*-glycolyneuraminic acid. *Journal of Experimental Medicine* **207**, 1637–46; Ghaderi D, Taylor RE, Padler-Karavani V *et al.* (2010). Implications of the presence of *N*-glycolylneuraminic acid in recombinant therapeutic glycoproteins. *Nature Biotechnology* **28**, 863–7; Varki A (2010). Uniquely human evolution of sialic acid genetics and biology. *Proceedings of the National Academy of Sciences of the USA* **107** (Suppl 2), 8939–46

46 Winter H, Langbein L, Krawczak M *et al.* (2001). Human type I hair keratin pseudogene *ψhHaA* has functional orthologs in the chimpanzee and gorilla: evidence for the inactivation of the human gene after the *Pan–Homo* divergence. *Human Genetics* **108**, 37–42

47 Kazantseva A, Goltsov A, Zinchenko R *et al.* (2006). Human hair growth deficiency is linked to a genetic defect in the phospholipase gene *LIPH*. *Science* **314**, 982–5

48 Stedman HH, Kozyak BW, Nelson A *et al.* (2004). Myosin gene mutation correlates with anatomical changes in the human lineage. *Nature* **428**, 415–18

49 Kim HL, Igawa T, Kawashima A *et al.* (2010). Divergence, demography and gene loss along the human lineage. *Philosophical Transactions of the Royal Society Series B: Biological Sciences* **365**, 2541–7

50 Zhu J, Sanborn JZ, Diekhans M *et al.* (2007). Comparative genomics search for losses of long-established genes on the human lineage. *PLoS Computational Biology* **3**, e247

51 Annilo T and Dean M (2004). Degeneration of an ATP-binding cassette transporter gene, *ABCC13*, in different mammalian lineages. *Genomics* **84**, 34–46

52 Martinez-Arias R, Calafell F, Mateu E *et al.* (2001). Sequence variability of a human pseudogene. *Genome Research* **11**, 1071–85; Wafaei JR and Choy FYM

(2005). Glucocerebrosidase recombinant allele: molecular evolution of the glucocerebrosidase gene and pseudogene in primates. *Blood Cells, Molecules and Diseases* **35**, 277–85

53 Apoil P-A, Roubinet F, Despiau S *et al.* (2000). Evolution of α2-fucosyltransferase genes in primates: relation between an intronic *Alu*-Y element and red cell expression of ABH antigens. *Molecular Biology and Evolution* **17**, 337–51; Saunier K, Barreaud J-P, Eggen A *et al.* (2001). Organization of the bovine α2-fucosyltransferase gene cluster suggests that the *Sec1* gene might have been shaped through a nonautonomous L1-retrotransposition event within the same locus. *Molecular Biology and Evolution* **18**, 2083–91

54 Oda M, Satta Y, Takenaka O and Takahata N (2002). Loss of urate oxidase activity in hominoids and its evolutionary implications. *Molecular Biology and Evolution* **19**, 640–53

55 Keebaugh AC and Thomas JW (2010). The evolutionary fate of the genes encoding the purine catabolic enzymes in hominoids, birds and reptiles. *Molecular Biology and Evolution* **27**, 1359–69

56 Ivell R, Pusch W, Balvers M *et al.* (2000). Progressive inactivation of the haploid gene for the sperm-specific endozepine-like peptide (ELP) through primate evolution. *Gene* **255**, 335–45

57 Goldberg A, Wildman DE, Schmidt TR *et al.* (2003). Adaptive evolution of cytochrome *c* oxidase subunit VIII in anthropoid primates. *Proceedings of the National Academy of Sciences of the USA* **100**, 5873–8

58 Koike C, Fung JJ, Geller DA *et al.* (2002). Molecular basis of evolutionary loss of the α1,3-galactosyltransferase gene in higher primates. *Journal of Biological Chemistry* **277**, 10114–20; Koike C, Uddin M, Wildman DE *et al.* (2007). Functionally important glycosyltransferase gain and loss during catarrhine primate emergence. *Proceedings of the National Academy of Sciences of the USA* **104**, 559–64 (Supplementary Information, Data Set 1)

59 Wigglesworth KM, Racki WJ, Mishra R *et al.* (2011). Rapid recruitment and activation of macrophages by anti-Gal/αGal liposome interaction accelerates wound healing. *Journal of Immunology* **186**, 4422–32

60 Morisset M, Richard C, Astier C *et al.* (2012). Anaphylaxis to pork kidney is related to IgE antibodies specific for galactose-alpha-1,3-galactose. *Allergy* **67**, 699–704; Wolver SE, Sun DR, Commins SP and Schwartz LB (2013). A peculiar case of anaphylaxis: no more steak? The journey to discovery of a newly recognized allergy to galactose-alpha-1,3-galactose found in mammalian meat. *Journal of General Internal Medicine* **28**, 322–5

61 Cai X and Patel S (2010). Degeneration of an intracellular ion channel in the primate lineage by relaxation of selective constraints. *Molecular Biology and Evolution* **27**, 2352–9

62 Ohta Y and Nishikimi M (1999). Random nucleotide substitutions in primate non-functional gene for L-gulono-γ-lactone oxidase, the missing enzyme in L-ascorbic acid biosynthesis. *Biochimica et Biophysica Acta* **1472**, 408–11; Inai Y, Ohta Y and Nishikimi M (2003). The whole structure of the human non-functional L-gulono-γ-lactone oxidase gene – the gene responsible for scurvy – and the evolution of repetitive sequences thereon. *Journal of Nutritional Science and Vitaminology* **49**, 315–19

63 Zhu *et al.* (2007), ref. 50

64 Lachapelle MY and Drouin G (2011). Inactivation dates of the human and guinea pig vitamin C genes. *Genetica* **139**, 199–207 (Supplementary Figure 1)

65 Cui J, Yuan X, Wang L *et al.* (2011). Recent loss of vitamin C biosynthesis ability in bats. *PLoS ONE* **6**, e27114

66 Drouin G, Godin JR and Page B (2011). The genetics of vitamin C loss in vertebrates. *Current Genomics* **12**, 371–8

67 Johnson RJ, Gaucher EA, Sautin YY *et al.* (2008). The planetary biology of ascorbate and uric acid and their relationship with the epidemic of obesity and cardiovascular disease. *Medical Hypotheses* **71**, 22–31; Johnson RJ, Andrews P, Benner SA and Oliver W (2010). The evolution of obesity: insights from the mid-Miocene. *Transactions of the American Clinical and Climatological Association* **121**, 295–305

68 Fernandez E, Torrents D, Zorzano A *et al.* (2005). Identification and functional characterization of a novel low-affinity aromatic-preferring amino acid transporter (arpAT). *Journal of Biological Chemistry* **280**, 19364–72; Casals F, Ferrer-Admetlla A, Chillaron J *et al.* (2008). Is there selection for the pace of successive inactivation of the *arpAT* gene in primates? *Journal of Molecular Evolution* **67**, 23–8

69 Naidu S, Peterson ML and Spear BT (2010). *Alpha-fetoprotein-related gene (ARG)*: a member of the albumin gene family that is no longer functional in primates. *Gene* **449**, 95–102

70 Zhang Z and Gerstein M (2003). The human genome has 49 cytochrome c pseudogenes, including a relic of a primordial gene that still functions in mouse. *Gene* **312**, 61–72; sequences are from http://bioinfo.mbb.yale.edu/genome/pseudogene/human-cyc/; Pierron D, Opazo JC, Heiske M *et al.* (2011). Silencing, positive selection and parallel evolution: busy history of primate cytochromes *c*. *PLoS ONE* **6**, e26269 (Supporting Information, Figure S3)

71 Zhang ZD, Frankish A, Hunt T *et al.* (2010). Identification and analysis of unitary pseudogenes: historic and contemporary gene losses in humans and other primates. *Genome Biology* **11**, R26

72 Liman ER (2006). Use it or lose it: molecular evolution of sensory signalling in primates. *Pflugers Archiv: European Journal of Physiology* **453**, 125–31; Niimura Y and Nei M (2006). Evolutionary dynamics of olfactory and other chemosensory receptor genes in vertebrates. *Journal of Human Genetics* **51**, 505–17; Rouquier S and Giorgi D (2007). Olfactory receptor gene repertoires in mammals. *Mutation Research* **616**, 95–102; Nei M, Niimura Y and Nozawa M (2008). The evolution of animal chemosensory receptor repertoires: roles of chance and necessity. *Nature Reviews Genetics* **9**, 951–63

73 Mundy NI (2006). Genetic basis of olfactory communication in primates. *American Journal of Primatology* **68**, 559–67

74 Liman ER and Innan H (2003). Released selective pressure on an essential component of pheromone transduction in primate evolution. *Proceedings of the National Academy of Sciences of the USA* **100**, 3328–32; Zhang J and Webb DM (2003). Evolutionary deterioration of the vomeronasal pheromone transduction pathway in catarrhine primates. *Proceedings of the National Academy of Sciences of the USA* **100**, 8337–41

75 Yu L, Jin W, Wang J-X *et al.* (2010). Characterization of *TRPC2*, an essential genetic component of VNS chemoreception provides insights into the evolution of pheromonal olfaction in secondary-adapted marine mammals. *Molecular Biology and Evolution* **27**, 1467–77

76 Zhang and Webb (2003), ref. 74; Mundy NI and Cook S (2003). Positive selection during the diversification of class I vomeronasal receptor-like (V1RL) genes, putative pheromone receptor genes, in human and primate evolution. *Molecular Biology and Evolution* **20**, 1805–10

77 Young JM, Kambere M, Trask B and Lane RP (2005). Divergent V1R repertoires in five species: amplification in rodents, decimation in primates, and a surprisingly small repertoire in dogs. *Genome Research* **15**, 231–40; Young JM, Massa HF, Hsu L and Trask BJ (2010). Extreme variability among mammalian V1R receptors. *Genome Biology* **20**, 10–18

78 Young JM and Trask B (2007). V2R gene families degenerated in primates, dog and cow, but expanded in opossum. *Trends in Genetics* **23**, 212–5; Suarez R, Fernandez-Aburto P, Manger PR and Mpodozis J (2011). Deterioration of the Gαo vomeronasal pathway in sexually dimorphic mammals. *PLoS ONE* **6**, e23436

79 Go Y and Niimura Y (2008). Similar numbers but different repertoires of olfactory receptor genes in humans and chimpanzees. *Molecular Biology and Evolution* **25**, 1897–907; Matsui A, Go Y and Niimura Y (2010). Degeneration of olfactory receptors gene repertories in primates: no direct link to full trichromatic vision. *Molecular Biology and Evolution* **27**, 1192–200

80 Gilad Y, Man O, Paabo S and Lancet D (2003). Human-specific loss of olfactory receptor genes. *Proceedings of the National Academy of Sciences of the USA* **100**, 3324–7

81 Gilad Y, Man O and Glusman G (2005). A comparison of the human and chimpanzee olfactory receptor gene repertoires. *Genome Research* **15**, 224–30

82 Olender T, Waszak SM, Viavant M *et al.* (2012). Personal receptor repertoires: olfaction as a model. *BMC Genomics* **13**, 414; Pierron D, Cortés NG, Letellier T and Grossman LI (2013). Current relaxation of selection on the human genome: tolerance of deleterious mutations on olfactory receptors. *Molecular Phylogenetics and Evolution* **66**, 558–64

83 Freitag J, Ludwig G, Andreini I *et al.* (1998). Olfactory receptors in aquatic and terrestrial vertebrates. *Journal of Comparative Physiology A* **183**, 635–50; Kishida T, Kubota S, Shirayama Y and Fukami H (2007). The olfactory receptor gene repertoires in secondary-adapted marine vertebrates: evidence for reduction of the functional proportions in cetaceans. *Biology Letters* **3**, 428–30; Niimura Y and Nei M (2007). Extensive gains and losses of olfactory receptor genes in mammalian evolution. *PLoS ONE* **2**, e708; Hayden S, Bekaert M, Crider TA *et al.* (2010). Ecological adaptation determines functional mammalian olfactory subgenomes. *Genome Research* **20**, 1–9

84 Young JM, Waters H, Dong C *et al.* (2007). Degeneration of the olfactory guanylyl cyclase D gene during primate evolution. *PLoS ONE* **2** e884

85 Wooding S, Bufe B, Gassi C *et al.* (2006). Independent evolution of bitter-taste sensitivity in humans and chimpanzees. *Nature* **440**, 930–4; Lalueza-Fox C, Gigli E, de la Rasilla M *et al.* (2009). Bitter taste perception in Neanderthals through the analysis of the *TAS2R38* gene. *Biology Letters* **5**, 809–11

86 Fischer A, Gilad Y, Man O and Paabo S (2004). Evolution of bitter taste receptors in humans and apes. *Molecular Biology and Evolution* **22**, 432–6; Perry CM, Erkner A and le Coutre J (2004). Divergence of T2R chemosensory receptor families in humans, bonobos, and chimpanzees. *Proceedings of the National Academy of Sciences of the USA* **101**, 14830–4

87 Li X, Li W, Wang H *et al.* (2005). Pseudogenization of a sweet-receptor gene accounts for cats' indifference towards sugar. *PLoS Genetics* **1**, 27–35; Li X, Li W, Wang H *et al.* (2006). Cats lack a sweet taste receptor. *Journal of Nutrition* **136**, 1932S–4S; Jiang P, Josue J, Li X *et al.* (2012). Major taste loss in carnivorous mammals. *Proceedings of the National Academy of Sciences of the USA* **109**, 4956–61

88 Zhao H, Zhou Y, Pinto CM *et al.* (2010). Evolution of the sweet taste receptor gene *Tas1R2* in bats. *Molecular Biology and Evolution* **27**, 2642–50

89 Zhao H, Yang J-R, Xu H and Zhang J (2010). Pseudogenization of the umami taste receptor gene *Tas1r1* in the giant panda coincided with its dietary switch

to bamboo. *Molecular Biology and Evolution* **27**, 2669–73; Jiang *et al.* (2012), ref. 87; Sato JJ and Wolsan M (2012). Loss or major reduction of umami taste sensation in pinnipeds. *Naturwissenschaften* **99**, 655–9

90 Renard C, Chardon P and Vaiman N (2003). The phylogenetic history of the MHC class I gene families in pigs, including a fossil gene predating mammalian radiation. *Journal of Molecular Evolution* **57**, 420–34

91 Brawand D, Wahli W and Kaessmann H (2008). Loss of egg yolk genes in mammals and the origin of lactation and placentation. *PLoS Biology* **6**, e63

92 Davit-Béal T, Tucker AS and Sire J-Y (2009). Loss of teeth and enamel in tetrapods: fossil record, genetic data and morphological adaptations. *Journal of Anatomy* **214**, 477–501

93 Deméré TA, McGowen MR, Berta A and Gatesy J (2008). Morphological and molecular evidence for a stepwise evolutionary transition from teeth to baleen in mysticete whales. *Systematic Biology* **57**, 15–37; Meredith RW, Gatesy J, Murphy WJ *et al.* (2009). Molecular decay of the tooth gene enamelin (*ENAM*) mirrors the loss of enamel in the fossil record of placental mammals. *PLoS Genetics* **5**, e1000634

94 Meredith *et al.* (2011), Chapter 2, ref. 102

95 Louchart A and Viriot L (2011). From snout to beak: the loss of teeth in birds. *Trends in Ecology and Evolution* **26**, 663–73

96 Harris MP, Hasso SM, Ferguson MWJ and Fallon JF (2006). The development of archosaurian first-generation teeth in a chicken mutant. *Current Biology* **16**, 371–7

97 Al-Hashimi N, Lafont A-G, Delgado S *et al.* (2010). The enamelin genes in lizard, crocodile and frog, and the pseudogene in the chicken provide new insights on enamelin evolution in tetrapods. *Molecular Biology and Evolution* **27**, 2078–94

98 Sire J-Y, Delgado SC and Girondot M (2008). Hen's teeth with enamel cap: from dream to impossibility. *BMC Evolutionary Biology* **8**, 246; McKnight DA and Fisher LW (2009). Molecular evolution of dentin phosphoprotein among toothed and toothless animals. *BMC Evolutionary Biology* **9**, 299; Meredith RW, Gatesy J and Springer S (2013). Molecular decay of enamel matrix protein genes in turtles and other edentulous amniotes. *BMC Evolutionary Biology* **13**, 20

99 Kawasaki K, Lafont A-G and Sire J-Y (2011). The evolution of milk casein genes from tooth genes before the origin of mammals. *Molecular Biology and Evolution* **28**, 2053–61

100 Esnault C, Maestre J and Heidmann T (2000). Human LINE retrotransposons generate processed pseudogenes. *Nature Genetics* **24**, 363–7

101 Awano H, Malueka RG, Yagi M *et al.* (2010). Contemporary retrotransposition of a novel non-coding gene induces exon-skipping in dystrophin mRNA. *Journal of Human Genetics* **55**, 785–90; see also Tabata A, Sheng J-S, Ushikai

M *et al.* (2008). Identification of 13 novel mutations including a retrotransposal insertion in *SLC25A13* gene and frequency of 30 mutations found in patients with citrin deficiency. *Journal of Human Genetics* **53**, 534–45

102 Zhang Z, Carriero N and Gerstein M (2004). Comparative analysis of processed pseudogenes in the mouse and human genomes. *Trends in Genetics* **20**, 62–7; Zhang Z and Gerstein M (2004). Large-scale analysis of pseudogenes in the human genome. *Current Opinion in Genetics and Development* **14**, 328–35; Pavlicek A, Gentles AJ Paces J *et al.* (2006). Retrotransposition of processed pseudogenes: the impact of RNA stability and translational control. *Trends in Genetics* **22**, 69–73

103 Nachman MW and Crowell SL (2000). Estimate of the mutations rate per nucleotide in humans. *Genetics* **156**, 297–304; Booth HAF and Holland PWH (2004). Eleven daughters of *NANOG*. *Genomics* **84**, 229–38; Fairbanks DJ and Maughan PJ (2006). Evolution of the *NANOG* pseudogene family in the human and chimpanzee genomes. *BMC Evolutionary Biology* **6**, 12

104 Fairbanks DJ, Fairbanks AD, Ogden TH *et al.* (2012). *NANOGP8*: evolution of a human-specific retro-oncogene. *G3: Genes, Genomes, Genetics* **2**, 1447–57

105 Ejima Y and Yang L (2003). *Trans* mobilization of genomic DNA as a mechanism for retrotransposon-mediated exon shuffling. *Human Molecular Genetics* **12**, 1321–8

106 Devor EJ, Dill-Devor RM, Magee HJ and Waziri R (1998). Serine hydroxymethyltransferase pseudogene *SHMT-ps1*: a unique genetic marker of the order Primates. *Journal of Experimental Zoology* **282**, 150–6; Friedberg F and Rhoads AR (2000). Calculation and verification of the ages of retroprocessed pseudogenes. *Molecular Phylogenetics and Evolution* **16**, 127–30; Devor EJ (2001). Use of molecular beacons to verify that the serine hydroxymethyltransferase pseudogene *SHMT-ps1* is unique to the order Primates. *Genome Biology* **2**, research0006.1; Gotter AL and Reppert SM (2001). Analysis of human *Per4*. *Molecular Brain Research* **92**, 19–26; Devor EJ and Moffat-Wilson K (2004). An ancient RNASE H1 splice junction mutant preserved in a 19-million-year-old genetic fossil in ape genomes. *Journal of Heredity* **95**, 257–61

107 Devor EJ (2006). Primate microRNAs *miR-220* and *miR-492* lie within processed pseudogenes. *Journal of Heredity* **97**, 186

108 Locke *et al.* (2011) (Supplemental Section S10), Chapter 1, ref. 37

109 Schmitz J, Churakov G, Zischler H and Brosius J (2004). A novel class of mammalian-specific tailless retropseudogenes. *Genome Research* **14**, 1911–15

110 Perreault J, Noel J-F, Briere F *et al.* (2005). Retropseudogenes derived from the human Ro/SS-A autoantigen-associated hY RNAs. *Nucleic Acids Research* **33**, 2032–41

111 Janecka JE, Miller W, Pringle TH *et al.* (2007). Molecular and genomic data identify the closest living relatives of primates. *Science* **318**, 792–4

112 Poux C, van Rheede T, Madsen O and de Jong WW (2002). Sequence gaps join mice and men: phylogenetic evidence from deletions in two proteins. *Molecular Biology and Evolution* **19**, 2035–7; de Jong WW, van Dijk MAM, Poux C *et al.* (2003). Indels in protein-coding sequences of Euarchontoglires constrain the rooting of the eutherian tree. *Molecular Phylogenetics and Evolution* **28**, 328–40; Springer MS, Stanhope MJ, Madsen MJ and de Jong WW (2004). Molecules consolidate the placental mammal tree. *Trends in Ecology and Evolution* **19**, 430–8

113 Murphy *et al.* (2007), Chapter 2, ref. 84

114 van Rheede T, Bastiaans T, Boone DN *et al.* (2006). The platypus is in its place: nuclear genes and indels confirm the sister group relation of Monotremes and Therians. *Molecular Biology and Evolution* **23**, 587–97

115 Wetterbom A, Sevov M, Cavelier L and Bergstrom TF (2006). Comparative genomic analysis of human and chimpanzee indicates a key role for indels in primate evolution. *Journal of Molecular Evolution* **63**, 682–90

116 Chaisson MJ, Raphael BJ and Pevzner PA (2006). Microinversions in mammalian evolution. *Proceedings of the National Academy of Sciences of the USA* **103**, 19824–9

117 Perelman P, Johnson WE, Roos C *et al.* (2011). A molecular phylogeny of living primates. *PLoS Genetics* 7, e1001342 (Supporting Information, Table S5)

118 McLean CY, Reno PL, Pollen AA *et al.* (2011). Human-specific loss of regulatory DNA and the evolution of human-specific traits. *Nature* **471**, 216–19

119 See ref. 32

120 Bekpen C, Marques-Bonet T, Alkan C *et al.* (2009). Death and resurrection of the human *IRGM* gene. *PLoS Genetics* **5**, e1000403; Bekpen C, Xavier RJ and Eichler EE (2010). Human *IRGM* gene 'to be or not to be'. *Seminars in Immunopathology* **32**, 437–44

121 Poliseno L, Salmena L, Zhang J *et al.* (2010). A coding-independent function of gene and pseudogene mRNAs regulates tumour biology. *Nature* **465**, 1033–8; Han YJ, Ma SF, Yourek G *et al.* (2011). A transcribed pseudogene of *MYLK* promotes cell proliferation. *FASEB Journal* **25**, 2305–12; Pink RC, Wicks K, Caley DP *et al.* (2011). Pseudogenes: pseudo-functional or key regulators in health and disease? *RNA* **17**, 792–8

122 Rebollo R, Romanish MT and Mager DL (2012). Transposable elements: an abundant and natural source of regulatory sequences for host cells. *Annual Review of Genetics* **46**, 21–42

4 THE ORIGINS OF NEW GENES

1 Vogt N, Lefevre S-H, Apoiu F *et al.* (2004). Molecular structure of double-minute chromosomes bearing amplified copies of the epidermal growth factor receptor gene in gliomas. *Proceedings of the National Academy of Sciences of the USA* **101**, 11368–73; Storlazzi CT, Lonoce A, Guastadisegni AC *et al.* (2010). Gene amplification as double minutes or homogeneously staining regions in solid tumors: origin and structure. *Genome Research* **20**, 1198–206

2 Isoda T, Ford AM, Tomizawa D *et al.* (2009). Immunologically silent cancer clone transmission from mother to offspring. *Proceedings of the National Academy of Sciences of the USA* **106**, 17882–5

3 Tannock IF and Hill RP, *The Basic Science of Oncology*, 3rd edn (New York: McGraw-Hill, 1998), 156

4 Rabes HM (2001). Gene rearrangements in radiation-induced thyroid carcinogenesis. *Medical and Pediatric Oncology* **36**, 574–82; Zhu Z, Ciampi R, Nikiforova MN *et al.* (2006). Prevalence of RET/PTC rearrangements in thyroid papillary carcinomas: effects of the detection methods and genetic heterogeneity. *Journal of Clinical Endocrinology and Metabolism* **91**, 3603–10; Kaye FJ (2009). Mutation-associated fusion cancer genes in solid tumours. *Molecular Cancer Therapeutics* **8**, 1399–408; Clark JP and Cooper CS (2010). ETS gene fusions in prostate cancer. *Nature Reviews Urology* **6**, 429–39; Palanisamy N, Ateeq B, Kalyana-Sundaram S *et al.* (2010). Rearrangements of the RAF kinase pathway in prostate cancer, gastric cancer, and melanoma. *Nature Medicine* **16**, 793–8; Sasaki T, Rodig SJ, Chirieac LR and Janne PA (2010). The biology and treatment of EML4-ALK non-small cell lung cancer. *European Journal of Cancer* **46**, 1773–80; Zitzelsberger H, Bauer V, Thomas G and Unger K (2010). Molecular rearrangements in papillary thyroid carcinomas. *Clinica Chimica Acta* **411**, 301–8

5 Iafrate AJ, Feuk L, Rivera MN *et al.* (2004). Detection of large-scale variation in the human genome. *Nature Genetics* **36**, 949–51; Sebat J, Lakshmi B, Troge J *et al.* (2004). Large-scale copy number polymorphism in the human genome. *Science* **305**, 525–8

6 Conrad DF, Pinto D, Redon R *et al.* (2010). Origins and functional impact of copy number variation in the human genome. *Nature* **464**, 704–12; Sudmant PH, Kitzman JO, Antonacci F *et al.* (2010). Diversity of human copy number variation and multicopy genes. *Science* **330**, 641–6

7 Nozowa M, Kawahara Y and Nei M (2007). Genomic drift and copy number variation of sensory receptor genes. *Proceedings of the National Academy of Sciences of the USA* **104**, 20421–6; Iskow RC, Gokcumen O and Lee C (2012).

Exploring the role of copy number variants in human adaptation. *Trends in Genetics* **28**, 245–57

8 Stults DM, Killen MW, Pierce HH *et al.* (2008). Genomic architecture and inheritance of ribosomal RNA gene clusters. *Genome Research* **18**, 13–18; Stults DM, Killen MW, Williamson EP *et al.* (2009). Human rRNA gene clusters are recombinational hotspots in cancer. *Cancer Research* **69**, 9096–104

9 Perry GH, Dominy NJ, Claw KG *et al.* (2007). Diet and the evolution of human amylase gene copy number variation. *Nature Genetics* **39**, 1256–60; Mandel AL, Peyrot des Gachons C, Plank KL *et al.* (2010). Individual differences in AMY1 copy number, salivary α-amylase levels, and the perception of oral starch. *PLoS ONE* **5**, e13352; Mandel AL and Breslin PAS (2012). High endogenous salivary amylase activity is associated with improved glycemic homeostasis following starch ingestion in adults. *Journal of Nutrition* **142**, 853–8

10 Bailey JA and Eichler EE (2006). Primate segmental duplications: crucibles of evolution, diversity and disease. *Nature Reviews Genetics* **7**, 552–64; Marques-Bonet T, Girirajan S and Eichler EE (2009). The origins and impact of primate segmental duplications. *Trends in Genetics* **25**, 443–54; Carvalho CMV, Zhang F and Lupski JR (2010). Genomic disorders: a window into human gene and genome evolution. *Proceedings of the National Academy of Sciences of the USA* **107** (Suppl. 1), 1765–71

11 Wimmer R, Kirsch S, Rappold GA and Schempp W (2002). Direct evidence for the *Homo–Pan* clade. *Chromosome Research* **10**, 55–61

12 Keller MP, Seifried BA and Chance PF (1999). Molecular evolution of the CMT1A region: a human- and chimpanzee-specific repeat. *Molecular Biology and Evolution* **16**, 1019–26

13 Delabre C, Nakauchi H, Bontrop R *et al.* (1993). Duplication of the *CD8* β-chain gene as a marker of the human–gorilla–chimpanzee clade. *Proceedings of the National Academy of Sciences of the USA* **90**, 7049–53

14 Saglio G, Storlazzi CT, Giugliano E *et al.* (2002). A 76-kb duplication maps close to the BCR gene on chromosome 22 and the ABL gene on chromosome 9: possible involvement in the genesis of the Philadelphia chromosome translocation. *Proceedings of the National Academy of Sciences of the USA* **99**, 9882–7; Albano F, Anelli L, Zagaria A *et al.* (2010). Genomic segmental duplications on the basis of the t(9;22) rearrangement in chronic myeloid leukemia. *Oncogene* **29**, 2509–16

15 Munch C, Kirsch S, Fernandez AMG and Schempp W (2008). Evolutionary analysis of the highly dynamic *CHEK2* duplicon in anthropoids. *BMC Evolutionary Biology* **8**, 269

16 Cheng Z, Ventura M, She X *et al.* (2005). A genome-wide comparison of recent chimpanzee and human segmental duplications. *Nature* **437**, 88–93; Linardopoulou EV, Williams EM, Fan Y *et al.* (2005). Human subtelomeres are hot spots of interchromosomal recombination and segmental duplication. *Nature* **437**, 94–100

17 Perry GH, Tchinda J, McGrath SD *et al.* (2006). Hotspots for copy number variation in chimpanzees and humans. *Proceedings of the National Academy of Sciences of the USA* **103**, 8006–11; Sudmant *et al.* (2010), ref. 6

18 Johnson ME, NISC Comparative Sequencing Program, Cheng Z *et al.* (2006). Recurrent duplication-driven transposition of DNA during hominoid evolution. *Proceedings of the National Academy of Sciences of the USA* **103**, 17626–31; She X, Liu G, Ventura M *et al.* (2006). A preliminary comparative analysis of primate segmental duplications shows elevated substitution rates and a great-ape expansion of intrachromosomal duplications. *Genome Research* **16**, 576–83

19 Marques-Bonet T, Kidd JM, Ventura M *et al.* (2009). A burst of segmental duplications in the genome of the African great ape ancestor. *Nature* **457**, 877–81

20 Ji X and Zhao S (2008). DA and Xiao – two giant and composite LTR-retrotransposon-like elements identified in the human genome. *Genomics* **91**, 249–58; Li X, Slife J, Patel N and Zhao S (2009). Stepwise evolution of two giant composite LTR-retrotransposon-like elements DA and Xiao. *BMC Evolutionary Biology* **9**, 128

21 Itoh T, Toyoda A, Taylor TD *et al.* (2005). Identification of large ancient duplications associated with human gene deserts. *Nature Genetics* **37**, 1041–3

22 Conrad B and Antonarakis SE (2007). Gene duplication: a drive for phenotypic diversity and cause of genetic disease. *Annual Review of Genomics and Human Genetics* **8**, 17–35; Kaessmann H (2010). Origins, evolution, and phenotypic impact of new genes. *Genome Research* **20**, 1313–26

23 Fortna A, Kim Y, MacLaren E *et al.* (2004). Lineage-specific gene duplication and loss in human and great ape evolution. *PLoS Biology* **2**, e207; Dumas L, Kim YH, Karimpour-Fard A *et al.* (2007). Gene copy number variation spanning 60 million years of human and primate evolution. *Genome Research* **17**, 1266–72; Armengol G, Knuutila S, Lozano J-J *et al.* (2010). Identification of human-specific gene duplications relative to other primates by array CGH and quantitative PCR. *Genomics* **95**, 203–9

24 Johnson ME, Viggiano L, Bailey JA *et al.* (2001). Positive selection of a gene family during the emergence of humans and African apes. *Nature* **413**, 514–19

25 Giannuzzi G, Siswara P, Malig M *et al.* (2012). Evolutionary dynamism of the LRRC37 gene family. *Genome Research* **23**, 46–59

26 Vandepoele K, van Roy N, Staes K *et al.* (2005). A novel gene family NBPF: intricate structure generated by gene duplications during primate evolution. *Molecular Biology and Evolution* **22**, 2265–74; Popesco MC, MacLaren EJ, Hopkins J *et al.* (2006). Human lineage-specific amplification, selection, and neuronal expression of DUF1220 domains. *Science* **313**, 1304–7

27 Hayakawa T, Angata T, Lewis AL *et al.* (2005). A human-specific gene in microglia. *Science* **309**, 1693; Wang X, Chow R, Deng L *et al.* (2011). Expression of Siglec-11 by human and chimpanzee ovarian stromal cells, with uniquely human ligands: implications for human ovarian physiology and pathology. *Glycobiology* **21**, 1038–48; Wang X, Mitra N, Cruz P *et al.* (2012). Evolution of Siglec-11 and Siglec-16 genes in hominins. *Molecular Biology and Evolution* **29**, 2073–86

28 Suh A, Kriegs JO, Brosius J and Schmitz J (2011). Retroposon insertions and the chronology of avian sex chromosome evolution. *Molecular Biology and Evolution* **28**, 2993–7

29 Bellott DW, Skaletsky H, Pyntikova T *et al.* (2010). Convergent evolution of chicken Z and human X chromosomes by expansion and gene acquisition. *Nature* **466**, 612–16

30 Rozen S, Skaletsky H, Marszalek JD *et al.* (2003). Abundant gene conversion between arms of palindromes in human and ape Y chromosomes. *Nature* **423**, 873–6; Skaletsky H, Kuroda-Kawaguchi T, Minx PJ *et al.* (2003). The male-specific region of the human Y chromosome is a mosaic of discrete sequence classes. *Nature* **423**, 825–37

31 Sin H-S, Kim D-S, Murayama M *et al.* (2010). Human endogenous retrovirus K14C drove genomic diversification of the Y chromosome during primate evolution. *Journal of Human Genetics* **55**, 717–25

32 Warburton PE, Giordano J, Cheung F *et al.* (2004). Inverted repeat structure of the human genome: the X-chromosome contain a preponderance of large, highly homologous inverted repeats that contain testes genes. *Genome Research* **14**, 1861–9 (Supplemental Research, Data Supplement 3)

33 Bhowmick BK, Satta Y and Takahata N (2007). The origin and evolution of human ampliconic gene families and ampliconic structure. *Genome Research* **17**, 441–50

34 Yu Y-H, Lin Y-W, Yu J-F *et al.* (2008). Evolution of the *DAZ* gene and the *AZFc* region on primate Y chromosomes. *BMC Evolutionary Biology* **8**, 96; Hughes JF, Skaletsky H and Page DC (2012). Sequencing of rhesus macaque Y chromosome clarifies origins and evolution of the *DAZ* (*Deleted in AZoospermia*) genes. *BioEssays* **34**, 1035–44

35 Vitek WS, Pagidas K, Gu G *et al.* (2012). Xq;autosome translocation in POF: Xq27.2 deletion resulting in haploinsufficiency for SPANX. *Journal of Assisted Reproduction and Genetics* **29**, 63–6

36 Kouprina N, Mullokandov M, Rogozin IM *et al.* (2004). The *SPANX* gene family of cancer testis-specific antigens: rapid evolution and amplification in African great apes and hominids. *Proceedings of the National Academy of Sciences of the USA* **101**, 3077–82; Kouprina N, Noskov VN, Pavlicek A *et al.* (2007). Evolutionary diversification of *SPANX-N* sperm protein gene structure and expression. *PLoS ONE* **4**, e359

37 Kouprina N, Pavlicek A, Noskov VN *et al.* (2005). Dynamic structure of the *SPANX* gene cluster mapped to the prostate cancer susceptibility locus *HPCX* at Xq27. *Genome Research* **15**, 1477–86

38 Hansen MA, Nielsen JE, Retelska D *et al.* (2008). A shared promoter region suggests a common ancestor for the human *VCX/Y*, *SPANX*, and *CSAG* gene families and the murine *CYPT* family. *Molecular Reproduction and Development* **75**, 219–29

39 Ruault M, Ventura M, Galtier N *et al.* (2003). *BAGE* genes generated by juxtacentromeric reshuffling in the Hominidae lineage are under selective pressure. *Genomics* **81**, 391–9

40 Artamonova II and Gelfand MS (2004). Evolution of the exon–intron structure and alternative splicing of the *MAGE-A* family of cancer/testis antigens. *Journal of Molecular Evolution* **59**, 620–31; Katsura Y and Satta Y (2011). Evolutionary history of the cancer immunity antigen *MAGE* gene family. *PLoS ONE* **6**, e20365; Zhao Q, Caballero OL, Simpson AJG and Strausberg RL (2012). Differential evolution of *MAGE* genes based on expression pattern and selection pressure. *PLoS ONE* **7**, e48240

41 Liu Y, Zhu Q and Zhu N (2008). Recent duplication and positive selection of the *GAGE* gene family. *Genetica* **133**, 31–5; Killen MW, Taylor TL, Stults DM *et al.* (2011). Configuration and rearrangement of the human *GAGE* gene clusters. *American Journal of Translational Research* **3**, 234–42

42 Paulding CA, Ruvolo M and Haber DA (2003). The *Tre2* (*USP6*) oncogene is a hominoid-specific gene. *Proceedings of the National Academy of Sciences of the USA* **100**, 2507–11; information on the junction sequence was provided by Dr Paulding, personal communication

43 Kuryshev VY, Vorobyov E, Zink D *et al.* (2006). An anthropoid-specific segmental duplication on human chromosome 1q22. *Genomics* **88**, 143–51

44 Perez-Maya AA, Rodriguez-Sanchez IP, de Jong P *et al.* (2012). The chimpanzee *GH* locus: composition, organisation, and evolution. *Mammalian Genome* **23**, 387–98

45 Scally *et al.* (2012), Chapter 1, ref. 36

46 Maston GA and Ruvolo M (2002). Chorionic gonadotropin has a recent origin within primates and an evolutionary history of selection. *Molecular Biology and Evolution* **19**, 320–35; Hallast P, Rull K and Laan M (2007). The evolution and genomic landscape of *CGB1* and *CGB2* genes. *Molecular and Cellular Endocrinology* **260–2**, 2–11; Hallast P, Saarela J, Palotie A and Laan M (2008). High divergence in primate-specific duplicated regions: human and chimpanzee *chorionic gonadotropin beta* genes. *BMC Evolutionary Biology* **8**, 195; Nagirnaja L, Rull K, Uuskula L *et al.* (2010). Genomics and genetics of gonadotropin beta-subunit genes: unique FSHB and duplicated *LHB/CGB* loci. *Molecular and Cellular Endocrinology* **329**, 4–16; Parrott AM, Sriram G, Liu Y and Mathews MB (2011). Expression of type II chorionic gonadotropin genes supports a role in the male reproductive system. *Molecular and Cellular Biology* **31**, 287–99

47 Bo M and Boime I (1992). Identification of the transcriptionally active genes of the chorionic gonadotropin β gene cluster *in vivo. Journal of Biological Chemistry* **267**, 3179–84; Hallast *et al.* (2007), Parrott *et al.* (2011), ref. 46

48 Parrott *et al.* (2011), Parrott and Mathews (2011), Chapter 2, ref. 111

49 Samuelson LC, Phillips RS and Swanberg LJ (1996). Amylase gene structures in primates: retroposon insertions and promoter evolution. *Molecular Biology and Evolution* **13**, 767–79

50 Irwin DM, Biegel JM and Stewart C-B (2011). Evolution of the mammalian lysozyme gene family. *BMC Evolutionary Biology* **11**, 166

51 Sorrentino S (2010). The eight human 'canonical' ribonucleases: molecular diversity, catalytic properties, and special biological actions of the enzyme proteins. *FEBS Letters* **584**, 2194–200

52 Narita Y, Oda S, Takenaka O and Kageyama T (2010). Lineage-specific duplication and loss of pepsinogen genes in hominoid evolution. *Journal of Molecular Evolution* **70**, 313–24

53 Lawrence MG, Stephens CR, Need EF *et al.* (2012). Long terminal repeats act as androgen-responsive enhancers for the PSA-kallikrein locus. *Endocrinology* **153**, 3199–210; Marques PI, Bernardino R, Fernandes T *et al.* (2012). Birth-and-death of *KLK3* and *KLK2* in primates: evolution driven by reproductive biology. *Genome Biology and Evolution* **4**, 1331–8

54 Zhang YE, Landback P, Vibranovsky MD and Long M (2011). Accelerated recruitment of new brain development genes into the human genome. *PLoS Biology* **9**, e1001179

55 Vandepoele *et al.* (2005), ref. 26

56 Vandepoele K and van Roy F (2007). Insertion of a HERV(K) LTR in the intron of *NBPF3* is not required for its transcriptional activity. *Virology* **362**, 1–5; Abrarova N, Simonova L, Vinogradova T and Sverdlov E (2011). Different

transcription activity of HERV-K LTR-containing and LTR-lacking genes of the *KIAA1245/NBPF* gene subfamily. *Genetica* **139**, 733–41

57 O'Bleness MS, Dickens CM, Dumas LJ *et al.* (2012). Evolutionary history and domain organisation of DUF1220 protein domains. *G3: Genes, Genomes, Genetics* **2**, 977–86

58 Dumas LJ, O'Bleness MS, Davis MJ *et al.* (2012). DUF1220-domain copy number implicated in human brain-size pathology and evolution. *American Journal of Human Genetics* **91**, 444–54

59 Dennis MY, Nuttle X, Sudmant PH *et al.* (2012). Evolution of human-specific neural *SRGAP2* genes by incomplete segmental duplication. *Cell* **149**, 912–22

60 Charrier C, Joshi K, Coutinho-Budd J *et al.* (2012). Inhibition of SRGAP2 function by its human-specific paralogs induces neoteny during spine maturation. *Cell* **149**, 923–35

61 Deeb SS, Jorgensen AL, Battisti L *et al.* (1994). Sequence divergence of the red and green visual pigments in great apes and humans. *Proceedings of the National Academy of Sciences of the USA* **91**, 7262–6

62 Dulai KS, von Dornum M, Mollon JD and Hunt DM (1999). The evolution of trichromatic color vision by opsin gene duplication in New World and Old World primates. *Genome Research* **9**, 629–38; for a review, see Jacobs GH (2008). Primate colour vision: a comparative perspective. *Visual Neuroscience* **25**, 619–33

63 Ueyama H, Torii R, Tanabe S *et al.* (2004). An insertion/deletion *TEX28* polymorphism and its application to analysis of red/green visual pigment arrays. *Journal of Human Genetics* **49**, 548–57

64 Goodman M (1999). The genomic record of humankind's evolutionary roots. *American Journal of Human Genetics* **64**, 31–9

65 Fitch DHA, Bailey WJ, Tagle DA *et al.* (1991). Duplication of the γ-globin gene mediated by L1 long interspersed repetitive elements in an early ancestor of simian primates. *Proceedings of the National Academy of Sciences of the USA* **88**, 7396–400

66 Opazo JC, Hoffmann FG and Storz JF (2008). Differential loss of embryonic globin genes during the radiation of placental mammals. *Proceedings of the National Academy of Sciences of the USA* **105**, 12950–5; Neumann R, Lawson VE and Jeffreys AJ (2010). Dynamics and processes of copy number instability in human γ-globin genes. *Proceedings of the National Academy of Sciences of the USA* **107**, 8304–9

67 Wolf R, Ruzicka T and Yuspa SH (2011). Novel S100A7 (psoriasin)/S100A15 (koebnerisin) subfamily: highly homologous but distinct in regulation and function. *Amino Acids* **41**, 789–96

68 Kulski JK, Lim CP, Dunn DS and Bellgard M (2003). Genomic and phylogenetic analysis of the S100A7 (psoriasin) gene duplications within the region of the S100 gene cluster on human chromosome 1q21. *Journal of Molecular Evolution* **56**, 397–406

69 Lehrer RI and Lu W (2012). α-Defensins in human innate immunity. *Immunological Reviews* **245**, 84–112

70 Semple CA, Gautier P, Taylor K and Dorin JR (2006). The changing of the guard: molecular diversity and rapid evolution of β-defensins. *Molecular Diversity* **10**, 575–84; Hollox EJ, Barber JCK, Brookes AJ and Armour JAL (2008). Defensins and the dynamic genome: what we can learn from structural variation at human chromosome band 8p23.1. *Genome Research* **18**, 1686–97; Fode P, Jespersgaard C, Hardwick RJ *et al.* (2011). Determination of beta-defensin genomic copy number in different populations: a comparison of three methods. *PLoS ONE* **6**, e16768

71 Whittington CM, Papenfuss AT, Bansal P *et al.* (2008). Defensins and the convergent evolution of the platypus and reptile venom genes. *Genome Research* **18**, 986–94

72 Lehrer and Lu (2012), ref. 69; Das S, Nikolaidis N, Goto H *et al.* (2010). Comparative genomics and evolution of the alpha-defensin multigene family in primates. *Molecular Biology and Evolution* **27**, 2333–43

73 Nguyen TX, Cole AM and Lehrer RI (2003). Evolution of primate θ-defensins: a serpentine path to a sweet tooth. *Peptides* **24**, 1647–54

74 Penberthy WT, Chari S, Cole AL and Cole AM (2011). Retrocyclins and their activity against HIV-1. *Cellular and Molecular Life Sciences* **68**, 2231–42; Doss M, Ruchala P, Tecle T *et al.* (2012). Hapivirins and diprovirins: novel θ-defensin analogs with potent activity against influenza A virus. *Journal of Immunology* **188**, 2759–68; for a review see Lehrer RI, Cole AM and Selsted ME (2012). θ-Defensins: cyclic peptides with endless potential. *Journal of Biological Chemistry* **287**, 27014–19

75 Venkataraman N, Cole AL, Ruchala P *et al.* (2009). Reawakening retrocyclins: ancestral human defensins active against HIV-1. *PLoS Biology* **7**, e95

76 Parham P, Abi-Rached L, Matevosyan L *et al.* (2010). Primate-specific regulation of natural killer cells. *Journal of Medical Primatology* **39**, 194–212; Han K, Lou DI and Sawyer SL (2011). Identification of a genomic reservoir for new *TRIM* genes in primate genomes. *PLoS Genetics* **7**, e1002388; Munk C, Willemsen A and Bravo IG (2012). An ancient history of gene duplications, fusions and losses in the evolution of APOBEC3 mutators in mammals. *BMC Evolutionary Biology* **12**, 71

77 Shiina T, Tamiya G, Oka A *et al.* (1999). Molecular dynamics of MHC genesis unraveled by sequence analysis of the 1,796,938-bp HLA class 1 region.

Proceedings of the National Academy of Sciences of the USA **96**, 13282–7; Anzai T, Shiina T, Kimura N *et al.* (2003). Comparative sequencing of human and chimpanzee MHC class I regions unveils insertions/deletions as the major path to genomic divergence. *Proceedings of the National Academy of Sciences of the USA* **100**, 7708–13

78 Fukami-Kobayashi K, Shiina T, Anzai T *et al.* (2005). Genomic evolution of MHC class I region in primates. *Proceedings of the National Academy of Sciences of the USA* **102**, 9230–4

79 Kulski JK, Anzai T and Inoko H (2005). ERVK9, transposons and the evolution of MHC class I duplicons within the alpha-block of the human and chimpanzee. *Cytogenetic and Genome Research* **110**, 181–92

80 Kulski JK, Anzai T, Shiina T and Inoko H (2004). Rhesus macaque class I duplicon structures, organization, and evolution within the alpha block of the major histocompatibility complex. *Molecular Biology and Evolution* **21**, 2079–91; Sawai H, Kawamoto Y, Takahata N and Satta Y (2004). Evolutionary relationships of major histocompatibility complex class I genes in simian primates. *Genetics* **166**, 1897–907; Kulski *et al.* (2005), ref. 79

81 Doxiadis GG, Hoof I, de Groot N and Bontrop RE (2012). Evolution of HLA-DRB genes. *Molecular Biology and Evolution* **29**, 3843–53

82 Das S, Nozawa M, Klein J and Nei M (2008). Evolutionary dynamics of the immunoglobulin heavy chain variable region genes in vertebrates. *Immunogenetics* **60**, 47–55; Pramanik S, Cui X, Wang H-Y *et al.* (2011). Segmental duplication as one of the driving forces underlying the diversity of the human immunoglobulin heavy chain variable gene region. *BMC Genomics* **12**, 78

83 Das S (2009). Evolutionary origin and genomic organisation of micro-RNA genes in immunoglobulin lambda variable region gene family. *Molecular Biology and Evolution* **26**, 1179–89; Than NG, Romero R, Goodman M *et al.* (2009). A primate subfamily of galectins expressed at the maternal–fetal interface that promote immune cell death. *Proceedings of the National Academy of Sciences of the USA* **106**, 9731–6

84 Nowick K, Hamilton AT, Zhang H and Stubbs L (2010). Rapid sequence and expression divergence suggest selection for novel function in primate-specific KRAB-ZNF genes. *Molecular Biology and Evolution* **27**, 2606–17; Nowick K, Fields C, Gernat T *et al.* (2011). Gain, loss and divergence in primate zinc finger genes: a rich resource for evolution of gene regulatory differences between species. *PLoS ONE* **6**, e21553

85 Shen S, Lin L, Cai JJ *et al.* (2011). Widespread establishment and regulatory impact of Alu exons in human genes. *Proceedings of the National Academy of Sciences of the USA* **108**, 2837–42; Thomas JH and Schneider SE (2011).

Coevolution of retroelements and tandem zinc finger genes. *Genome Research* **21**, 1800–12

86 Bosch N, Caceres M, Cardone MF *et al.* (2007). Characterisation and evolution of the novel gene family *FAM90A* in primates originated by multiple duplication and rearrangement events. *Human Molecular Genetics* **16**, 2572–82; Bosch N, Escaramis G, Mercader JM *et al.* (2008). Analysis of the multi-copy gene family *FAM90A* as a copy number variant in different ethnic backgrounds. *Gene* **420**, 113–17

87 Matsunami M, Sumiyama K and Saitou N (2010). Evolution of conserved, non-coding sequences within the vertebrate Hox clusters through the two-round whole-genome duplications revealed by phylogenetic footprinting analysis. *Journal of Molecular Evolution* **71**, 427–36

88 Zhong Y-F and Holland PWH (2011). The dynamics of vertebrate homeobox gene evolution: gains and losses of genes in mouse and human lineages. *BMC Evolutionary Biology* **11**, 169; Zhong Y-F and Holland PWH (2011). Correction: The dynamics of vertebrate homeobox gene evolution: gain and loss of genes in mouse and human lineages. *BMC Evolutionary Biology* **11**, 204

89 Niu A-L, Wang Y-Q, Zhang H *et al.* (2011). Rapid evolution and copy number variation of primate RHOXF2, an X-linked homeobox gene, involved in male reproduction and possibly brain function. *BMC Evolutionary Biology* **11**, 298

90 Leidenroth A, Clapp J, Mitchell LM *et al.* (2012). Evolution of *DUX* gene macrosatellites in placental mammals. *Chromosoma* **121**, 489–97

91 Ding W, Lin L, Chen B and Dai J (2006). L1 elements, processed pseudogenes and retrogenes in mammalian genomes. *IUBMB Life* **58**, 677–85; Kaessmann H, Vinckenbosch N and Long M (2009). RNA-based gene duplication: mechanistic and evolutionary insights. *Nature Reviews Genetics* **10**, 19–31

92 Parker HG, VonHoldt BM, Quignon P *et al.* (2009). An expressed *Fgf4* retrogene is associated with breed-defining chondrodysplasia in domestic dogs. *Science* **325**, 995–8

93 Marques AC, Dupanloup I, Vinckenbosch N *et al.* (2005). Emergence of young human genes after a burst of retroposition in primates. *PLoS Biology* **3**, e357; Vinckenbosch N, Dupanloup I and Kaessmann H (2006). Evolutionary fate of retroposed gene copies in the human genome. *Proceedings of the National Academy of Sciences of the USA* **103**, 3220–5; Fablet M, Bueno M, Potrzebowski L and Kaessmann H (2010). Evolutionary origin and functions of retrogene introns. *Molecular Biology and Evolution* **26**, 2147–56

94 Baertsch R, Diekhans M, Kent WJ *et al.* (2008). Retrocopy contribution to the evolution of the human genome. *BMC Genomics* **9**, 466

95 Burki F and Kaessmann H (2004). Birth and adaptive evolution of a hominoid gene that supports high neurotransmitter flux. *Nature Genetics* **36**, 1061–3

96 Rosso L, Marques AC, Reidhert AS and Kaessmann H (2008). Mitochondrial targeting adaption of the hominoid-specific glutamate dehydrogenase driven by positive Darwinian selection. *PLoS Genetics* **4**, e1000150; Kotzamani D and Plaitakis A (2012). Alpha helical structures in the leader sequence of human *GLUD2* glutamate dehydrogenase responsible for mitochondrial import. *Neurochemistry International* **61**, 463–9

97 Spanaki C, Zaganas I, Kleopa KA and Plaitakis A (2010). Human *GLUD2* glutamate dehydrogenase is expressed in neural and testicular supporting cells. *Journal of Biological Chemistry* **285**, 16748–56; Zaganas I, Spanaki C and Plaitakis A (2012). Expression of human *GLUD2* glutamate dehydrogenase in human tissues: functional implications. *Neurochemistry International* **61**, 455–62

98 Kanavouras K, Mastorodemos V, Borompokas M *et al.* (2007). Properties and molecular evolution of human *GLUD2* (neural and testicular tissue-specific) glutamate dehydrogenase. *Journal of Neuroscience Research* **85**, 3398–406; Spanaki S, Zaganas I, Kounoupa Z and Plaitakis A (2012). The complex regulation of human *GLUD1* and *GLUD2* glutamate dehydrogenases and its implications in nerve tissue biology. *Neurochemistry International* **61**, 470–81

99 Plaitakis A, Latsoudis H, Kanavouras K *et al.* (2010). Gain-of-function variant in *GLUD2* glutamate dehydrogenase modifies Parkinson's disease onset. *European Journal of Human Genetics* **18**, 336–41

100 Rosso L, Marques AC, Weier M et *al* (2008). Birth and rapid subcellular adaptation of a hominoid-specific CDC14 protein. *PLoS Biology* **6**, e140

101 Babushok DV, Ohshima K, Ostertag EM *et al.* (2007). A novel testis ubiquitin-binding protein gene arose by exon shuffling in hominoids. *Genome Research* **17**, 1129–38; Ohshima K and Igarashi K (2010). Inference for the initial stage of domain shuffling: tracing the evolutionary fate of the *PIPSL* retrogene in hominoids. *Molecular Biology and Evolution* **27**, 2522–33

102 Zingler N, Willhoeft U, Brose HP *et al.* (2005). Analysis of 5′ junctions of human LINE-1 and Alu retrotransposons suggests an alternative model for 5′-end attachment requiring microhomology-mediated end joining. *Genome Research* **15**, 780–9; Kojima KK (2010). Different integration site structures between L1 protein-mediated retrotransposition in *cis* and retrotransposition in *trans*. *Mobile DNA* **1**, 17

103 Kojima KK and Okada N (2009). mRNA retrotransposition coupled with 5′ inversion as a possible source of new genes. *Molecular Biology and Evolution* **26**, 1405–20

104 Lee Y, Ise T, Ha D *et al.* (2006). Evolution and expression of chimeric POTE-actin genes in the human genome. *Proceedings of the National Academy of Sciences of the USA* **103**, 17885–90

105 Liu XF, Bera TK, Liu LJ and Pastan I (2009). A primate-specific POTE-actin fusion protein plays a role in apoptosis. *Apoptosis* **14**, 1237–44; Bera TK, Walker DA, Sherins RJ and Pastan I (2012). POTE protein, a cancer-testis antigen, is highly expressed in spermatids in human testis and is associated with apoptotic cells. *Biochemical and Biophysical Research Communications* **417**, 1271–4

106 Vinckenbosch *et al.* (2006), ref. 93 (Supporting Data Set 1); adapted by Wood AJ, Roberts RG, Monk D *et al.* (2007). A screen for retrotransposed imprinted genes reveals an association between X chromosome homology and maternal germ-line methylation. *PLoS Genetics* **3**, 192; see also Svensson O, Arvestad L and Lagergren J (2006). Genome-wide survey for biologically functional pseudogenes. *PLoS Computational Biology* **2**, e46

107 Luo C, Lu X, Stubbs L and Kim J (2006). Rapid evolution of recently retroposed transcription factor *YY2* in mammalian genomes. *Genomics* **87**, 348–55; Kim JD, Faulk C and Kim J (2007). Retroposition and evolution of the DNA-binding motifs of YY1, YY2 and REX1. *Nucleic Acids Research* **35**, 3442–52

108 Chen L, Tioda T, Coser KR *et al.* (2011). Genome-wide analysis of YY2 versus YY1 target genes. *Nucleic Acids Research* **38**, 4011–26

109 Guallar D, Perez-Palacios R, Climent M *et al.* (2012). Expression of endogenous retroviruses is negatively regulated by the pluipotency marker *REX1/Zpf42*. *Nucleic Acids Research* **40**; 8993–9007

110 Wood *et al.* (2007), ref. 106; McCole RB, Loughran NB, Chahal M *et al.* (2011). A case-by-case evolutionary analysis of four imprinted retrogenes. *Evolution* **65**, 1413–27

111 Bradley J, Baltus A, Skaletsky H *et al.* (2004). An X-to-autosome retrogene is required for spermatogenesis in mice. *Nature Genetics* **36**, 872–6; Rohozinski J, Lamb DL and Bishop CE (2006). *UTP14c* is a recently acquired retrogene associated with spermatogenesis and fertility in man. *Biology of Reproduction* **74**, 644–51

112 Yang F, Skaletsky H and Wang PJ (2007). *Ubl4b*, an X-derived retrogene, is specifically expressed in post-meiotic germ cells in mammals. *Gene Expression Patterns* **7**, 131–6

113 McLysaght A (2008). Evolutionary steps of sex chromosomes are reflected in retrogenes. *Trends in Genetics* **24**, 478–81; Potrzebowski L, Vinckenbosch N, Marques AC *et al.* (2008). Chromosomal gene movements reflect the recent origin and biology of therian sex chromosomes. *PLoS Biology* **6**, e80

114 Potrzebowski L, Vinckenbosch N and Kaessmann H (2010). The emergence of new genes on the young therian X. *Trends in Genetics* **26**, 1–4

115 Ciomborowska J, Rosikiewicz W, Szklarczyk D *et al.* (2013). 'Orphan' retrogenes in the human genome. *Molecular Biology and Evolution* **30**, 384–96

116 Weber MJ (2006). Mammalian small nucleolar RNAs are mobile genetic elements. *PLoS Genetics* **2**, e205; Luo Y and Li S (2007). Genome-wide analyses of retrogenes derived from the human box H/ACA snoRNAs. *Nucleic Acids Research* **35**, 559–71

117 Kapitonov *et al.* (2004), Chapter 2, ref. 1

118 Collier LS, Carlson CM, Ravimohan S *et al.* (2005). Cancer gene discovery in solid tumours using transposon-based somatic mutagenesis in the mouse. *Nature* **436**, 272–6; Dupuy AJ, Akagi K, Largaespada DA *et al.* (2005). Mammalian mutagenesis using a highly mobile somatic *Sleeping Beauty* transposon system. *Nature* **436**, 221–6; Rad R, Rad L, Wang W *et al.* (2010). *PiggyBac* transposon mutagenesis: a tool for cancer gene discovery in mice. *Science* **330**, 1104–7; Starr TK, Scott PM, Marsh BM *et al.* (2011). A *Sleeping Beauty* transposon-mediated screen identifies murine susceptibility genes for adenomatous polyposis coli (*Apc*)-dependent intestinal tumorigenesis. *Proceedings of the National Academy of Sciences of the USA* **108**, 5765–70

119 Pace JK, II and Feschotte C (2007). The evolutionary history of human DNA transposons: evidence for intense activity in the primate lineage. *Genome Research* **17**, 422–32

120 Pritham EJ and Feschotte C (2007). Massive amplification of rolling-circle transposons in the lineage of the bat *Myotis lucifugus*. *Proceedings of the National Academy of Sciences of the USA* **104**, 1895–900; Ray DA, Pagan HJT, Thompson ML and Stevens RD (2007). Bats with *hATs*: evidence for DNA transposon activity in genus *Myotis*. *Molecular Biology and Evolution* **24**, 632–9

121 Volff J-N (2006). Turning junk into gold: domestication of transposable elements and the creation of new genes in eukaryotes. *BioEssays* **28**, 913–22

122 Piriyapongsa J and Jordan IK (2007). A family of microRNA genes from miniature inverted-repeat transposable elements. *PLoS ONE* **2**, e203; Yuan Z, Sun X, Jiang D *et al.* (2010). Origin and evolution of a placental-specific microRNA family in the human genome. *BMC Evolutionary Biology* **10**, 346

123 Piriyapongsa J, Marino-Ramirez L and Jordan IK (2007). Origin and evolution of human microRNAs from transposable elements. *Genetics* **176**, 1323–37; Borchert GM, Holton NW, Williams JD *et al.* (2011). Comprehensive analysis of microRNA genomic loci identifies pervasive repetitive-element origins. *Mobile Genetic Elements* **1**, 8–17

124 Cordaux R, Udit S, Batzer MA and Feschotte C (2006). Birth of a chimeric gene by capture of the transposase gene from a mobile element. *Proceedings of the National Academy of Sciences of the USA* **103**, 8101–6

125 Shaleen M, Williamson E, Nickoloff J *et al.* (2010). Metnase/SETMAR: a domesticated primate transposase that enhances DNA repair, replication and decatenation. *Genetica* **138**, 559–66; Wray J, Williamson EA, Chester S *et al.* (2010). The transposase domain protein Metnase/SETMAR suppresses chromosomal translocations. *Cancer Genetics and Cytogenetics* **200**, 184–90

126 Hromas R, Williamson EA, Fnu S *et al.* (2012). Chk1 phosphorylation of metnase enhances DNA repair but inhibits replication fork restart. *Oncogene* **31**, 4245–54

127 Almeida LM, Silva IT, Silva WA, Jr *et al.* (2007). The contribution of transposable elements to *Bos taurus* gene structure. *Gene* **390**, 180–9

128 Markljung E, Jiang L, Jaffe JD *et al.* (2009). ZBED6, a novel transcription factor derived from a domesticated DNA transposon regulates IGF2 expression and muscle growth. *PLoS Biology* **7**, e1000256; Andersson L, Andersson G, Hjalm G *et al.* (2010). ZBED6: the birth of a new transcription factor in the common ancestor of placental mammals. *Transcription* **1**, 144–8

129 Casola C, Hucks D and Feschotte F (2008). Convergent domestication of *pogo*-like transposases into centromere-binding proteins in fission yeast and mammals. *Molecular Biology and Evolution* **25**, 29–41

130 Kojima KK and Jurka J (2011). Crypton transposons: identification of new diverse families and ancient domestication events. *Mobile DNA* **2**, 12

131 Sinzelle L, Kapitonov VV, Grzela DP *et al.* (2008). Transposition of a reconstructed *Harbinger* element in human cells and functional homology with two transposon-derived cellular genes. *Proceedings of the National Academy of Sciences of the USA* **105**, 4715–20; Smith JJ, Sumiyama J and Amemiya CT (2012). A living fossil in the genome of a living fossil: *Harbinger* transposons in the coelacanth genome. *Molecular Biology and Evolution* **29**, 985–93

132 Fugmann SD (2010). The origins of the *Rag* genes: from transposition to V(D)J recombination. *Seminars in Immunology* **22**, 10–16

133 Tautz D and Domazet-Loso T (2011). The evolutionary origin of orphan genes. *Nature Reviews Genetics* **12**, 692–702

134 Li Y, Qian Y-P, Yu X-J *et al.* (2004). Recent origin of a hominid-specific splice form of neuropsin, a gene involved in learning and memory. *Molecular Biology and Evolution* **21**, 2111–15

135 Lu Z-X, Peng J and Su B (2007). A human-specific mutation leads to the origin of a novel splice form of neuropsin (*KLK8*), a gene involved in learning and memory. *Human Mutation* **28**, 978–84

136 Gianfrancesco F, Esposito T, Casu G *et al.* (2004). Emergence of talanin protein associated with human uric acid nephrolithiasis in the Hominidae lineage. *Gene* **339**, 131–8

137 Knowles DG and McLysaght A (2009). Recent *de novo* origin of human protein-coding genes. *Genome Research* **19**, 1752–9

138 Wu DD, Irwin DM and Zhang Y-P (2011). *De novo* origin of human protein-coding genes. *PLoS Genetics* **7**, e1002379

139 Tay S-K, Blythe J and Lipovich L (2009). Global discovery of primate-specific genes in the human genome. *Proceedings of the National Academy of Sciences of the USA* **106**, 12019–24; Toll-Riera M, Bosch N, Bellora N *et al.* (2009). Origin of primate orphan genes: a comparative genomics approach. *Molecular Biology and Evolution* **26**, 603–12 (Supplementary Information)

140 Hu HY, He L, Forminykh K *et al.* (2012). Evolution of the human-specific microRNA miR-941. *Nature Communications* **3**, 114; Iwama H, Kato K, Imachi A *et al.* (2012). Human microRNAs originated from two periods at accelerated rates in mammalian evolution. *Molecular Biology and Evolution* **30**, 613–26

141 Duret L, Chureau C, Samain S *et al.* (2006). The *XIST* gene evolved in eutherians by pseudogenization of a protein-coding gene. *Science* **312**, 1653–5; Yen ZC, Meyer IM, Karalic S and Brown CJ (2007). A cross-species comparison of X-chromosome inactivation in Eutheria. *Genomics* **90**, 453–63 (Supplementary Data, Figure 5)

142 Elisaphenko EA, Kolesnikov NN, Shevchenko AI *et al.* (2008). A dual origin of the *XIST* gene from a protein-coding gene and a set of transposable elements. *PLoS ONE* **3**, e2521

143 Romito A and Rougeulle C (2011). Origin and evolution of the long non-coding genes in the X-inactivation centre. *Biochimie* **93**, 1935–42

144 Gontan C, Achame EM, Demmers J *et al.* (2012). RNF-12 initiates X-chromosome inactivation by targeting REX1 for degradation. *Nature* **485**, 386–90

145 He S, Liu S and Zhu H (2011). The sequence, structure and evolutionary features of HOTAIR in mammals. *BMC Evolutionary Biology* **11**, 102; see also Cartault F, Munier P, Benko E *et al.* (2012). Mutation in a primate-conserved retrotransposon reveals a non-coding RNA as a mediator of infantile encephalopathy. *Proceedings of the National Academy of Sciences of the USA* **109**, 4980–5; Novikova IV, Hennelly SP and Sanbonmatsu KY (2012). Structural architecture of the long non-coding RNA, steroid receptor RNA activator. *Nucleic Acids Research* **40**, 5034–51

146 Kelley DR and Rinn JL (2012). Transposable elements reveal a stem cell-specific class of long noncoding RNAs. *Genome Biology* **13**, R107

147 Xie C, Zhang YE, Chen J-Y *et al.* (2012). Hominoid-specific *de novo* protein-coding genes originating from long non-coding RNAs. *PLoS Genetics* **8**, e1002942

148 Matsunami *et al.* (2010), ref. 87

149 Lek M, Quinlan KGR and North KN (2009). The evolution of skeletal muscle performance: gene duplication and divergence of human sarcomeric α-actinins. *BioEssays* **32**, 17–25

150 Huminiecki L and Heldin C-H (2010). 2R and the modeling of vertebrate signal transduction engine. *BMC Biology* **8**, 146; Manning G and Scheeff E (2010). How the vertebrates were made: selective pruning of a double-duplicated genome. *BMC Biology* **8**, 144

151 Hoffmann FG and Opazo JC (2011). Evolution of the relaxin/insulin-like gene family in placental mammals: implications for its early evolution. *Journal of Molecular Evolution* **72**, 72–9; Arroyo JI, Hoffmann FG and Opazo JC (2012). Gene turnover and differential retention in the relaxin/insulin-like gene family in primates. *Molecular Phylogenetics and Evolution* **63**, 768–76

152 Hellstein U, Aspden JL, Rio DC and Rokhsar DS (2011). A segmental duplication generates a functional intron. *Nature Communications* **2**, 454

153 Kordis D (2011). Extensive intron gain in the ancestor of placental mammals. *Biology Direct* **6**, 59; Kordis D and Kokosar J (2012). What can domesticated genes tell us about the intron gain in mammals? *International Journal of Evolutionary Biology* **2012**, Article ID 278981

154 Wu DD, Irwin DM and Zhang YP (2008). Molecular evolution of the keratin-associated protein gene family in mammals, role in the evolution of human hair. *BMC Evolutionary Biology* **8**, 241; similarly, Vandeburgh W and Bossuyt F (2012). Radiation and functional diversification of alpha keratins during early vertebrate evolution. *Molecular Biology and Evolution* **29**, 995–1004

EPILOGUE: WHAT REALLY MAKES US HUMAN

1 Check Hayden E (2009). Darwin 200: The other strand. *Nature* **457**, 776–9; Varki A, Geschwind DH and Eichler EE (2010). Explaining human uniqueness: genome interactions with environment, behaviour, and culture. *Nature Reviews Genetics* **9**, 749–63

2 Lai Y, Di Nardo A, Nakatsuji T *et al.* (2009). Commensal bacteria regulate toll-like receptor 3-dependent inflammation after skin injury. *Nature Medicine* **15**, 1377–82

3 Fukata M, Chen A, Klepper A *et al.* (2006). Cox-2 is regulated by Toll-like receptor-4 (TLR4) signaling: role in proliferation and apoptosis in the intestine. *Gastroenterology* **131**, 862–77

4 Abreu MT (2010). Toll-like receptor signalling in the intestinal epithelium: how bacterial recognition shapes intestinal function. *Nature Reviews Immunology* **10**, 131–44; van Baarlen P, Troost F, van der Meer C *et al.* (2011). Human mucosal in vivo transcriptome responses to three lactobacilli indicate how probiotics may modulate cellular pathways. *Proceedings of the National Academy of Sciences of the USA* **108** (Suppl. 1), 4562–9; Wells JM, Rossi O, Meijerink M and van Baarlen P (2011). Epithelial crosstalk at the microbiota–mucosal interface. *Proceedings of the National Academy of Sciences of the USA* **108** (Suppl. 1), 4607–14

5 Bouskra D, Brezillon C, Berard M *et al.* (2008). Lymphoid tissue genesis by commensals through NOD1 regulates intestinal homeostasis. *Nature* **456**, 507–10

6 Guarner F, Bourdet-Sicard R, Brandtzaeg P *et al.* (2006). Mechanisms of disease: the hygiene hypothesis revisited. *Nature Clinical Practice Gastroenterology and Hepatology* **3**, 275–84; Elliott DE, Summers RW and Weinstock JV (2007). Helminths as governors of immune-mediated inflammation. *International Journal for Parasitology* **37**, 457–64; Liu Z, Liu Q, Bleich D *et al.* (2010). Regulation of type 1 diabetes, tuberculosis, and asthma by parasites. *Journal of Molecular Medicine* **88**, 27–38; Scher JU and Abramson SB (2011). The microbiome and rheumatoid arthritis. *Nature Reviews Rheumatology* **7**, 569–78

7 Maslowski KM and Mackay CR (2011). Diet, gut microbiota and immune responses. *Nature Immunology* **12**, 5–9; Pennisi E (2011). Newsfocus: girth and the gut. *Science* **332**, 32–3; Rooks MG and Garrett WS (2011). Sharing the bounty. *Scientist* **25**(8), 38

8 Mazmanian SK, Round JL and Kasper DL (2008). A microbial symbiosis factor prevents intestinal inflammatory disease. *Nature* **453**, 620–5; Mazmanian S (2009). The microbial health factor. *Scientist* **23**(8), 34; Round JL and Mazmanian SK (2009). The gut microbiota shapes intestinal immune responses during health and disease. *Nature Reviews Immunology* **9**, 313–23

9 Atarashi K, Tanoue T, Shima T *et al.* (2011). Induction of colonic regulatory T cells by indigenous *Clostridium* species. *Science* **331**, 337–41

10 Wen L, Ley RE, Volchkov PY *et al.* (2008). Innate immunity and intestinal microbiota in the development of type 1 diabetes. *Nature* **455**, 1109–13; Cook A (2009). Review series on helminths, immune modulation and the hygiene hypothesis: how might infection modulate the onset of type 1 diabetes? *Immunology* **126**, 12–17; Hubner MP, Stocker JT and Mitre E (2009). Inhibition of type 1 diabetes in filaria-infected non-obese diabetic mice is associated with a T helper type 2 shift and induction of FoxP3+ regulatory T cells. *Immunology* **127**, 512–22; Zaccone P, Burton OT, Gibbs S *et al.* (2010). Immune modulation by *Schistosoma mansoni* antigens in NOD mice: effects on both innate

and adaptive immune systems. *Journal of Biomedicine and Biotechnology* **2010**, 795210

11 Cho I, Yamanishi S, Cox L *et al.* (2012). Antibiotics in early life alter the murine colonic microbiome and adiposity. *Nature* **488**, 621–6

12 Reddy A and Fried B (2009). An update on the use of helminths to treat Crohn's and other autoimmune diseases. *Parasitology Research* **104**, 217–21; Jouvin M-H and Kinet J-P (2012). *Trichuris suis* ova: testing a helminth-based therapy as an extension of the hygiene hypothesis. *Journal of Allergy and Clinical Immunology* **130**, 3–10

13 Palmer R (2011). Fecal matters. *Nature Medicine* **17**, 150–2; Borody TJ and Khoruts A (2012). Fecal microbiota transplantation and emerging applications. *Nature Reviews Gastroenterology and Hepatology* **9**, 88–96; Damman CJ, Miller SI, Surawicz CM and Zisman TL (2012). The microbiome and inflammatory bowel disease: is there a therapeutic role for fecal microbiota transplantation? *American Journal of Gastroenterology* **107**, 1452–9

14 Correale J and Farez M (2007). Association between parasite infection and immune responses in multiple sclerosis. *Annals of Neurology* **61**, 97–108; Correale J and Farez MF (2012). Does helminth activation of toll-like receptors modulate immune responses in multiple sclerosis patients? *Frontiers in Cellular and Infection Microbiology* **2**, 112

15 von Mutius E and Vercelli D (2010). Farm living: effects on childhood asthma and allergy. *Nature Reviews Immunology* **10**, 861–8

16 Wickens K, Black PN, Stanley TV *et al.* (2008). A differential effect of 2 probiotics in the prevention of eczema and atopy: a double-blind, randomized, placebo-controlled trial. *Journal of Allergy and Clinical Immunology* **122**, 788–94; Pelucchi C, Chatenoud L, Turati F *et al.* (2012). Probiotics supplementation during pregnancy or infancy for the prevention of atopic dermatitis: a meta-analysis. *Epidemiology* **23**, 402–14

17 Heijtz RD, Wang S, Anuar F *et al.* (2011). Normal gut microbiota modulates brain development and behaviour. *Proceedings of the National Academy of Sciences of the USA* **108**, 3047–52; Collins SM, Surette M and Bercik B (2012). The interplay between the intestinal microbiota and the brain. *Nature Reviews Microbiology* **10**, 735–42

18 Rook GAW and Lowry CA (2008). The hygiene hypothesis and psychiatric disorders. *Trends in Immunology* **29**, 150–8; Blaser MJ and Falkow S (2009). What are the consequences of the disappearing human microbiota? *Nature Reviews Microbiology* **7**, 887–94; Rook GAW (2009). Review series on helminths, immune modulation and the hygiene hypothesis: the broader implications of the hygiene hypothesis. *Immunology* **126**, 3–11

19 Claesson MJ, Jeffery IB, Conde S *et al.* (2012). Gut microbiota composition correlates with diet and health in the elderly. *Nature* **488**, 178–84; Hughes V (2012). Microbiome: cultural differences. *Nature* **492**, S14–15

20 Khaitovich P, Enard W, Lachmann M and Paabo S (2006). Evolution of primate gene expression. *Nature Reviews Genetics* **7**, 693–702

21 Somel M, Franz H, Yan Z *et al.* (2009). Transcriptional neoteny in the human brain. *Proceedings of the National Academy of Sciences of the USA* **106**, 5743–8

22 Maguire EA, Gadian DG, Johnsrude IS *et al.* (2000). Navigation-related structural change in the hippocampi of taxi drivers. *Proceedings of the National Academy of Sciences of the USA* **97**, 4398–403

23 Munte TF, Altenmuller E and Jancke L (2002). The musician's brain as a model of neuroplasticity. *Nature Reviews Neuroscience* **3**, 473–8; Draganski B, Gaser C, Busch V *et al.* (2004). Changes in grey matter induced by training. *Nature* **427**, 311–12; Draganski B, Gaser C, Kempermann G *et al.* (2006). Temporal and spatial dynamics of brain structure changes during extensive learning. *Journal of Neuroscience* **26**, 6314–17; Jancke L, Koeneke S, Hoppe A *et al.* (2009). The architecture of the golfer's brain. *PLoS ONE* **4**, e4785; Zatorre RJ, Fields RD and Johansen-Berg H (2011). Plasticity in grey and white: neuroimaging in brain structure during learning. *Nature Neuroscience* **15**, 528–36

24 Sameroff A (2010). A unified theory of development: a dialectic integration of nature and nurture. *Child Development* **81**, 6–22

25 Stiles J (2011). Brain development and the nature versus nurture debate. *Progress in Brain Research* **189**, 3–22

26 Morishita H and Hensch TK (2008). Critical period revisited: impact on vision. *Current Opinion in Neurobiology* **18**, 101–7

27 Bedny M, Konkle T, Pelphrey K *et al.* (2010). Sensitive period for a multimodal response in human visual motion area MT/MST. *Current Biology* **20**, 1900–6

28 Leppanen JM and Nelson CA (2009). Tuning the developing brain to social signals of emotions. *Nature Reviews Neuroscience* **10**, 37–47

29 de Villiers-Sidani E, Simpson KL, Lu Y-F *et al.* (2008). Manipulating critical period closure across different sectors of the primary auditory cortex. *Nature Neuroscience* **11**, 957–65

30 Kuhl PL (2010). Brain mechanisms in early language acquisition. *Neuron* **67**, 713–27; Perani D, Saccuman MC, Scifo P *et al.* (2012). Neural networks at birth. *Proceedings of the National Academy of Sciences of the USA* **108**, 16056–61

31 Baumeister RF and Masicampo EJ (2011). Conscious thought is for facilitating social and cultural interactions: how mental simulations serve the animal–culture interface. *Psychological Review* **117**, 945–71

32 Dehaene S, Pegado F, Braga F *et al.* (2010). How learning to read changes the cortical networks for vision and language. *Science* **330**, 1359–64

33 Maggi S, Irwin LJ, Siddiqi A and Hertzman C (2010). The social determinants of early child development: an overview. *Journal of Paediatrics and Child Health* **46**, 627–35

34 Bradshaw GA, Schore AN, Brown JL *et al.* (2005). Elephant breakdown. *Nature* **433**, 807

35 Bock J and Braun K (2011). The impact of perinatal stress on the functional maturation of prefronto-cortical synaptic circuits: implications for the pathophysiology of ADHD? *Progress in Brain Research* **189**, 155–69; Mogi K, Nagasawa M and Kikusui T (2011). Developmental consequences and biological significance of mother–infant bonding. *Progress in Neuro-Psychopharmacology and Biological Psychiatry* **35**, 1232–41; Kolb B, Mychasiuk R, Muhammad A *et al.* (2012). Experience and the developing prefrontal cortex. *Proceedings of the National Academy of Sciences of the USA* **109** (Suppl 2), 17186–93

36 Makinodan M, Rosen KM, Ito S and Corfas G (2012). A critical period for social experience-dependent oligodendrocyte maturation and myelination. *Science* **337**, 1357–60

37 Perry BD (2002). Childhood experience and the expression of genetic potential: what childhood neglect tells us about nature and nurture. *Brain and Mind* **3**, 79–100; McGowan PO, Sasaki A, D'Alessio AC *et al.* (2009). Epigenetic regulation of the glucocorticoid receptor in human brain associates with childhood abuse. *Nature Neuroscience* **12**, 342–8; Neigh GN, Gillespie CF and Nemeroff CB (2009). The neurobiological toll of child abuse and neglect. *Trauma, Violence and Neglect* **10**, 389–410; Heim C, Shugart M, Craighead WE and Nemeroff CB (2010). Neurobiological and psychiatric consequences of child abuse and neglect. *Developmental Psychobiology* **52**, 671–90; Teicher MH, Anderson CM and Polcari A (2012). Childhood maltreatment is associated with reduced volume in the hippocampal subfields CA3, dentate gyrus, and subiculum. *Proceedings of the National Academy of Sciences of the USA* **109**, E563–72

38 Pollak SD, Nelson CA, Schlaak MF *et al.* (2010). Neurodevelopmental effects of early deprivation in postinstitutionalized children. *Child Development* **81**, 224–36; see also Majer M, Nater UM, Lin J-MS *et al.* (2010). Association of childhood trauma with cognitive function in healthy adults: a pilot study. *BMC Neurology* **10**, 61

39 Chugani HT, Behen ME, Muzik O *et al.* (2001). Local brain functional activity following early deprivation: a study of postinstitutional Romanian orphans.

NeuroImage **14**, 1290–301; see also Maheu FS, Dozier M, Guyer AE *et al.* (2010). A preliminary study of medial temporal lobe function in youths with a history of caregiver deprivation and emotional neglect. *Cognitive, Affective and Behavioral Neuroscience* **10**, 34–49

40 Eluvathingal TJ, Chugani HT, Behen ME *et al.* (2006). Abnormal brain connectivity in children after early severe socioemotional deprivation: a diffusion tensor imaging study. *Pediatrics* **117**, 2093–100; Sheridan MA, Fox NA, Zeanah CH *et al.* (2012). Variation in neural development as a result of exposure to institutionalization early in childhood. *Proceedings of the National Academy of Sciences of the USA* **109**, 12927–32

41 Teicher MH, Dumont NL, Ito Y *et al.* (2004). Childhood neglect is associated with reduced corpus callosum area. *Biological Psychiatry* **56**, 80–5

42 Benetti S, McCrory E, Arulanantham S *et al.* (2010). Attachment style, affective loss and grey matter volume: a voxel-based morphometry study. *Human Brain Mapping* **31**, 1482–9

43 Frodl T, Reinhold E, Koutsouleris N *et al.* (2010). Interaction of childhood stress with hippocampus and prefrontal cortex volume reduction in major depression. *Journal of Psychiatric Research* **44**, 799–807

44 Spinelli S, Chefer S, Suomi SJ *et al.* (2009). Early-life stress induces long-term morphologic changes in primate brain. *Archives of General Psychiatry* **66**, 658–65; Arabadzisz D, Diaz-Heijtz R, Knuesel I *et al.* (2010). Primate early-life stress leads to long-term mild hippocampal decreases in glucocorticoid receptor expression. *Biological Psychiatry* **11**, 1106–9; Feng X, Wang L, Yang S *et al.* (2011). Maternal separation produces lasting changes in cortisol and behaviour in rhesus monkeys. *Proceedings of the National Academy of Sciences of the USA* **108**, 14312–17; Cole SW, Conti G, Arevalo JMG *et al.* (2012). Transcriptional modulation of the developing immune system by early-life social adversity. *Proceedings of the National Academy of Sciences of the USA* **109**, 20578–83

45 Sallet J, Mars RB, Noonan MP *et al.* (2011). Social network size affects neural circuits in macaques. *Science* **334**, 697–700

46 Hackman DA, Farah MJ and Meaney MJ (2010). Socioeconomic status and the brain: mechanistic insights from human and animal research. *Nature Reviews Neuroscience* **11**, 651–9; Maggi *et al.* (2010), ref. 33

47 Perry (2002), ref. 37; Venderwert RE, Marshall PJ, Nelson CE, III *et al.* (2010). Timing of intervention affects brain electrical activity in children exposed to severe psychosocial neglect. *PLoS ONE* **5**, e11415

48 Messer N, *Selfish Genes and Christian Ethics* (London: SCM Press, 2007), 113, 128

49 Fonagy P and Luyten P (2009). A developmental, mentalization-based approach to the understanding and treatment of borderline personality disorder. *Development and Psychopathology* **21**, 1355–81

50 Colvert E, Rutter M, Kreppner JH *et al.* (2008). Do theory of mind and executive function deficits underlie the adverse outcomes associated with profound early deprivation? Findings from the English and Romanian adoptees study. *Journal of Abnormal Child Psychology* **36**, 1057–68; Lewis-Morrarty E, Dozier M, Bernard K *et al.* (2012). Cognitive flexibility and theory of mind outcomes among foster children: preschool follow-up results of a randomized clinical trial. *Journal of Adolescent Health* **51** (Suppl. 2), S17–22

51 Pyers JE and Senghas A (2009). Language promotes false-belief understanding. *Psychological Science* **20**, 805–12

52 Kobayashi C, Glover GH and Temple E (2007). Cultural and linguistic effects on neural bases of 'theory of mind' in American and Japanese children. *Brain Research* **1164**, 95–107; Kobayashi C, Glover GH and Temple E (2008). Switching language switches mind: linguistic effects on developmental neural bases of 'theory of mind'. *Social Cognitive and Affective Neuroscience* **3**, 62–70

53 Kapogiannis D, Barbey AK, Su M *et al.* (2009). Cognitive and neural foundations of religious belief. *Proceedings of the National Academy of Sciences of the USA* **106**, 4876–81

54 Jeeves M, in Jeeves (ed.), *Rethinking Human Nature* (2011), see Prologue, ref. 36

55 Rochat P (2007). Intentional action arises from early reciprocal exchanges. *Acta Psychologica* **124**, 8–25

56 Steeves HP (2003). Humans and animals at the divide: the case of feral children. *Between the Species* **III** (an online journal for the study of philosophy and animals, www.cla.calpoly.edu/bts/)

57 Villa-Vicencio C and De Gruchy J, *Doing Ethics in Context: South African Perspectives* (Claremont: David Philip and Maryknoll, NY: Orbis Books, 1994), 196; De Gruchy J, *Christianity and Democracy* (Cambridge: Cambridge University Press, 1995), 188–92

58 Boyd R, Richerson PJ and Henrich J (2011). The cultural niche: why social learning is essential for human adaptation. *Proceedings of the National Academy of Sciences of the USA* **108** (Suppl. 2), 10918–25

59 Perry JC, Sigal JJ, Boucher S and Pare N (2006). Seven institutionalized children and their adaptation in late adulthood: the children of Duplessis (les enfants de Duplessis). *Psychiatry* **69**, 283–301

60 Cozza CJ (2006). Case studies of the orphans of Duplessis: the power of stories. *Psychiatry* **69**, 325–7

61 Broom BC (2010). A reappraisal of the role of 'mindbody' factors in chronic urticaria. *Postgraduate Medical Journal* **86**, 365–70

62 Baumeister and Masicampo (2011), ref. 31

63 McAdams DP and Olson BD (2010). Personality development: continuity and change over the life course. *Annual Review of Psychology* **61**, 517–42

64 McLean KC, Pasupathi M and Pals JL (2007). Selves creating stories creating selves: a process model of self-development. *Personality and Social Psychology Review* **11**, 262–78; see also Fivush R, Habermas T, Waters TEA and Zaman W (2011). The making of autobiographical memory: intersections of culture, narratives and identity. *International Journal of Psychology* **46**, 321–45

65 Mauron A (2001). Is the genome the secular equivalent of the soul? *Science* **291**, 831–2

66 Wright NT, *The New Testament and the People of God* (London: SPCK, 1992), 32, 38, 40–5, 116

67 Birch BC and Rasmussen LL, *Bible and Ethics in the Christian Life* (Minneapolis: Augsburg, 1989), 106–7

68 Hauerwas S, *Truthfulness and Tragedy* (Notre Dame: University of Notre Dame Press, 1977), 71–8, 223–5; *Vision and Virtue* (Notre Dame: University of Notre Dame Press, 1981), 68–77

69 Lk 6:27–8; Mt 5:44

70 Lk 10:25–37

71 Rolston H, III, *Genes, Genesis and God* (Cambridge: Cambridge University Press, 1999), 248

72 Mk 10:43–4; also 9:35; Mt 23:11

73 Paabo S (2001). The human genome and our view of ourselves. *Science* **291**, 1219–20

74 Ayala FJ, *Darwin's Gift to Science and Religion* (Washington, DC: Joseph Henry Press, 2007), 177–8

Index